HEAT TREATMENTS FOR POSTHARVEST PEST CONTROL: THEORY AND PRACTICE

HEAT TREATMENTS FOR POSTHARVEST PEST CONTROL: THEORY AND PRACTICE

Edited by

Juming Tang

Elizabeth Mitcham

Shaojin Wang

and

Susan Lurie

www.cabi.org

CABI is a trading name of CAB International

CABI Head Office
Nosworthy Way
Wallingford
Oxon OX10 8DE
UK

Tel: +44 (0)1491 832111
Fax: +44 (0)1491 833508
E-mail: cabi@cabi.org
Website: www.cabi.org

CABI North American Office
875 Massachusetts Avenue
7th Floor
Cambridge, MA 02139
USA

Tel: +1 617 395 4056
Fax: +1 617 354 6875
E-mail: cabi-nao@cabi.org

A catalogue record for this book is available from the British Library, London, UK.

A catalogue record for this book is available from the Library of Congress, Washington, DC.

ISBN 978 1 84593 252 7

Typeset by Columns Design Ltd, Reading, UK
Printed and bound in the UK by Cromwell Press, Trowbridge

Contents

Preface   ix

Contributors   xi

Reviewers   xiii

1   **Introduction**   1
         J.D. Hansen and J.A. Johnson
         1.1   History and Purpose of Quarantine and   1
                Phytosanitation Requirements
         1.2   Review of Treatments   4
         1.3   Survey of Heat Treatments   8
         1.4   Heat Treatments for Microbial Control   16
         1.5   Tolerance of Commodities to Heat Treatments   16
         1.6   Conclusions   16
         1.7   References   17

2   **Fundamental Heat Transfer Theory for Thermal Treatments**   27
         J. Tang and S. Wang
         2.1   Introduction   27
         2.2   Conventional Heat Transfer Theory   28
         2.3   Dielectric Heating   41
         2.4   Case Studies to Demonstrate the Differences
                between Conventional and Dielectric Heating   47
         2.5   Closing Remarks   51
         2.6   References   53

3   **Temperature Measurement**   56
         J. Tang and S. Wang
         3.1   Introduction   56

3.2   Principles and Properties   56
3.3   Sensor Calibration, Precision and Response Time   68
3.4   Application of Temperature Sensors   70
3.5   Temperature Control   76
3.6   Closing Remarks   76
3.7   References   77

**4   Physiological Responses of Agricultural Commodities to   79
       Heat Treatments**
       S. Lurie and E.J. Mitcham
4.1   Introduction   79
4.2   Effects on Physiology   79
4.3   Types of Heat Damage   83
4.4   Responses of Dried Commodities to Heat Treatment   90
4.5   Factors Affecting Response to Heat Treatment   92
4.6   Conclusions   97
4.7   References   97

**5   Experimental and Simulation Methods of Insect Thermal   105
       Death Kinetics**
       S. Wang, J. Tang and J.D. Hansen
5.1   Introduction   105
5.2   Experimental Methods for Obtaining Thermal Kinetic
       Response Information   107
5.3   Insect Mortality Models   112
5.4   Model Comparisons   124
5.5   Model Applications   126
5.6   Closing Remarks   128
5.7   References   128

6   Biology and Thermal Death Kinetics of Selected Insects   133
       J. Tang, S. Wang and J.A. Johnson
6.1   Introduction   133
6.2   Biology and Economic Impact of Target Species   134
6.3   Thermal Death Data   144
6.4   Influence of Life Stages and Species on Thermal
       Mortality   149
6.5   Activation Energies for Thermal Kill of Insect Pests   154
6.6   Preconditioning Effects on Thermotolerance of Pests   155
6.7   Effect of Heating Rates in Thermal Treatments   157
6.8   Closing Remarks   158
6.9   References   159

7   Thermal Control of Fungi in the Reduction of Postharvest Decay 162
       E. Fallik and S. Lurie
7.1   Introduction   162
7.2   Responses of Fungi to Thermal Heat: in Vitro Studies   162
7.3   Methods of Thermal Treatment   165
7.4   Conclusions   174

7.5	Acknowledgements	177
7.6	References	177

8 Disinfestation of Stored Products and Associated Structures Using Heat — **182**
S.J. Beckett, P.G. Fields and Bh. Subramanyam

8.1	Introduction	182
8.2	The Use of Heat for Insect Management	190
8.3	Effects of High Temperatures on Stored-product Insects	193
8.4	Heat Tolerance in Stored-product Insects	194
8.5	Survey of Current Thermal Kinetic Data: Empirical Methods and Common Models	195
8.6	Current Status of Research and Development in Heat Disinfestation of Stored Products	212
8.7	Heat Disinfestation of Structures	220
8.8	Conclusions	228
8.9	References	229

9 Considerations for Phytosanitary Heat Treatment Research — **238**
G.J. Hallman

9.1	Introduction	238
9.2	Source of Research Organisms	239
9.3	Rearing Conditions	240
9.4	Methods of Infesting Commodities for Disinfestation Research	241
9.5	Determination of Disinfestation Policy	245
9.6	Commodity Conditioning	246
9.7	Commercial Possibilities	246
9.8	Conclusions and Recommendations	247
9.9	References	248

10 Heat with Controlled Atmospheres — **251**
E.J. Mitcham

10.1	Introduction	251
10.2	Mode of Action of Controlled Atmospheres on Insects	252
10.3	Effects of Controlled Atmospheres on Commodities	254
10.4	Effects of Heat and Controlled Atmospheres on Arthropod Pests	255
10.5	Commodity Response to High-temperature Controlled Atmospheres	260
10.6	Synergistic Effects of Heat and Controlled Atmospheres	262
10.7	Promising Treatments	264
10.8	Summary	264
10.9	References	265

**11   The Influence of Heat Shock Proteins on Insect Pests and   269
        Fruits in Thermal Treatments**
        S. Lurie and E. Jang
        11.1   Introduction   269
        11.2   Heat Shock Proteins   271
        11.3   Heat Shock Responses and Heat Shock Proteins in
                Plant Tissue   273
        11.4   Heat Shock Responses and Heat Shock Proteins in
                Insects   276
        11.5   Discussion   282
        11.6   References   285

12   Thermal Treatment Protocol Development and Scale-up   291
        J. Tang, S. Wang and J.W. Armstrong
        12.1   Introduction   291
        12.2   Strategies for Thermal Treatment Development   291
        12.3   Systematic Development of RF   295
                Treatment for In-shell Walnuts
        12.4   Developing RF Treatments for Fresh
                Fruits   304
        12.5   Conclusions   308
        12.6   References   308

13   Commercial Quarantine Heat Treatments   311
        J.W. Armstrong and R.L. Mangan
        13.1   Introduction   311
        13.2   Definitions and Concepts   312
        13.3   Quarantine Heat Treatments   313
        13.4   Quarantine Treatment Protocols   315
        13.5   Quarantine Security Statistics   316
        13.6   Developing Quarantine Heat Treatments   318
        13.7   Commodity Quality   325
        13.8   Experimental Heat Treatment Equipment   329
        13.9   Heat Treatment Research   330
        13.10   Commercial Heat Treatment Equipment and
                Facilities   332
        13.11   Approved Commercial Heat Treatments   334
        13.12   Summary   334
        13.13   References   336

Index   **341**

Preface

With increasing globalization facilitated by World Trade Organization agreements and other treaties, the international trade of agricultural commodities has increasingly become an important part of the global economy. Unlike manufactured goods, agricultural commodities are natural carriers of exotic pests. Unintended introduction of these pests to new areas may cause major losses to native crops. Importing countries or regions have thus imposed quarantine or phytosanitary measures against several economically important insect pests.

Methyl bromide has been the most effective pest control fumigant for host commodities structures; however, it is now listed as an ozone-depleting chemical under the Montreal Protocol of 1992 and other related international agreements. As such, there are severe restrictions against its use, creating an urgent need to find environmentally friendly and effective alternatives. Thermal treatments have drawn increasing attention because they are environmentally friendly, leave no chemical residues and have already been successfully used for some commodities structures.

The purpose of this book is to provide fundamental and up-to-date published information on thermal treatments managing postharvest pests associated with agricultural commodities structures. Specific topics of this book include: (i) regulatory issues for quarantine and phytosanitory treatments; (ii) basic information on temperature measurement, heat transfer and thermal death kinetics of insects; (iii) biological responses of agricultural commodities and insect pests; (iv) biological responses of plants, insects and pathogens to heat; and (v) an introduction to current and potential quarantine treatments based on hot air, hot water and radio frequency energy.

The book is divided into 13 chapters written by leading experts with extensive research experience in their relevant fields. The contributors are affiliated with research universities or government research institutes in the USA, Israel, Australia and Canada. The contributors of this book bring expertise from three different disciplines that are highly relevant to this subject, namely, engineering, entomology and plant physiology. From this perspective, it is the most comprehensive book yet on the subject of thermal treatments for pest control.

Through a multidisplinary book, we hope to establish a sound basis for further development and implementation of thermal control methods for pests of harvested agricultural commodities structures. It should serve as an important resource for readers who are interested in knowledge, methods and strategies used in the development of environmentally friendly pest control processes based on thermal energy. This book may also be suited for readers worldwide at different levels in academia, industry and government.

The editorial team thanks the contributors for sharing their valuable experiences and knowledge. Each chapter has been reviewed by two internal or two external reviewers. We, therefore, thank those reviewers (see the list of reviewers) for helping us maintain a high standard throughout the book's content. We appreciate grant support from the US Department of Initiative for Future Agriculture and Food Systems (IFAFS), National Research Initiative (NRI) and Cooperative State Research, Education, Extension Service (CSREES) Methyl Bromide Transitions programs, and from other grant agencies, including BARD (USA–Israel Binational Agricultural Research and Development), California Department of Food and Agriculture (CDFA), Washington State University IMPACT Centre and commodity commissions from the States of California and Washington. Those grants provided opportunities for several book contributors to work closely over the past 8 years on numerous multidisciplinary research projects and allowed us to develop understanding, collaboration and friendship across disciplinary, institutional and geographic borders.

Juming Tang
Elizabeth Mitcham
Shaojin Wang
Susan Lurie

Contributors

John W. Armstrong, *Research Leader, USDA-Agricultural Research Service (ARS), Pacific Basin Agricultural Research Center, PO Box 4459, Hilo, HI 96720, USA; email: jarmstrong@pbarc.ars.usda.gov*

Stephen Beckett, *Entomologist, Commonwealth Scientific and Industrial Research Organization (CSIRO) Entomology, PO Box 1700, Canberra, ACT 2601, Australia; e-mail: stephen.beckett@csiro.au*

Elazar Fallik, *Professor, Department of Postharvest Science, Agriculture Research Organization (ARO), The Volcani Center, PO Box 6, Bet-Dagan 50250, Israel; e-mail: efallik@volcani.agri.gov.il*

Paul Fields, *Professor, Agriculture & Agrifood Canada, Cereal Research Centre 195, Dafoe Road, Winnipeg, Manitoba, Canada, R3T 2M9; e-mail: pfields@agr.gc.ca*

Guy J. Hallman, *Research Entomologist, USDA-ARS, 2413 E. Business 83, Weslaco, TX 78596, USA; e-mail: ghallman@weslaco.ars.usda.gov*

James D. Hansen, *Research Entomologist, USDA-ARS, Yakima Agricultural Research Laboratory, 5230 Konnowac Pass Road, Wapato, WA 98951, USA; e-mail: jimbob@yarl.ars.usda.gov*

Eric Jang, *Research Entomologist, USDA-ARS, Pacific Basin Agricultural Research Center, PO Box 4459, Hilo, HI 96720, USA; e-mail: ejang@pbarc.ars.usda.gov*

Judy Johnson, *Research Entomologist, USDA-ARS, San Joaquin Valley Agricultural Sciences Center, 9611 S. Riverbend Ave., Parlier, CA 93648, USA; e-mail: jjohnson@fresno.ars.usda.gov*

Susan Lurie, *Senior Scientist, Department of Postharvest Science, ARO, The Volcani Center, PO Box 6, Bet-Dagan 50250, Israel; e-mail: slurie43@volcani.agri.gov.il*

Robert L. Mangan, *Research Leader, USDA-ARS, Subtropical Agricultural Research Service CQFIRU, 2413 E. Highway 83 Building 200, Weslaco, TX 78596, USA; e-mail: rmangan@weslaco.ars.usda.gov*

Elizabeth Mitcham, *Professor, Department of Plant Sciences, Mail Stop 2, University of California, One Shields Avenue, Davis, CA 95616, USA; e-mail: ejmitcham@ucdavis.edu*

Bhadriraju Subramanyam, *Professor, Kansas State University, Department of Grain Science and Industry, Kansas State University, Manhattan, KS 66506, USA; e-mail: bhs@wheat.ksu.edu*

Juming Tang, *Professor, WSU IMPACT Fellow, Department of Biological Systems Engineering, Washington State University, 213 L.J. Smith Hall, Pullman, WA 99164-6120, USA; e-mail: jtang@wsu.edu*

Shaojin Wang, *Assistant Research Professor, Department of Biological Systems Engineering, Washington State University, 213 L.J. Smith Hall, Pullman, WA 99164-6120, USA; e-mail: shaojin_wang@wsu.edu*

Reviewers

Dr J.W. Armstrong, *USDA-ARS, US Pacific Basin Agricultural Research Center, PO Box 4459, Hilo, Hawaii 96720, USA.*

Dr F. Arthur, *USDA-ARS, 1515 College Ave., Manhattan, KS 66502, USA.*

Dr C. Burks, *USDA-ARS, San Joaquin Valley Agricultural Science, 9611 South Riverbend Ave., Parlier, CA 93648, USA.*

Dr M. Casada, *USDA-ARS, Grain Marketing and Production Research Center, 1515 College Avenue, Manhattan, KS 66502, USA.*

Dr Y. Chen, *USDA-ARS, Bee Research Laboratory, 10300 Baltimore Avenue, Beltsville, MD 20705, USA.*

Dr W.S. Conway, *Produce Quality and Safety Laboratory, USDA-ARS, Beltsville Agricultural Research Center, Beltsville, MD 20705, USA.*

Dr D.E. Evans, *28 Clissold Street, Mollymook, NSW 2539, Australia.*

Dr P. Follett, *USDA-ARS, Pacific Basin Agricultural Research Center, PO Box 4459, Hilo, HI 96720, USA.*

Dr S. Garczynski, *USDA-ARS Yakima Agricultural Research Laboratory, 5230 Konnowac Pass Road, Wapato, WA 98951-9651, USA.*

Dr G. Hallman, *USDA-ARS Crop Quality and Research Laboratory, 2413 E. Highway 83, Building 200, Weslaco, TX 78596, USA.*

Dr J. Johnson, *USDA-ARS, San Joaquin Valley Agricultural Sciences Center, 9611 S. Riverbend Ave., Parlier, CA 93648, USA.*

Dr Y. Liu, *USDA-ARS, 1636 E. Alisal Street, Salinas, CA 93905, USA.*

Dr S. Lurie, *Department of Postharvest Science, ARO, The Volcani Center, PO Box 6, Bet-Dagan 50250, Israel.*

Dr E. Mitcham, *Department of Plant Sciences, Mail Stop 2, University of California, One Shields Avenue, Davis, CA 95616-8780, USA.*

Dr L.G. Neven, *USDA-ARS, Yakima Agricultural Research Laboratory, 5230 Konnowac Pass Road, Wapato, WA 98951, USA.*

Dr D. Obenland, *USDA-ARS, San Joaquin Valley Agricultural Science, 9611 South Riverbend Ave., Parlier, CA 93648, USA.*

1 Introduction

J.D. HANSEN[1] AND J.A. JOHNSON[2]

[1]USDA-ARS Yakima Agricultural Research Laboratory, Wapato, Washington, USA; e-mail: jimbob@yarl.ars.usda.gov ; [2]USDA-ARS San Joaquin Valley Agricultural Sciences Center, Parlier, California, USA; e-mail: jjohnson@fresno.ars.usda.gov

1.1 History and Purpose of Quarantine and Phytosanitation Requirements

Introduction

Heat has had a variety of uses since primitive times, such as in cooking and food preservation, but its use was limited as a pest control method for stored products until the modern era. Heat can be generated by various ways: chemical oxidation, combustion, electrical resistance and electromagnetic exposure, and heat treatments have been devised that take advantage of each.

The manner in which heat is produced affects both products and their pests, and the success of a given treatment depends on its ability to control insects without causing product damage. This chapter will briefly review the history and development of product treatments in general and heat treatments in particular that meet pest phytosanitation and quarantine security requirements. Terms used in the regulatory processes will be defined, and a short review of the current status of heat treatments will be presented.

Treatments

Postharvest treatments are a recent development. For most of human history, insect pests in stored products have been tolerated. However, two events led to the development of postharvest methods to eliminate insects. The first relates to the distribution of pest insects, which historically was limited by their biology and geophysical forces. With increased travel by humans – particularly wide-ranging exploration and commerce – insects can now be transported into new areas. If the insect has potential to do damage, even though it is a minor pest in its native range, measures that prevent its introduction and establishment are needed. This concern has resulted in a series of procedures and regulations pertaining to quarantine insects.

The second event that led to the development of postharvest methods to eliminate insects involved the rapid evolution of agricultural technology to store surplus commodities until future consumption or transport to distant markets for greater economic gain (see Table 1.1, which illustrates the huge global market for two particular commodities, fruit and nuts). To prevent damage during storage by various insects – particularly beetles and moths, phytosanitation procedures were developed to reduce and control the populations of these pests. Because of the complexities of modern commerce, the distinction between quarantine and phytosanitation is often unclear.

Quarantine Regulations

Quarantine statutes involve the treatment or shipment of an agricultural commodity from one location to another. They can be imposed by various bodies: (i) local governmental jurisdiction, such as the Washington State (USA) quarantine against intrastate movement of apples from areas where fruits have been infested with the apple maggot, *Rhagoletis pomonella* (Walsh) (Diptera:

Table 1.1. World export of fruit and nut products.

Commodity	Four-year average (2000–2003)	
	Quantity (metric tonnes)	Value (US$1,000)
Fruits		
Cherries	155,769	388,011
Dates	494,863	260,179
Figs (fresh and dried)	88,270	141,338
Grapefruit and Pomelos	1,037,373	475,102
Grapes	2,806,023	2,701,020
Lemons and limes	1,737,864	796,061
Mangoes	714,655	437,758
Olives	36,813	34,761
Oranges	4,783,691	2,024,224
Papayas	209,718	132,112
Peaches and Nectarines	1,222,043	1,007,580
Pineapples	1,294,431	589,146
Plums	434,231	351,183
Prunes (dried plums)	160,846	267,899
Raisins	632,459	607,248
Strawberries	492,654	789,275
Tangerines	2,564,985	1,559,091
Nuts (shelled and in-shell)		
Almonds	393,268	1,099,125
Hazelnuts	224,843	557,993
Pistachios	211,212	697,965
Walnuts	186,885	460,667

Tephritidae); (ii) within national boundaries, such as the quarantine against movement of cherries from the US Pacific Northwest states to California (CDFA, 1998) because of possible infestation by cherry fruit fly, *Rhagoletis cingulata* (Loew) (Diptera: Tephritidae); or (iii) between nations, such as the quarantine of possible host materials of the codling moth, *Cydia pomonella* (L.) (Lepidoptera: Tortricidae), in exports to Japan (MAFF–Japan, 1950).

To facilitate the implementation of such quarantines, groups of nations within a region have formed organizations that provide similar regulations against a common potential pest. Laws involving the importation of commodities that may contain the Caribbean fruit fly, *Anastrepha suspense* (Loew) (Diptera: Tephritidae), imposed by nations in the European and Mediterranean Plant Protection Organization (EPPO), are good examples.

Other international regulatory organizations include the Caribbean Plant Protection Commission, Inter-African Phytosanitary Council and the North American Plant Protection Organization. Individual countries have their own plant protection agencies, such as the Australian Quarantine Inspection Service, Ministry of Agriculture, Forestry and Fisheries (Japan) and Animal and Plant Health Inspection Service (USA). Regulation within a country often falls on a local or state government body, such as the US California Department of Food and Agriculture and Washington State Department of Agriculture.

All of the above entities may write their own regulations to meet immediate concerns and generate lists of quarantine pests. For example, the EPPO has a list for quarantine pests not present (A1 pests) and another for quarantine pests present but not widely distributed and officially under control (A2 pests). The state of California has a similar rating system, with 'A' pests requiring that infested commodities be rejected or treated, and 'Q' pests temporarily requiring the same action as A pests until a permanent rating can be determined. To assure exclusion of the quarantine pest, these lists include all potential hosts.

Enforcement of quarantine regulations relies primarily on international trade agreements and standards that may include the use of specific treatments on select commodities to assure quarantine security. Hence, international commerce in commodities potentially containing quarantine pests is strictly regulated.

Phytosanitation

Phytosanitation generally does not involve regulation as extensive as that applied to quarantine. Rather, economic considerations direct its application. It benefits the producer to promote commodities free of pests, both to restrain internal costs and to promote an attractive product. The burden of phytosanitation usually falls on the fruit or produce packer or marketer, and treatments may be applied repeatedly.

Probit-9 and Quarantine Security

The standard of quarantine security for many importing countries is probit-9, or 99.996832% mortality of the pest population (Robertson *et al.*, 1994b). In practical terms, probit-9 means that only 32 individuals can survive out of 1 million treated pests. Originally suggested by Baker (1939), the use of probit-9 was directed at those not familiar with sophisticated experimentation and data analysis. Baker used moisture-saturated hot air to control the Mediterranean fruit fly, *Ceratitis capitata* (Wiedemann) (Diptera: Tephritidae), on Hawaiian kamani nuts shipped to the American mainland, and intended that the criteria used to calculate probit-9 mortality be successful adult emergence from a single batch of insects.

Probit-9 security levels are normally applied to quarantine treatments of commodities that have a high and frequently unknown level of pest infestation. In practice, this condition is often ignored. For example, cherries imported to Japan from the Pacific Northwest of the USA must be treated with quarantine procedures that have demonstrated probit-9 efficacy on the last codling moth instar, even though intense postharvest inspections demonstrate its rare occurrence. Between 1978 and 1996, one potential codling moth larva was identified from out of 4.9×10^8 individually inspected fruits from the Pacific Northwest region of the USA (Wearing *et al.*, 2001).

In developing a new quarantine procedure, it is very difficult to produce the probit-9 level by treating 1 million insects at one time. Instead, a probit model is developed based on a dose-response curve from much smaller populations, and then projected out to the probit-9 level to determine the quarantine treatment. A statistical test may be included to determine whether the original data fit a probit distribution. Confirmatory tests are then performed, during which the proposed quarantine treatment is tested against a large number of pest insects.

Some countries require a specific number of test insects to be treated during confirmatory studies: 30,000 test insects with no survivors are commonly used. Although this approach is flawed by several severe statistical problems, described in detail elsewhere (Landolt *et al.*, 1984; Chew, 1994; Robertson *et al.*, 1994a, b), many regulatory bodies continue to base their quarantine programs on probit-9 security. A more thorough discussion of the mathematical approaches for determining efficacy is given in Chapter 5, this volume.

1.2 Review of Treatments

General Overview

Quarantine and phytosanitation treatments can be sorted into several categories: chemical, biological, irradiation and physical. Briefly, chemical approaches include a wide assortment of distinct compounds: (i) fumigants (hydrogen cyanide, phosphine, carbon disulphide, halogenated hydrocarbons such as ethylene dibromide and methyl bromide, and acrylonitrile); (ii) dips (insecticides and soaps); and (iii) controlled or modified atmospheres (often used in combination with a temperature treatment).

Fumigants are the most prevalent, but are losing support because of health, environmental and safety concerns (Stark, 1994; Yokoyama, 1994). The postharvest alternatives to methyl bromide fumigation will be discussed later in this chapter. Controlled atmospheres (CA), reviewed by Carpenter and Potter (1994) and Hallman (1994), involve changing the composition of gases, usually substituting oxygen with nitrogen or carbon dioxide. Insecticides used in dips and sprays include soaps, organophosphorus compounds, organochlorines and insect growth regulators, and these have been reviewed by Hansen and Hara (1994) and Heather (1994). Approval of chemical approaches varies between different jurisdictions.

Biological approaches include the use of natural enemies or microbial agents (generally limited to phytosanitary treatments of stored grains or other durables), host-plant resistance (particularly among cultivars) and pest-free zones, which may be geographical or temporal. For pest-free zones to be effective, the host–pest relationship must be well known, and site manipulation – such as the removal of alternative hosts as well as intensive sampling – may be required. Armstrong (1994a), Greany (1994), Riherd et al. (1994), Moore et al. (2000) and Schöller and Flinn (2000) have all reviewed different biological methods.

Irradiation, including exposure to gamma and X-rays, is an old technology where dose mortality schedules have been established for several pests on a variety of commodities. Safety, expense, environmental concerns and consumer acceptance interfere with its widespread use. For further information, consult Moy (1985), Burditt (1994) and Nation and Burditt (1994).

Physical methods of quarantine and phytosanitation treatments include mechanical, ultrasound, vacuum and temperature extremes. Mechanical systems use brushes and water sprays (Walker et al., 1996; Whiting et al., 1998; Prusky et al., 1999). Ultrasound produces alternating high and low pressure waves that cause cavitation at the cellular level (Sala et al., 1999) to reduce surface pests on fruit (Hansen, 2001).

Vacuum treatments, applied primarily to durable commodities, are gaining interest due to the development of flexible, low-cost treatment containers (Navarro et al., 2001). Alone, vacuum treatments appear to cause mortality in insects by creating a low-oxygen atmosphere and, in low-moisture environments, may cause rapid dehydration (Navarro and Calderon, 1979). A vacuum microwave (MW) grain dryer was found to rapidly disinfest stored grain of insects (Tilton and Vardell, 1982a, b).

Cold temperature treatments, particularly cold storage, have practical application for many commercial operations (Armstrong, 1994b; Gould, 1994), such as holding commodities before marketing – as well as pest control. Thermal treatments must be precise because of the narrow margin between efficacy and commodity tolerance to the temperature, particularly in fresh fruit and vegetables. Heat treatments have been accepted for commodities entering the USA (USDA, 2005a) and for interstate shipments (see Table 1.2) (USDA,

Table 1.2. Quarantine heat treatments in the USA (from USDA, 2005b, c).

Commodity	Target pests	Treatment schedule
Hot water immersion		
Limes	Mealybugs and other surface pests	49°C or above for 20 min
Longan (lychee from Hawaii)	*Ceratitis capitata, Bactrocera dorsalis*	49°C or above for 20 min
Mango	*Ceratitis capitata, Anastrepha* spp., *Anastrepha ludens*	46°C for 65–110 min, depending on origin, size and shape of fruit
High-temperature forced air		
Citrus from Mexico, infested areas in the USA	*Anastrepha* spp.	Raise centre of fruit to 44°C over 90 min, hold at 44°C for 100 min
Citrus from Hawaii	*Ceratitis capitata, Bactrocera dorsalis, B. cucurbitae*	47.2°C (fruit centre) for at least 4 h total treatment time
Mango from Mexico	*Anastrepha ludens, A. oblique, A. serpentina*	Until seed surface reaches 48°C
Papaya from Chile, Belize and Hawaii	*Ceratitis capitata, Bactrocera dorsalis, B. cucurbitae*	47.2°C (fruit centre) for at least 4 h total treatment time
Rambutan from Hawaii	*Ceratitis capitata, Bactrocera dorsalis*	Raise centre of fruit to 47.2°C over 60 min, hold at 47.2°C for 20 min
Vapour heat treatments		
Bell pepper, aubergine, papaya, pineapple, squash, tomato, courgette from Hawaii	*Ceratitis capitata, Bactrocera dorsalis, B. cucurbitae*	44.4°C (fruit centre) for 8.75 h (heating rate variable)
Clementine, orange, grapefruit, mango from Mexico	*Anastrepha* spp.	Raise centre of fruit to 43.3°C over 8 h, hold at 43.3°C for 6 h
Clementine from Mexico (alternate treatment)	*Anastrepha* spp.	Raise centre of fruit to 43.3°C over 6 h, hold at 43.3°C for 4 h; fruit should be heated rapidly during first 2 h
Lychee from Hawaii	*Ceratitis capitata, Bactrocera dorsalis*	Raise centre of fruit to 47.2°C over 60 min, hold at 47.2°C for 20 min
Mango from Philippines	*Bactrocera* spp.	Raise centre of fruit to 46°C over 4 h, hold at 46°C for 10 min
Mango from Taiwan	*Bactrocera dorsalis*	Raise centre of fruit to 46.5°C, hold at 46.5°C for 30 min
Papaya	*Ceratitis capitata, Bactrocera dorsalis, B. cucurbitae*	Raise centre of fruit to 47.2°C over 4 h
Yellow pitaya from Colombia	*Ceratitis capitata, Anastrepha fraterculus*	Raise centre of fruit to 46°C over 4 h, hold at 46°C for 20 min
Rambutan from Hawaii	*Ceratitis capitata, Bactrocera dorsalis*	Raise centre of fruit to 47.2°C over 60 min, hold at 47.2°C for 20 min

2004). Reviews of heat treatments can be found in Stout and Roth (1983), Armstrong (1994b), Hallman and Armstrong (1994), Sharp (1994a, b) and USDA (2005b). Heat treatments will be discussed in more detail later in this chapter.

Methyl Bromide Fumigation and the Montreal Protocol

For many years, ethylene dibromide was the preferred fumigant against quarantine pests, but was withdrawn in 1984 because of its carcinogenic properties. Since the demise of ethylene dibromide, methyl bromide has become the most popular quarantine treatment (Gaunce *et al.*, 1981; Yokoyama *et al.*, 1990; Moffit *et al.*, 1992). In addition to quarantine uses, methyl bromide has been used extensively for phytosanitary treatments of durables such as dried fruit, nuts, beans, processed foods and pet foods, as well as treatment of processing facilities.

However, methyl bromide was identified by the US Environmental Protection Agency (EPA) under the Federal Clean Air Act (Anon., 1990) and by the Montreal Protocol (Anon., 1995) as having high ozone depletion potential. The EPA mandated the removal of methyl bromide from the chemical register and the phase-out of its production and import into the USA by 31 December 2005.

Similar legislation has occurred in other industrialized countries (see Fig. 1.1). Although methyl bromide fumigation for postharvest quarantine treatments is exempt from the ban, price increases from reduced production, unreliable sources and future restrictions under international agreements (USEPA, 2001) suggest that methyl bromide may become unavailable for quarantine applications as well. Consequently, alternatives to methyl bromide are needed for all postharvest applications.

Practical Alternatives to Methyl Bromide Fumigation

Practical alternatives to methyl bromide must be safe and efficacious, must not reduce product quality, storage life or marketability, must be environmentally acceptable and economically feasible. Some applications require a relatively rapid treatment, either to treat large volumes of product or to meet the needs of specific markets. Although many alternatives have been suggested, most require much longer treatment times, extensive changes to the way processors handle product or substantial capital expenditure.

Alternatives to methyl bromide are easiest to adopt when they serve as drop-in replacements, requiring little change to existing plant infrastructure or

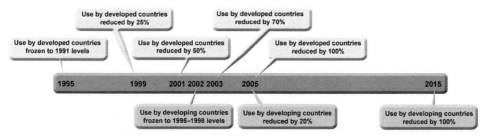

Fig. 1.1. Timeline for the reduction of methyl bromide use globally.

processing procedures. Because other fumigants are considered to be the only such alternatives, they are most likely to be adopted when available. However, issues such as extended treatment times, reduced efficacy and increased costs for some alternative fumigants make them less suitable for certain applications. New fumigants may be developed, but testing and registration of new compounds is a lengthy and expensive process.

Several alternatives have been proposed that do not rely on fumigation for postharvest insect control. These range from single treatments – such as modified atmospheres or vacuum – to combining existing sub-efficacious treatments so that the net efficacy reaches acceptable levels. Related to this is the Systems Approach for quarantine applications, where existing commercial operations, along with treatments that can easily be added without disruption, accumulatively reach the quarantine security level, but not necessarily probit-9. Also, intensive inspection is added to assure compliance.

The industry usually favours this procedure because, with no or few added steps, costs remain low while quality is not adversely affected. This has become the standard for many importing countries, along with a phytosanitation certificate indicating that the commodity has passed the required inspections. Regulatory agencies of both the exporting and importing countries oversee this process. Jang and Moffit (1994) provide a good review of the Systems Approach.

Heat treatments appear as likely candidates for methyl bromide alternatives because: (i) the technology is well established; (ii) they can be applied to a wide range of pests; (iii) there are usually no harmful residues or chemical byproducts; (iv) heat can be generated from a variety of sources; and (v) the costs can usually be controlled. The major impediment is the maintenance of product quality. Dried fruit, nuts and other durables – as well as tropical and subtropical fruit and vegetables – are good candidates for thermal treatments because of their heat tolerances (McDonald and Miller, 1994). However, successful treatments have been developed for temperate fruits as well. Other alternatives to methyl bromide fumigation, in addition to temperature methods, are reviewed by Fields and White (2002) and Vincent et al. (2003).

1.3 Survey of Heat Treatments

Early Applications of Heat for Pest Control

Since ancient times, heat in the form of solar energy or fire has been used to control insect pests. Early civilizations killed insect pests in stored grains with the heat of the sun (Cotton, 1963). To control locusts, the Chinese Kingdom of Shang (c.1520–1030 BC) appointed anti-locust officials, who used bonfires to burn collected locusts or repel them (Nevo, 1996). This practice was carried into the 20th century in the USA, when it was suggested for control of chinch bugs (Headlee, 1911) and grasshoppers (Milliken, 1916), a practice that continues to be studied for US rangelands (Vermeire et al., 2004).

In more recent times, the most extensive use of heat treatment has been to control grain insects. For this purpose, heat has been used both as a commodity treatment and as a structural treatment for mills and processing facilities. Thermal treatments against the Angoumois grain moth, *Sitotroga cerealella* (Oliver) (Lepidoptera: Gelechiidae), were used in France for stored grains as early as 1792 (Fields and White, 2002), and later the French used devices known as 'insect mills' for heating infested grain (Dean, 1913).

In the USA as early as 1835, heated rooms were used to control *Sitophilus* spp. (Coleoptera: Curculionidae) in wheat (Oosthuizen, 1935). During the first half of the 20th century, research on heat treatments was conducted on a variety of insects, including (i) the khapra beetle, *Trogoderma granarium* Everts (Coleoptera: Dermestidae) (Husain and Bhasin, 1921); (ii) the red flour beetle, *Tribolium castaneum* (Herbst) F. (Coleoptera: Tenebrionidae) (Grossman, 1931); (iii) the confused flour beetle, *Tribolium confusum* Jacqueline du Val (Coleoptera: Tenebrionidae) (Oosthuizen, 1935); (iv) the Angoumois grain moth (Grossman, 1931; Harukawa, 1941); (v) grain weevils, *Sitophilus* spp. (Back and Cotton, 1924; Grossman, 1931; Tsuchiya, 1943); and (vi) others.

Early work with heat treatments using electromagnetic energy (radio frequency or microwaves) centered primarily on stored-product insects. The lethal thermal effects on insects caused by exposure to radio frequencies were first reported by Lutz (1927) and Headlee and Burdette (1929). Hadjinicolaou (1931) and Whitney (1932) attributed the death of a variety of stored-product pests exposed to high-frequency radio waves to internal heat generated within the body of each insect. Davis (1933) described how radio frequency energy could kill stored-product pests. Mouromtseff (1933) and Ulrey (1936) worked on designing oscillators for use against grain pests. Kuznetzova (1937) showed that the mature larvae of the granary weevil were more resistant to high-frequency electric currents than were other life stages and noted that adults outside the grain were more susceptible than those within.

Heat has been used to disinfest mills and processing plants for some time. Chittenden (1897) recommended using steam on flour mill machinery to control contamination by the Mediterranean flour moth, *Anagast kuehniella* (Zeller) (Lepidoptera: Pyralidae), and recommended 52–60°C for a few hours to kill other grain insects. Dean (1913) noted that the heating of mills in Kansas and other Midwestern states demonstrated the efficacy of this method against all life stages of common mill insects. Pepper and Strand (1935) described how heating the structure of a grain mill to 66°C could control grain pests within 24 h.

Since the late 19th century, heat treatments in the form of hot water dips, vapour heat or hot forced air have been used to treat numerous fresh commodities (Hallman and Armstrong, 1994; Sharp, 1994a). In 1909, one of the earliest attempts at postharvest treatment of a horticultural commodity used immersion of fruit in hot water to control tarsonemid mites (Cohen, 1967).

Vapour heat was first used in Mexico in 1913 to control the Mexican fruit fly, *Anastrepha ludens* (Loew) (Diptera: Tephritidae). Procedures for using vapour heat were also developed to control the Mediterranean fruit fly in Florida citrus (Latta, 1932) and several types of fruits and vegetables in

California (Mackie, 1931). A similar approach was adopted with Texas citrus for fruit fly control (Hawkins, 1932).

Recent Progress in Heat Treatments

Since these early attempts to control insects by thermal methods, the applicable technologies have progressed in mechanical design and theory. Advances in instrumentation now provide accurate measurements of temperature and other treatment variables, and refined techniques have improved precision and replication. The following provides a short review of each of the current thermal methods.

Hot water

Hot (> 40°C) water baths and dips are the simplest form of heat treatment. Because of the aqueous medium, treatments are anaerobic, with a rapid energy transfer. Although the procedure has a long history, hot water treatments for quarantine applications have only recently been approved, primarily for fruit flies in tropical and subtropical fruits (Sharp, 1986; Sharp *et al.*, 1988, 1989a, b, c; Sharp and Picho-Martinez, 1990).

Couey *et al.* (1985) and Couey and Hayes (1986) combined ethylene dibromide fumigation with hot water baths to control tephritid fruit flies in Hawaiian papayas. Morgan and Crocker (1986) recommended 49°C water baths for 15 min to control weevils in acorns. For other pests, McLaren *et al.* (1997, 1999) described a commercial-scale 2-min hot water bath (50°C) to control the New Zealand flower thrips, *Thrips obscuratus* (Crawford) (Thysanoptera: Thripidae), on apricots, peaches and nectarines. Other examples of the use of hot water for commercial treatments are found in Chapter 13, this volume, and a discussion of thermal death studies using hot water baths in the laboratory is in Chapter 5, this volume.

Vapour

Like water baths, vapour heat treatments use moisture (as saturated air) to transfer thermal energy, usually involving air movement. As presented earlier in this chapter, vapour heat treatment is an old process. It was developed by Baker (1952) for citrus and is now widely used for papaya, pineapple, bell pepper, eggplant, tomato and zucchini (USDA, 2005c). Sinclair and Lindgren (1955) modified vapour heat treatments for California citrus and avocados, and Seo *et al.* (1974) applied this technology for treatment against the oriental fruit fly, *Bactrocera dorsalis* (Hendel) (Diptera: Tephritidae). In Florida, McCoy *et al.* (1994) presented a vapour heat treatment against the eggs of the Fuller rose beetle, *Asynonychus godmani* (Crotch) (Coleoptera: Curculionidae). For more information on vapour heat treatments, see Hallman and Armstrong (1994) and Chapter 13 of this volume.

Forced hot air

Forced hot air treatment is similar to vapour heat treatment, but does not have the moisture component and is a more recent development (Armstrong et al., 1989). Improvements in temperature and moisture monitoring and air delivery have advanced forced hot air treatments (Hallman and Armstrong, 1994). Forced hot air treatments are being devised for commodities normally subjected to vapour heat treatment and have also been applied to new commodities. Their disadvantages are the long treatment durations and sophisticated equipment needed for operation. Also, not all horticultural commodities are suitable, such as avocado (Kerbel et al., 1987).

Heat treatments for durable commodities

Heat has been effective against stored-product pests of durable products such as grains, nuts, dried fruits, wood products and museum artefacts. Most recent work has been carried out on grain pests, and there exists a considerable body of laboratory work on the thermal sensitivity of those (Fields, 1992; Saxena et al., 1992; Mahroof et al., 2003, 2005; Boina and Subramanyam, 2004). Dry heat (82.2°C for 7 min) has been recommended for disinfection of milled products of the khapra beetle, *Trogoderma granarium* Everts (Coleoptera: Dermestidae) – an important quarantine grain pest (Stout and Roth, 1983).

In order to improve the efficiency and reduce the cost, continuous flow, fluidized-bed heating systems for grain have been examined in Australia (Dermott and Evans, 1978; Evans et al., 1983). Commercial heat treatments (usually at 56°C) to control insects in imported wood packaging have been approved in Japan, China, New Zealand, Australia, Europe, North America and most of South America (Anon., 2005). More information on heat treatments for grain may be found in Chapter 8, this volume.

Heat treatment for structures

The application of dry heat to structures to eliminate pests is an attractive alternative to fumigation, and has been used successfully for decades. Sheppard (1984), Heaps (1988, 1996) and Heaps and Black (1994) recount the measures used in commercial mills and food plants when using heat to control stored-grain pests residing in the infrastructures. More recent work has better defined the treatment parameters (Mahroof et al., 2003; Akdoğan et al., 2005). Detailed information on structural heat treatments for control of storage insects can be found in Chapter 8, this volume.

In addition to stored-product pests, structural heat treatments have been applied to a variety of other structural pests, including roaches (Forbes and Ebeling, 1987), termites (Forbes and Ebeling, 1987; Lewis and Haverty, 1996; Zeichner et al., 1998), ants (Forbes and Ebeling, 1987), and powderpost beetles (Ebeling et al., 1989). Ebeling (1994) reviewed the thermal parameters and equipment needed to control structural pests. Quarles (1994) noted that thermal treatments were as effective as conventional fumigations for structural pest control and that their costs were decreasing.

High-temperature controlled atmospheres

Another anaerobic procedure combines forced hot air with an oxygen-poor environment, achieved by replacing oxygen with nitrogen or using high concentrations of carbon dioxide. The mechanism of control is to increase respiratory demands for the target pest – as during heating – yet restrict the amount of oxygen available, leading to metabolic arrest and death. Besides exchanging gases, sophisticated instrumentation may be used like those in the forced hot air system.

The greatest advancement and application of high-temperature controlled atmosphere (HTCA) treatments has been against stored-product pests. Many studies on controlled atmosphere (CA) targeting stored-product pests have noted the relationship between temperature and mortality (Harein and Press, 1968; AliNiazee, 1972; Storey, 1975, 1977; Banks and Annis, 1977; Bailey and Banks, 1980; Soderstrom et al., 1986; Delate et al., 1990; Wang et al., 2001). High-temperature controlled atmosphere treatments have now replaced methyl bromide fumigation, particularly in Europe, for pest control of stored food items, spices, grain in silos and ships, furniture and floorboards (Bergwerff and Vroom, 2003).

Studies on HTCA treatments for horticultural crops are of more recent origin. Early experimental units were based on forced hot air design (Gaffney and Armstrong, 1990; Gaffney et al., 1990; Sharp et al., 1991; Neven and Mitcham, 1996). Many HTCA studies for horticultural pests were for control of the light brown apple moth, Epiphyas postvittana (Walker) (Lepidoptera: Tortricidae), and other pests of New Zealand apples (Whiting et al., 1991, 1995, 1999a, b; Dentener et al., 1992; Lay-Yee et al., 1997; Chervin et al., 1998; Whiting and Hoy, 1998).

Neven and Mitcham (1996) discussed the development of an HTCA treatment known as the Controlled Atmosphere Temperature Treatment System (CATTS). Applications that show promise include codling moth on fresh sweet cherries (Shellie et al., 2001; Neven, 2005) and nectarines and peaches (Obenland et al., 2005). Neven (2004) provided further discussion of CATTS and other HTCA treatments for fresh commodities. More information on heat with controlled atmospheres may be found in Chapter 10, this volume.

Solar energy

Very little has been done using solar energy to control insects. Although the approach is simple and has potential for long-term storage in rural areas and developing countries, there are problems with temperature control in terms of the regulation of consistent temperatures. Most work with solar heat treatments of commodities has targeted bruchid pests of seeds using a variety of solar heating methods including plastic bags, corrugated metal and wooden racks (Murdock and Shade, 1991; Kitch *et al.*, 1992; Chinwada and Giga, 1996; Ntoukam *et al.*, 1997; Arogba *et al.*, 1998; Songa and Rono, 1998; Ugwu *et al.*, 1999; Chauhan and Ghaffar, 2002).

Solar heating has been used to control other stored product and fruit pests, including: (i) the hide beetle in dried mullet; (ii) the Indianmeal moth in peaches; (iii) the merchant grain beetle in oatmeal (Nakayama *et al.*, 1983); (iv) the larger grain borer in maize cobs (McFarlane, 1989); and (v) nitidulid beetles in figs (Shorey *et al.*, 1989). Solar heating has also been suggested for museum artefacts (Baskin, 2001; Brokerhof, 2003; Pearce, 2003).

Infrared

Infrared is strongly emitted by hot substances and readily absorbed by living tissue, making it a logical choice for thermal treatment. Considerable research has been carried out using infrared heat to control internal pests of grains (rice weevils, lesser grain borers and Angoumois grain moths) using short exposures to temperatures of 56–68°C (Schroeder and Tilton, 1961; Tilton and Schroeder, 1963; Kirkpatrick *et al.*, 1972).

Kirkpatrick and Tilton (1972) measured mortalities of 12 species of stored product beetles in soft winter wheat and obtained > 99.5% mortality for all when treated at 65°C for less than 1 min. Kirkpatrick (1975) treated wheat infested with rice weevils and lesser grain borers in bulk with infrared and obtained > 93% mortality after 24 h exposure to 43.3°C. More recently, Subramanyam (2004) reported mortality from flameless catalytic infrared heaters on adults of the sawtoothed grain beetle, rice weevil and red flour beetle.

Microwave

The microwave (MW) region of the electromagnetic spectrum is from 1 to 100 GHz, between infrared and FM radio, and it is close to radio frequency range (see Fig. 1.2). Microwaves have been applied to a wide range of products, from soil and museum artefacts to fresh fruits. However, the most predominant efforts with current MW technology are in the control of pests of grain and stored products (Nelson 1973, 1996; Roseberg and Bögl, 1987; Nelson *et al.*, 1998; Wang and Tang, 2001).

A number of studies developing MW grain drying systems noted that insect disinfestation was an additional benefit (Hamid and Boulanger, 1970; Boulanger *et al.*, 1971; Tilton and Vardell, 1982a, b). Langlinais (1989)

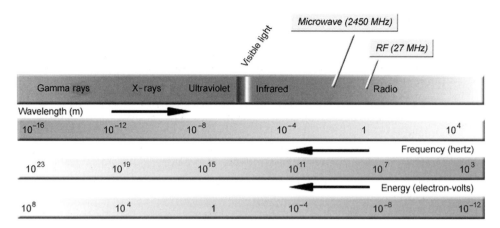

Fig. 1.2. The electromagnetic spectrum.

demonstrated that the confused flour beetle and flat grain beetle, *Cryptolestes pusillus* (Schnherr) (Coleoptera: Cucujidae), can be economically controlled by MW exposures. Other applications for MW in durable food commodities include treatment for pecan weevil in pecans (Nelson and Payne, 1982), red flour beetle, sawtoothed grain beetle, almond moth and Indianmeal moth in walnuts (Wilkin and Nelson, 1987) and almond moth in sun-dried figs (Baysal *et al.*, 1998).

Investigations into the MW applications of fresh horticultural commodities have been limited. In tests to control the mango weevil, *Cryptorhynchus mangiferae* (F.) (Coleoptera: Curculionidae), Seo *et al.* (1970) observed that MW exposures of 45 s resulted in cooked rind and pulp of treated mangoes. Sharp (1994b) described the equipment used to study the effects of MW on grapefruits infested with the Caribbean fruit fly, *Anastrepha suspense* (Loew) (Diptera: Tephritidae). Sharp *et al.* (1999) examined the effect of MW on mature larvae of the Caribbean fruit fly and concluded that rapid heating imposes serious constraints on the use of heat-induced mortality. Ikediala *et al.* (1999) improved treatment efficacy against the third-instar codling moth by adding 1–2 day cold storage after MW treatments.

MW energy has been explored in the treatment of wood and wood products (Hightower *et al.*, 1974; Burdette *et al.*, 1975; Mashek, 1998) and other non-food items (Reagan *et al.*, 1980; Hall, 1981). Lewis and Haverty (1996) compared MW treatments to five other methods of controlling the drywood termite and obtained 99% mortality 4 weeks after treatment.

To control the Asian longhorned beetle, *Anoplophora glabripennis* (Motschulsky) (Coleoptera: Cerambycidae), Fleming *et al.* (2003) found that MWs heated wood to the controlling temperature of 60°C within 5 min compared to 123 min with conventional heating, and recommended MW treatment to eradicate Asian longhorned beetles in solid wood-packing materials. Philbrick (1984) was concerned about seed viability and slight morphological changes in specimens when MWs were used for herbarium pest control. As with the other applications, MWs work best on dried subjects.

Radio frequency

Radio frequency (RF) waves are at the lower frequency range of the electromagnetic spectrum, with longer wavelength, and the most commonly accepted frequencies used for industrial purposes are 13.56 MHz, 27.12 MHz and 40.68 MHz (Tang *et al.*, 2000). RF energy generates internal heat by resistance from a very rapid change in molecular polarity and migration of charged ions.

The advantages of RF heating are: (i) it is very fast; (ii) it can penetrate deep into the target material because of its long wavelength; (iii) it can produce differential heating between the product and the pest; and (iv) it leaves no toxic residues.

Other than investigations with grain insects (Nelson and Kantack, 1966; Nelson, 1996) and recent progress described later in this volume, RF has not maintained the attention it received in the middle of the last century. Early reviews are provided by Ark and Parry (1940), Webber *et al.* (1946), Frings (1952) and Thomas (1952). Later commentaries are given by Whitney *et al.* (1961) and Watters (1962).

Recent pest control efforts using RF have targeted fresh fruits and nuts (Ikediala *et al.*, 2000; Mitcham *et al.*, 2004; Wang *et al.*, 2005b). Wang *et al.* (2005a) demonstrated the commercial feasibility of RF methodologies to control stored-product pests of walnuts at a walnut-packing house. In Chapters 12 and 13 respectively, this volume, the properties of electromagnetic energy treatments and their potential as commercial treatments will be discussed more thoroughly. A comparison of the heat treatment strategies described above is given in Table 1.3.

Table 1.3. Comparison of heat treatment strategies.

Strategy	First used	Commodity	Advantages	Disadvantages
Hot water	1925	Fruits, bulbs, ornamentals, seeds	Simplest, efficient	Surface heating first; fuel costs
Vapour heat	1913	Fruits, vegetables	Relatively simple	Expensive facilities; surface heating first; slow
Forced hot air	1989	Fruits, vegetables	Product quality retained	Expensive facilities; surface heating first; slow
Dry heat	1792	Structures, grains, fibres, museum artefacts, books	Simple; versatile, known technology	Surface heating first; slow
CATTS	1996	Experimental	Faster than other air methods	Surface heating first; complicated, expensive facilities
Solar	1983	Experimental, structures	Simple, inexpensive	Variable effects
Electromagnetic energy	1927	Experimental, grains, seeds, nuts	Very fast; internal heating first	Expensive facilities; variable effects

1.4 Heat Treatments for Microbial Control

Heat treatments also affect microbial populations on commodities and the incidence of natural decay. Heat treatments are applied to some commodities specifically for decay control, while in other cases the heat treatment is applied for the purpose of insect control and has a secondary effect on decay susceptibility. The secondary effect on decay could be beneficial, resulting in a lower incidence of decay, or detrimental, increasing decay susceptibility. The tolerance of the commodity to the heat treatment has the greatest influence on resultant decay susceptibility. In Chapter 7, this volume, the use of heat treatments for microbial control will be discussed.

1.5 Tolerance of Commodities to Heat Treatments

The tolerance of commodities to heat for insect or decay control must be carefully considered in development of such treatments. If the commodity quality is compromised significantly, the treatment will be unsuccessful. Commodity tolerance to the various methods of heating (water, air, RF, etc.) varies. In some cases, product quality and postharvest life is improved by heat treatment: ripening can be delayed, extending storage life; decay incidence may be reduced.

However, most heat treatments, particularly those designed for control of internal insect pests, cause some detrimental effects on product quality. There are some strategies for increasing the tolerance of commodities. The overall goal is to minimize detrimental effects while maintaining treatment efficacy. See Chapter 4, this volume, for detailed information about the tolerance of commodities to heat treatment and Chapter 11, this volume, for a discussion of induced heat tolerance.

1.6 Conclusions

Insects affect the storage, marketing and trade of food products, resulting in considerable economic loss. Methods to prevent or control insect infestations are needed to maintain food quality and allow free movement of produce. Although chemical fumigation is still the most commonly used control method, environmental and safety issues concerning fumigants provide a strong incentive for the development of alternative, non-chemical methods.

The wide variety of heat treatments currently available or in development show great potential as alternatives, but much work is needed to understand better the biological and physical processes involved. This book provides the foundation for advancing the application of postharvest heat treatments for effective pest control.

1.7 References

Akdoğan, H., Casada, M.E., Dowdy, A.K. and Subramanyam, Bh. (2005) A novel method for analyzing grain facility heat treatment data. *Journal of Stored Products Research* 41, 175–185.

AliNiazee, M.T. (1972) Susceptibility of the confused and red flour beetles to anoxia produced by helium and nitrogen at various temperatures. *Journal of Economic Entomology* 65, 60–64.

Anon. (1990) *Public Law 101–549*. Federal Clean Air Act enacted 15 November 1990. Washington, DC.

Anon. (1995) Montreal protocol on substances that deplete the ozone layer. *Report of Methyl Bromide Technical Options Committee: 1995 Assessment.* United Nations Environmental Program, Ozone Secretariat, Nairobi, Kenya.

Anon. (2005) ExO$_2$ by fumigation in a natural way – wood. *ExO$_2$* (The Netherlands, http://www.eco2.nl/UK/wood.htm).

Ark, P.A. and Parry, W. (1940) Application of high-frequency electrostatic fields in agriculture. *Quarterly Review of Biology* 15, 172–191.

Armstrong, J.W. (1994a) Commodity resistance to infestation by quarantine pests. In: Sharp, J.L. and Hallman, G.J. (eds) *Quarantine Treatments for Pests of Food Plants.* Westview Press, Boulder, Colorado, pp. 199–211.

Armstrong, J.W. (1994b) Heat and cold treatments. In: Paull, R.E. and Armstrong, J.W. (eds) *Insect Pests and Horticultural Products: Treatments and Responses.* CAB International, Wallingford, UK, pp. 103–119.

Armstrong, J.W., Hansen, J.D., Hu, B.K.S. and Brown, S.A. (1989) High-temperature, forced-air quarantine treatment for papayas infested with tephritid fruit flies (Diptera: Tephritidae). *Journal of Economic Entomology* 82, 1667–1674.

Arogba, S.S., Ugwu, F.M. and Abu, J.D. (1998) The effects of sun-drying surfaces and packaging materials on the storability of cowpea (*Vigna unguiculata*) seed. *Plant Foods for Human Nutrition* 53, 113–120.

Back, E.A. and Cotton, R.T. (1924) Relative resistance of the rice weevil, *Sitophilus oryza* L., and the granary weevil, *S. granaries* L., to high and low temperatures. *Journal of Agricultural Research* 28, 1043–1044.

Bailey, S.W. and Banks, H.J. (1980) A review of recent studies of the effects of controlled atmospheres on stored-product pests. In: Shejbal, J. (ed.) *Controlled Atmosphere Storage of Grains.* Elsevier, Amsterdam, pp. 101–118.

Baker, A.C. (1939) The basis for treatment of products where fruitflies are involved as a condition for entry into the USA. *USDA Circular* 551, 1–7.

Baker, A.C. (1952) The vapor-heat process. In: USDA (ed.) *Insects: the Yearbook of Agriculture.* US Government Printing Office, Washington, DC, pp. 401–404.

Banks, H.J. and Annis, P.C. (1977) Suggested procedures for controlled atmosphere storage of dry grain. *Commonwealth Scientific and Industrial Research Organization, Australia, Division of Entomology Technical Paper* 13, 1–23.

Baskin, B. (2001) Solar bagging: putting sunlight to work to eliminate insect infestations in mere hours. *Western Association for Art Conservation* 23 (2), 4 pp (http://palimpsest.stanford. edu/waac/wn/ wn23/wn23-2/wn23-207.html).

Baysal, T., Ural, A., Çakir, M. and Özen, Ç. (1998) Microwave application for the control of dried fig moth. *Acta Horticulturae* 480, 215–219.

Bergwerff, F. and Vroom, N. (2003) Exterminating insects and vermin in raw materials. *Grain & Feed Milling Technology* November/December, 28–29.

Boina, D. and Subramanyam, Bh. (2004) Relative susceptibility of *Tribolium confusum* life stages exposed to elevated temperatures. *Journal of Economic Entomology* 97, 2168–2173.

Boulanger, R.J., Boerner, W.M. and Hamid, M.A.K. (1971) Microwave and dielectric heating systems. *Milling* 153 (2), 18–21, 24–28.

Brokerhof, A.W. (2003) The solar tent – cheap and effective pest control in museums. *Australian Institute for the Conservation of Cultural Materials Bulletin* 28, 39–46.

Burdette, E.C., Hightower, N.C., Burns, C.P. and Cain, F.L. (1975) Microwave energy for wood products insect control. *Proceedings of the Microwave Power Symposium* 10, 276–281.

Burditt, A.K. (1994) Irradiation. In: Sharp, J.L. and Hallman, G.J. (eds) *Quarantine Treatments for Pests of Food Plants*. Westview Press, Boulder, Colorado, pp. 101–116.

Carpenter, A. and Potter, M. (1994) Controlled atmospheres. In: Sharp, J.L. and Hallman, G.J. (eds) *Quarantine Treatments for Pests of Food Plants*. Westview Press, Boulder, Colorado, pp. 171–198.

CDFA (California Department of Food and Agriculture) (1998) Cherry fruit fly. In: *Plant Quarantine Manual*. CDFA, Sacramento, California, pp. 305.1–305.3.

Chauhan, Y.S. and Ghaffar, M.A. (2002) Solar heating of seeds – a low cost method to control bruchid (*Callosobruchus* spp.) attack during storage of pigeonpea. *Journal of Stored Products Research* 38, 87–91.

Chervin, C.C., Jessup, A., Hamilton, A., Kreidl, S., Kulkarni, S. and Franz, P. (1998) Non-chemical disinfestations: combining the combinations – additive effects of the combination of three postharvest treatments on insect mortality and pome fruit quality. *Acta Horticulturae* 464, 273–278.

Chew, V. (1994) Statistical methods for quarantine treatment data analysis. In: Sharp, J.L. and Hallman, G.J. (eds) *Quarantine Treatments for Pests of Food Plants*. Westview Press, Boulder, Colorado, pp. 33–46.

Chinwada, P. and Giga, D.P. (1996) Sunning as a technique for disinfesting stored beans. *Postharvest Biology and Technology* 9, 335–342.

Chittenden, F.H. (1897) Some insects injurious to stored grain. *USDA Farmer's Bulletin* 45, 1–24.

Cohen, M. (1967) Hot-water treatment of plant material. *Ministry of Agriculture, Fisheries and Food Bulletin* 201, 1–39.

Cotton, R.T. (1963) *Pests of Stored Grain and Grain Products*. Burgess Publishing Co., Minneapolis, Minnesota, 318 pp.

Couey, H.M. and Hayes, C.F. (1986) Quarantine procedure for Hawaiian papaya using fruit selection and a two-stage hot-water immersion. *Journal of Economic Entomology* 79, 1307–1314.

Couey, H.M., Armstrong, J.W., Hylin, J.W., Thornburg, W., Nakamura, A.N., Linse, E.S., Ogata, J. and Vetro, R. (1985) Quarantine procedure for Hawaii papaya, using a hot-water treatment and high-temperature, low-dose ethylene dibromide fumigation. *Journal of Economic Entomology* 78, 879–884.

Davis, J.H. (1933) Radio waves kill insect pests. *Scientific American* May, 272–273.

Dean, G.A. (1913) Mill and stored-grain insects. *Kansas State Agricultural College, Agricultural Experiment Station Bulletin* No. 189, 63 pp.

Delate, K.M., Brecht, J.K. and Coffelt, J.A. (1990) Controlled atmosphere treatments for control of sweetpotato weevil (Coleoptera: Curculionidae) in stored tropical sweet potatoes. *Journal of Economic Entomology* 83, 461–465.

Dentener, P.R., Peetz, S.M. and Birtles, D.B. (1992) Modified atmospheres for the postharvest disinfestations of New Zealand persimmons. *New Zealand Journal of Crop and Horticultural Science* 20, 203–208.

Dermott, T. and Evans, D.E. (1978) An evaluation of fluidized-bed heating as a means of disinfesting wheat. *Journal of Stored Product Research* 14, 1–12.

Ebeling, W. (1994) The thermal pest eradication system for structural pest control. *The IPM Practitioner* 16 (2), 1–7.

Ebeling, W., Forbes, C.F. and Ebeling, S. (1989) Heat treatment for powerpost beetles. *The IPM Practitioner* 11 (9), 1–4.

Evans, D.E., Thorpe, G.R., and Dermott, T. (1983) The disinfestations of wheat in a continuous-flow fluidized bed. *Journal of Stored Products Research* 19, 125–137.

Fields, P.G. (1992) The control of stored product insects and mites with extreme temperatures. *Journal of Stored Products Research* 28, 89–118.

Fields, P.G. and White, N.D.G. (2002) Alternative to methyl bromide treatments for stored-product and quarantine insects. *Annual Review of Entomology* 47, 331–359.

Fleming, M.R., Hoover, K., Janowiak, J.J., Fang, Y., Wang, X., Liu, W., Wang, Y., Hang, X., Agrawal, D., Mastro, V.C., Lance, D.R., Shield, J.E. and Roy, R. (2003) Microwave irradiation of wood packing material to destroy the Asian longhorned beetle. *Forest Products Journal* 53, 46–52.

Forbes, C.F. and Ebeling, W. (1987) Update: use of heat for elimination of structural pests. *The IPM Practitioner* 9 (8), 1–5.

Frings, H. (1952) Factors determining the effects of radio-frequency electromagnetic fields on insects and materials they infest. *Journal of Economic Entomology* 45, 396–408.

Gaffney, J.J. and Armstrong, J.W. (1990) High-temperature forced-air research facility for heating fruits for insect quarantine treatments. *Journal of Economic Entomology* 83, 1954–1964.

Gaffney, J.J., Hallman, G.J. and Sharp, J.L. (1990) Vapour heat research unit for insect quarantine treatments. *Journal of Economic Entomology* 83, 1965–1971.

Gaunce, A.P., Madsen, H.F. and McMullen, R.D. (1981) Fumigation with methyl bromide to kill larvae and eggs of codling moth in Lambert cherries *Journal of Economic Entomology* 74, 154–157.

Gould, W.P. (1994) Cold storage. In: Sharp, J.L. and Hallman, G.J. (eds) *Quarantine Treatments for Pests of Food Plants*. Westview Press, Boulder, Colorado, pp. 119–132.

Greany, P.D. (1994) Plant host status and natural resistance. In: Paull, R.E. and Armstrong, J.W. (eds) *Insect Pests and Horticultural Products: Treatments and Responses.* CAB International, Wallingford, UK, pp. 37–46.

Grossman, E.F. (1931) Heat treatment for controlling the insect pest of stored corn. *University of Florida Agricultural Experiment Station Bulletin* 239, 1–24.

Hadjinicolaou, J. (1931) Effect of certain radio waves on insects affect certain stored products. *Journal of the New York Entomological Society* 39, 145–150.

Hall, D.W. (1981) Microwave: a method to control herbarium insects. *Taxon* 30, 818–819.

Hallman, G.J. (1994) Controlled atmospheres. In: Paull, R.E. and Armstrong, J.W. (eds) *Insect Pests and Horticultural Products: Treatments and Responses.* CAB International, Wallingford, UK, pp. 121–136.

Hallman, G.J. and Armstrong, J.W. (1994) Heated air treatments. In: Sharp, J.L. and Hallman, G.J. (eds) *Quarantine Treatments for Pests of Food Plants.* Westview Press, Boulder, Colorado, pp. 149–163.

Hamid, M.A.K. and Boulanger, R.J. (1970) Control of moisture and insect infestation by microwave power. *Milling* 152 (5), 25, 28, 30, 31.

Hansen, J.D. (2001) Ultrasound treatments to control surface pests of fruit. *HortTechnology* 11, 186–188.

Hansen, J.D. and Hara, A.H. (1994) A review of postharvest disinfestations of cut flowers and foliage with special reference to tropicals. *Postharvest Biology and Technology* 4, 193–212.

Harein, P.K. and Press, A.F. (1968) Mortality of stored-peanut insects exposed to mixtures of atmospheric gases at various temperatures. *Journal of Stored Products Research* 4, 77–82.

Harukawa, C. (1941) Heat as means of controlling Angoumois grain-moth. II. Velocity of the rise of wheat temperature during heating. *Berichte des Ohara Instituts für Landwirtschaftliche Forschungen* 8, 455–464.

Hawkins, L.A. (1932) Sterilization of citrus fruit by heat. *Texas Citriculture* 9, 7–8, 21–22.

Headlee, T.J. (1911) Burn the chinch bug in winter quarters. *Kansas State Agricultural College, Experiment Station Circular* No. 19, 8 pp.

Headlee, T.J. and Burdette, R.C. (1929) Some facts relative to the effect of high frequency radio waves on insect activity.

Journal of the New York Entomological Society 37, 59–64.

Heaps, J.W. (1988) Turn on the heat to control insects. *Dairy and Food Sanitation* 8, 416–418.

Heaps, J.W. (1996) Heat for stored product insects. *The IPM Practioner* 18 (5/6), 18–19.

Heaps, J.W. and Black, T. (1994) Using portable rented electric heaters to generate heat and control stored product insects. *Association of Operative Millers Bulletin* July, 6408–6411.

Heather, N.W. (1994) Pesticide quarantine treatments. In: Sharp, J.L. and Hallman, G.J. (eds) *Quarantine Treatments for Pests of Food Plants.* Westview Press, Boulder, Colorado, pp. 89–100.

Hightower, N.C., Burdette, E.C. and Burns, C.P. (1974) Investigation of the use of microwave energy for weed seed and wood products insect control. *Georgia Institute of Technology Final Technical Report* Projects E-230–901, 1–53.

Husain, M.A. and Bhasin, H.D. (1921) Preliminary observations on lethal temperatures for the larvae of *Trogoderma khapra*, a pest of stored wheat. *Proceedings of the Fourth Entomological Meeting, Pusa*, February 1921, Calcutta, pp. 240–248.

Ikediala, J.N., Tang, J., Neven, L.G. and Drake, S.R. (1999) Quarantine treatment of cherries using 915 MHz microwaves: temperature mapping, codling moth mortality and fruit quality. *Postharvest Biology and Technology* 16, 127–137.

Ikediala, J.N., Tang, J., Drake, S.R. and Neven, L.G. (2000) Dielectric properties of apple cultivars and codling moth larvae. *Transactions of the American Society of Agricultural Engineers* 43, 1175–1184.

Jang, E.B. and Moffit, H.R. (1994) Systems approaches to achieving quarantine security. In: Sharp, J.L. and Hallman, G.J. (eds) *Quarantine Treatments for Pests of Food Plants.* Westview Press, Boulder, Colorado, pp. 225–237.

Kerbel, E.L., Mitchell, F.G. and Mayer, G. (1987) Effect of postharvest heat treatments for insect control on the quality and market life of avocados. *HortScience* 22, 92–94.

Kirkpatrick, R.L. (1975) Infrared radiation for control of lesser grain borers and rice weevils in bulk wheat (Coleoptera: Bostrichidae and Curculionidae). *Journal of the Kansas Entomological Society* 48, 100–104.

Kirkpatrick, R.L. and Tilton, E.W. (1972) Infrared radiation to control adult stored-product Coleoptera. *Journal of the Georgia Entomological Society* 7, 73–75.

Kirkpatrick, R.L., Brower, J.H. and Tilton, E.W. (1972) A comparison of microwave and infrared radiation to control rice weevil (Coleoptera: Curculionidae) in wheat. *Journal of the Kansas Entomological Society* 45, 434–438.

Kitch, L.W., Ntoukam, G., Shade, R.E., Wolfson, J.L. and Murdock, L.L. (1992) A solar heater for disinfesting stored cowpeas on subsistence farms. *Journal of Stored Product Research* 28, 261–267.

Kuznetzova, E.A. (1937) Study of the action of the high frequency field on insects. *Review of Applied Entomology (Series A)* 25, 154.

Landolt, P.J., Chambers, D.L. and Chew, V. (1984) Alternative to the use of probit-9 mortality as a criterion for quarantine treatments of fruit fly-infested fruit. *Journal of Economic Entomology* 77, 285–287.

Langlinais, S.J. (1989) Economics of microwave treated rice for controlling weevils. *American Society of Agricultural Engineers Paper* 89-3544, 1–16.

Latta, R. (1932) The vapour-heat treatment as applied to the control of narcissus pests. *Journal of Economic Entomology* 25, 1020–1026.

Lay-Yee, M., Whiting, D.C. and Rose, K.J. (1997) Response of 'Royal Gala' and 'Granny Smith' apples to high-temperature controlled atmosphere treatments for control of *Epiphyas postvittana* and *Nysius huttoni*. *Postharvest Biology and Technology* 12, 127–136.

Lewis, V.R. and Haverty, M.I. (1996) Evaluation of six techniques for control of the western drywood termite (Isoptera:

Kalotermitidae) in structures. *Journal of Economic Entomology* 89, 922–934.

Lutz, F.E. (1927) A much abused but still cheerful cricket. *Journal of the New York Entomological Society* 35, 307–308.

Mackie, D.B. (1931) Heat treatments of California fruits from the standpoint of compatibility of the Florida process. *Monthly Bulletin of the California Department of Agriculture* 20, 211–218.

MAFF (Ministry of Agriculture, Forestry and Fisheries) – Japan (1950) Plant protection law enforcement regulation. *Ministerial Ordinance* No. 73, Annexed Table 1. MAFF, Tokyo.

Mahroof, R., Subramanyam, Bh. and Eustace, D. (2003) Temperature and relative humidity profiles during heat treatment of mills and its efficacy against *Tribolium castaneum* (Herbst) life stages. *Journal of Stored Product Research* 39, 555–569.

Mahroof, R., Subramanyam, Bh. and Flinn, P. (2005) Reproductive performance of *Tribolium castaneum* (Coleoptera: Tenebrionidae) exposed to the minimum heat treatment temperature as pupae and adults. *Journal of Economic Entomology* 98, 626–633.

Mashek, B.H. (1998) Appropriate technology for drywood termite control. *The IPM Practitioner* 20 (2), 7–8.

McCoy, C.W., Terranova, A.C., Miller, W.R. and Ismail, M.A. (1994) Vapour heat treatment for the eradication of Fuller rose beetle eggs on grapefruit and its effect on fruit quality. *Proceedings of the Florida Horticultural Society* 107, 235–240.

McDonald, R.E. and Miller, W.R. (1994) Quality and condition maintenance. In: Sharp, J.L. and Hallman, G.J. (eds) *Quarantine Treatments for Pests of Food Plants.* Westview Press, Boulder, Colorado, pp. 249–277.

McFarlane, J.A. (1989) Preliminary experiments on the use of solar cabinets for thermal disinfestations of maize cobs and some observations on heat tolerance in *Prostephanus truncates* (Horn) (Coleoptera: Bostrichidae). *Tropical Science* 29, 75–89.

McLaren, G.F., Fraser, J.A. and McDonald, R.M. (1997) The feasibility of hot water disinfestations of summerfruit. *Proceedings of the New Zealand Plant Protection Conference* 50, 425–430.

McLaren, G.F., Fraser, J.A. and McDonald, R.M. (1999) Non-chemical disinfestations of a quarantine pest on apricots. *Acta Horticulturae* 488, 687–690.

Milliken, F.B. (1916) Methods of controlling grasshoppers. *Kansas State Agricultural College, Agricultural Experiment Station Bulletin* No. 215, 28 pp.

Mitcham, E.J., Veltman, R.H., Feng, X., de Castro, E., Johnson, J.A., Simpson, T.L., Biasi, W.V., Wang, S. and Tang, J. (2004) Application of radio frequency treatments to control insects in in-shell walnuts. *Postharvest Biology and Technology* 33, 93–100.

Moffit, H.R., Drake, S.R., Toba, H.H. and Hartsell, P.L. (1992) Comparative efficacy of methyl bromide against codling moth (Lepidoptera: Tortricidae) larvae in 'Bing' and 'Rainier' cherries and confirmation of efficacy of a quarantine treatment for 'Rainier' cherries. *Journal of Economic Entomology* 85, 1855–1858.

Moore, D., Lord, J.C. and Smith, S.M. (2000) Pathogens. In: Subramanyam, B. and Hagstrum, D.W. (eds) *Alternatives to Pesticides in Stored-Product IPM.* Kluwer Academic Publishers, Boston, Massachusetts, pp. 193–227.

Morgan, D.L. and Crocker, R.L. (1986) Survival of weevil-infested live oak acorns treated by heated water. *Texas Agricultural Experiment Station* PR-4412 (August), 1–5.

Mouromtseff, I.E. (1933) Oscillator kills grain weevils in a few seconds. *Electrical World* 102, 667.

Moy, J.H. (ed.) (1985) *Radiation Disinfestations of Food and Agricultural Products.* University Hawaii Press, Honolulu, Hawaii, 424 pp.

Murdock, L.L. and Shade, R.E. (1991) Eradication of cowpea weevil (Coleoptera: Bruchidae) in cowpeas by solar heating. *American Entomologist* 37, 228–231.

Nakayama, T.O.M., Allen, J.M., Cummins, S. and Wang, Y.Y.D. (1983) Disinfestation of dried foods by focused solar energy. *Journal of Food Processing and Preservation* 7, 1–8.

Nation, J.L. and Burditt, A.K. (1994) Irradiation. In: Paull, R.E. and Armstrong, J.W. (eds) *Insect Pests and Horticultural Products: Treatments and Responses.* CAB International, Wallingford, UK, pp. 85–102.

Navarro, S. and Calderon, M. (1979) Mode of action of low atmospheric pressures on *Ephestia cautella* (Wlk.) pupae. *Experientia* 35, 620–621.

Navarro, S., Donahaye, J.E., Dias, R., Azrieli, A., Rindner, M., Phillips, T., Noyes, R., Villers, P., Debruin, T., Truby, R. and Rodriquez, R. (2001) Application of vacuum in a transportable system for insect control. In: Donahaye, E.J., Navarro, S. and Leesch, J.G. (eds) *Proceedings of the International Conference of Controlled Atmospheres and Fumigation in Stored Products*, Fresno, California, 29 October–3 November 2000. Executive Printing Services, Clovis, California, pp. 308–315.

Nelson, S.O. (1973) Insect-control studies with microwaves and other radiofrequency energy. *Bulletin of the Entomological Society of America* 19, 157–163.

Nelson, S.O. (1996) Review and assessment of radio-frequency and microwave energy for stored-grain insect control. *Transactions of the American Society of Agricultural Engineers* 39, 1475–1484.

Nelson, S.O. and Kantack, B.H. (1966) Stored-grain insect control studies with radio-frequency energy. *Journal of Economic Entomology* 59, 588–594.

Nelson, S.O. and Payne, J.A. (1982) Pecan weevil control by dielectric heating. *Journal of Microwave Power* 17, 51–55.

Nelson, S.O., Bartley, P.G. and Lawrence, K.C. (1998) RF and microwave dielectric properties of stored-grain insects and their implications for potential insect control. *Transactions of the American Society of Agricultural Engineers* 41, 685–692.

Neven, L.G. (2004) Hot forced air with con- trolled atmospheres for disinfestations of fresh commodities. In: Dis, R. and Jain, S.M.S.M. (eds) *Production Practices and Quality Assessment of Food Crops. Vol. 4: Post Harvest Treatments.* Springer, New York, pp. 297–315.

Neven, L.G. (2005) Combined heat and controlled atmosphere quarantine treatments for control of codling moth in sweet cherries. *Journal of Economic Entomology* 98, 709–715.

Neven, L.G. and Mitcham, E.J. (1996) CATTS (Controlled Atmosphere/Temperature Treatment System): novel tool for the development of quarantine treatments. *American Entomologist* 42, 56–59.

Nevo, D. (1996) The desert locust, *Schistocerca gregaria*, and its control in the land of Israel and the near east in antiquity, with some reflections on its appearance in Israel in modern times. *Phytoparasitica* 24, 7–32.

Ntoukam, G., Kitch, L.W., Shade, R.E. and Murdock, L.L. (1997) A novel method for conserving cowpea germplasm and breeding stocks using solar disinfestations. *Journal of Stored Product Research* 33, 175–179.

Obenland, D., Neipp, P., Mackey, B. and Neven, L. (2005) Peach and nectarine quality following treatment with high-temperature forced air combined with controlled atmosphere. *HortScience* 40, 1425–1430.

Oosthuizen, M.J. (1935) The effect of high temperature on the confused flour beetle. *Minnesota Technical Bulletin* 107, 1–45.

Pearce, A. (2003) Heat eradication of insect infestations: the development of a low cost, solar heated treatment unit. *Australian Institute for the Conservation of Cultural Materials Bulletin* 28, 3–8.

Pepper, J.H. and Strand, A.L. (1935) Superheating as a control for cereal-mill insects. *Montana State College Agricultural Experiment Station Bulletin* 297, 1–26.

Philbrick, C.T. (1984) Comments on the use of microwave as a method of herbarium insect control: possible drawbacks. *Taxon* 33, 73–74.

Prusky, D., Fuchs, Y., Kobiler, I., Roth, I.,

Weksler, A., Shalom, Y., Fallik, E., Zauberman, G., Pesis, E., Akerman, M., Ykutiely, O., Weisblum, A., Regev, R. and Artes, L. (1999) Effect of hot water brushing, prochloraz treatment and waxing on the incidence of black spot decay caused by *Alternaria alternata* in mango fruits. *Postharvest Biology and Technology* 15, 165–174.

Quarles, W. (1994) Pest control operators and heat treatment. *The IPM Practitioner* 16 (2), 8.

Reagan, B.M., Chiao-Cheng, J.-H. and Streit, N.J. (1980) Effects of microwave radiation on the webbing clothes moth, *Tineola bisselliella* (Humm.) and textiles. *Journal of Food Protection* 43, 658–663.

Riherd, C., Nguyen, R. and Brazzel, J.R. (1994) Pest free areas. In: Sharp, J.L. and Hallman, G.J. (eds) *Quarantine Treatments for Pests of Food Plants*. Westview Press, Boulder, Colorado, pp. 213–223.

Robertson, J.L., Preisler, H.K. and Frampton, E.R. (1994a) Statistical concept and minimum threshold concept. In: Paull, R.E. and Armstrong, J.W. (eds) *Insect Pests and Horticultural Products: Treatments and Responses*. CAB International, Wallingford, UK, pp. 47–65.

Robertson, J.L., Preisler, H.K., Frampton, E.R. and Armstrong, J.W. (1994b) Statistical analyses to estimate efficacy of disinfestations treatments. In: Sharp, J.L. and Hallman, G.J. (eds) *Quarantine Treatments for Pests of Food Plants*. Westview Press, Boulder, Colorado, pp. 47–65.

Roseberg, U. and Bögl, W. (1987) Microwave pasteurization, sterilization, blanching, and pest control in the food industry. *Food Technology* June, 92–99.

Sala, F.J., Burgos, J., Condón, S., Lopez, P. and Raso, J. (1999) Effect of heat and ultrasound on microorganisms and enzymes. In: Gould, G.W. (ed.) *New Methods of Food Preservation*. Chapman & Hall, New York, pp. 176–204.

Saxena, B.P., Sharma, P.R., Thappa, R.K. and Tikku, K. (1992) Temperature induced sterilization for control of three stored grain beetles. *Journal of Stored Products Research* 28, 67–70.

Schöller, M. and Flinn, P.W. (2000) Parasites and predators. In: Subramanyam, B.H. and Hagstrum, D.W. (eds) *Alternatives to Pesticides in Stored-Product IPM*. Kluwer Academic Publishers, Boston, Massachusetts, pp. 229–271.

Schroeder, H.W. and Tilton, E.W. (1961) Infrared radiation for the control of immature insects in kernels of rough rice. *USDA Agricultural Marketing Service* 445, 110.

Seo, S.T., Chambers, D.L., Komura, M. and Lee, C.Y.L. (1970) Mortality of mango weevils in mangoes treated by dielectric heating. *Journal of Economic Entomology* 63, 1977–1978.

Seo, S.T., Hu, B.K.S., Komura, M., Lee, C.Y. and Harris, E.J. (1974) *Dacus dorsalis* vapour heat treatment in papayas. *Journal of Economic Entomology* 67, 240–242.

Sharp, J.L. (1986) Hot-water treatment for control of *Anastrepha suspense* (Diptera: Tephritidae) in mangoes. *Journal of Economic Entomology* 79, 706–708.

Sharp, J.L. (1994a) Hot water immersion. In: Sharp, J.L. and Hallman, G.J. (eds) *Quarantine Treatments for Pests of Food Plants*. Westview Press, Boulder, Colorado, pp. 133–147.

Sharp, J.L. (1994b) Microwaves as a quarantine treatment to disinfest commodities. In: Champ, B.R., Highley, E. and Johnson, G.I. (eds) *Postharvest Handling of Tropical Fruits*. Australian Centre for International Agricultural Research Proceedings No. 56, Canberra, Australia, pp. 362–364.

Sharp, J.L. and Picho-Martinez, H. (1990) Hot-water quarantine treatment to control fruit flies in mangoes imported into the USA from Peru. *Journal of Economic Entomology* 83, 1940–1943.

Sharp, J.L., Ouye, M.T., Thalman, R., Hart, W., Ingle, S. and Chew, V. (1988) Submersion of 'Francis' mango in hot water as a quarantine treatment for the West Indian fruit fly and the Caribbean fruit fly (Diptera: Tephritidae) in mangoes. *Journal of Economic Entomology* 81, 1431–1436.

Sharp, J.L., Ouye, M.T., Hart, W., Ingle, S., Hallman, G., Gould, W. and Chew, V.

(1989a) Immersion of Florida mangoes in hot water as a quarantine treatment for Caribbean fruit fly (Diptera: Tephritidae). *Journal of Economic Entomology* 82, 186–188.

Sharp, J.L., Ouye, M.T., Ingle, S.J. and Hart, W.G. (1989b) Hot-water quarantine treatment for mangoes from Mexico infested with Mexican fruit fly and West Indian fruit fly (Diptera: Tephritidae). *Journal of Economic Entomology* 82, 1657–1662.

Sharp, J.L., Ouye, M.T., Ingle, S.J., Hart, W.G., Enkerlin, H., W.R., Celedonio H., H., Toledo A., J., Stevens, L., Quintero, E., Reyes F., J. and Schwarz, A. (1989c) Hot-water quarantine treatment for mangoes from the state of Chiapas, Mexico, infested with Mediterranean fruit fly and *Anastrepha serpentine* (Wiedemann) (Diptera: Tephritidae). *Journal of Economic Entomology* 82, 1663–1666.

Sharp, J.L., Gaffney, J.J., Moss, J.I. and Gould, W.P. (1991) Hot-air treatment device for quarantine research. *Journal of Economic Entomology* 84, 520–527.

Sharp, J.L., Robertson, J.L. and Preisler, H.K. (1999) Mortality of mature third-instar Caribbean fruit fly (Diptera: Tephritidae) exposed to microwave energy. *Canadian Entomologist* 131, 71–77.

Shellie, K.C., Neven, L.G. and Drake, S.R. (2001) Assessing 'Bing' sweet cherry tolerance to a heated controlled atmosphere for insect pest control. *HortTechnology* 11, 308–311.

Sheppard, K.O. (1984) Heat sterilization (superheating) as a control for stored-grain pests in a food plant. In: Baur, F.J. (ed.) *Insect Management for Food Storage and Processing.* American Association of Cereal Chemists, St Paul, Minnesota, pp. 194–200.

Shorey, H.H., Ferguson, L. and Wood, D.L. (1989) Solar heating reduces insect infestations in ripening and drying figs. *HortScience* 24, 443–445.

Sinclair, W.B. and Lindgren, D.L. (1955) Vapour heat sterilization of California citrus and avocado fruits against fruit-fly insects. *Journal of Economic Entomology* 48, 133–138.

Soderstrom, E.L., Mackey, B.E. and Brandl, D.G. (1986) Interactive effects of low-oxygen atmospheres, relative humidity, and temperature on mortality of two stored-product moths (Lepidoptera: Pyralidae). *Journal of Economic Entomology* 79, 1303–1306.

Songa, J.M. and Rono, W. (1998) Indigenous methods for bruchid beetle (Coleoptera: Bruchidae) control in stored beans (*Phaseolus vulgaris* L.). *International Journal of Pest Management* 44, 1–4.

Stark, J.D. (1994) Chemical fumigants. In: Paull, R.E. and Armstrong, J.W. (eds) *Insect Pests and Horticultural Products: treatments and responses.* CAB International, Wallingford, UK, pp. 69–84.

Storey, C.L. (1975) Mortality of adult stored-product insects in an atmosphere produced by an exothermic inert atmosphere generator. *Journal of Economic Entomology* 68, 316–318.

Storey, C.L. (1977) Effect of low oxygen atmospheres on mortality of red and confused flour beetles. *Journal of Economic Entomology* 70, 253–255.

Stout, O.O. and Roth, H.L. (1983) *International Plant Quarantine Manual.* FAO Plant Production and Protection Paper 50, FAO-UN, Rome, 195 pp.

Subramanyam, Bh. (2004) Hot technology for killing insects. *Milling Journal* First Quarter, 48–50.

Tang, J., Ikediala, J.N., Wang, S., Hansen, J.D. and Cavalieri, R.P. (2000). High-temperature, short-time thermal quarantine methods. *Postharvest Biology and Technology* 21, 129–145.

Thomas, A.M. (1952) Pest control by high-frequency electric fields – critical résumé. *Electrical Research Association Technical Report* W/T23, 7–40.

Tilton, E.W. and Schroeder, H.W. (1963) Some effects of infrared irradiation on the mortality of immature insects in kernels of rough rice. *Journal of Economic Entomology* 56, 727–730.

Tilton, E.W. and Vardell, H.H. (1982a) Combination of microwaves and partial vacuum for control of four stored-product insects in stored grain. *Journal of the*

Georgia Entomological Society 17, 106–112.

Tilton, E.W. and Vardell, H.H. (1982b) An evaluation of a pilot-plant microwave vacuum drying unit for stored-product insect control. *Journal of the Georgia Entomological Society* 17, 133–138.

Tsuchiya, T. (1943) Experimental studies on the resistance of the rice weevil, *Calandra oryzae* L., to heat. I. Resistance of adult weevil to heat. *Berichte des Ohara Instituts für Landwirtschaftliche Forschungen* 9, 170–190.

Ugwu, F.M., Ekwu, F.C. and Abu, J.O. (1999) Effect of different sun-drying surfaces on the functional properties, cooking and insect infestation of cowpea seeds. *Bioresource Technology* 69, 87–90.

Ulrey, D. (1936) New electronic tubes and new uses. *Physics* 7, 97–105.

USDA (USA Department of Agriculture) (2004) *USDA–Animal and Plant Health Inspection Service–Plant Protection and Quarantine Manual: Hawaii*. Riverdale, Maryland, 26 pp.

USDA (USA Department of Agriculture) (2005a) *USDA–Animal and Plant Health Inspection Service–Plant Protection and Quarantine Manual: Regulating the Importation of Fresh Fruits and Vegetables*. Riverdale, Maryland, 306 pp.

USDA (2005b) *USDA–Animal and Plant Health Inspection Service–Plant Protection and Quarantine Treatment Manual: Nonchemical Treatments*. Riverdale, Maryland, pp. 3.1.1–3.6.10.

USDA (2005c) *USDA–Animal and Plant Health Inspection Service–Plant Protection and Quarantine Treatment Manual: T106 – Vapour Heat*. Riverdale, Maryland, pp. 5.2.69–5.2.75.

USEPA (US Environmental Protection Agency) (2001) Protection of stratospheric ozone: process for exempting quarantine and pre-shipment applications of methyl bromide. *Rules and Regulations, Federal Register* 66 (139), 37752–37769.

Vermeire, L.T., Mitchell, R.B., Fuhlendorf, S.D. and Westerm D.B. (2004) Selective control of rangeland grasshoppers with prescribed fire. *Journal of Range Management* 57, 29–33.

Vincent, C., Hallman, G., Panneton, B. and Fleurat-Lessard, F. (2003) Management of agricultural insects with physical control methods. *Annual Review of Entomology* 48, 261–281.

Walker, G.P., Morse, J.G. and Arpaia, M.L. (1996) Evaluation of a high-pressure washer for postharvest removal of California red scale (Homoptera: Diaspididae) from citrus scale. *Journal of Economic Entomology* 89, 148–155.

Wang, J.J., Tsai, J.H., Zhao, Z.M. and Li, L.S. (2001) Interactive effects of temperature and controlled atmosphere at biologically relevant levels on development and reproduction of the psocid, *Liposcelis bostrychophila* Badoonel (Psocoptera: Liposcelididae). *International Journal of Pest Management* 47, 55–62.

Wang, S. and Tang, J. (2001) Radio frequency and microwave alternative treatments for insect control in nuts: a review. *International Agricultural Engineering Journal* 10, 105–120.

Wang, S., Monzon, M., Tang, J., Johnson, J.A. and Mitcham, E. (2005a) Large-scale radio frequency treatments for insect control in walnuts. *Proceedings of the Tenth Annual International Research Conference on Methyl Bromide Alternatives and Emissions Reduction*, San Diego, California, 31 October –3 November .

Wang, S., Yue, J., Tang, J. and Chen, B. (2005b) Mathematical modelling of heating uniformity for in-shell walnuts subjected to radio frequency treatments with intermittent stirrings. *Postharvest Biology and Technology* 35, 97–107.

Watters, F.L. (1962) Control of insects in foodstuffs by high-frequency electric fields. *Proceedings of the Entomological Society of Ontario* 92, 26–32.

Wearing, C.H., Hansen, J.D., Whyte, C., Miller, C.E. and Brown, J. (2001) The potential for spread of codling moth (Lepidoptera: Tortricidae) via commercial sweet cherry fruit: a critical review and risk assessment. *Crop Protection* 20, 465–488.

Webber, H.H., Wagner, R.P. and Pearson, A.G. (1946) High-frequency electric fields as lethal agents for insects. *Journal of Economic Entomology* 39, 487–498.

Whiting, D.C. and Hoy, L.E. (1997) High-temperature controlled atmosphere and air treatments to control obscure mealy bug (Hemiptera: Pseudococcidae) on apples. *Journal of Economic Entomology* 90, 546–550.

Whiting, D.C. and Hoy, L.E. (1998) Effect of temperature establishment time on the mortality of *Epiphyas postvittana* (Lepidoptera: Tortricidae) larvae exposed to a high-temperature controlled atmosphere. *Journal of Economic Entomology* 91, 287–292.

Whiting, D.C., Foster, S.P. and Maindonald, J.H. (1991) Effects of oxygen, carbon dioxide, and temperature on the mortality responses of *Epiphyas postvittana* (Lepidoptera: Tortricidae). *Journal of Economic Entomology* 84, 1544–1549.

Whiting, D.C., O'Connor, G.M., van den Heuvel, J. and Maindonald, J.H. (1995) Comparative mortalities of six tortricid (Lepidoptera) species to two high-temperature controlled atmospheres and air. *Journal of Economic Entomology* 88, 1365–1370.

Whiting, D.C., Hoy, L.E., Maindonald, J.H., Connolly, P.G. and McDonald, R.M. (1998) High-pressure washing treatments to remove obscure mealy bug (Homoptera: Pseudococcidae) and light brown apple moth (Lepidoptera: Tortricidae) from harvested apples. *Journal of Economic Entomology* 91, 1458–1463.

Whiting, D.C., Jamieson, L.E. and Connolly, P.G. (1999a) Effects of sublethal tebufenozide applications on the mortality responses of *Epiphyas postvittana* (Lepidoptera: Tortricidae) larvae exposed to a high-temperature controlled atmosphere. *Journal of Economic Entomology* 92, 445–452.

Whiting, D.C., Jamieson, L.E., Spooner, K.J. and Lay-Yee, M. (1999b) Combination high-temperature controlled atmosphere and cold storage as quarantine treatment against *Ctenopseustis obliquana* and *Epiphyas postvittana* on 'Royal Gala' apples. *Postharvest Biology and Technology* 16, 119–126.

Whitney, W.K., Nelson, S.O. and Walkden, H.H. (1961) Effects of high-frequency electric fields on certain species of stored-grain insects. *USDA Market. Research Report* 455, 1–52.

Whitney, W.R. (1932) Radiothermy. *General Electric Review* 35, 410–413.

Wilkin, D.R. and Nelson, G. (1987) Control of insects in confectionery walnuts using microwaves. *British Crop Protection Council Monograph* 37, 247–254.

Yokoyama, V.Y. (1994) Fumigation. In: Sharp, J.L. and Hallman, G.J. (eds) *Quarantine Treatments for Pests of Food Plants*. Westview Press, Boulder, Colorado, pp. 67–87.

Yokoyama, V.Y., Miller, G.T. and Hartsell, P.L. (1990) Evaluation of a methyl bromide quarantine treatment to control codling moth (Lepidoptera: Tortricidae) on nectarine cultivars proposed for export to Japan. *Journal of Economic Entomology* 83, 466–471.

Zeichner, B.C., Hoch, A.L. and Wood, Jr., D.F. (1998) Heat and IPM for cockroach control. *The IPM Practitioner* 20 (2), 1–6.

2 Fundamental Heat Transfer Theory for Thermal Treatments

J. TANG[1] AND S. WANG[2]

Department of Biological Systems Engineering, Washington State University, Pullman, Washington, USA; e-mails: [1]jtang@wsu.edu; [2]shaojin_wang@wsu.edu

2.1 Introduction

Thermal treatment methods using hot water, vapour, hot air, microwave (MW) and radio frequency (RF) energy have been investigated extensively as alternatives to methyl bromide fumigation for disinfestation of agricultural commodities (Sharp et al., 1991; Yokoyama et al., 1991; Moffitt et al., 1992; Sokhansanj et al., 1992, 1993; Neven, 1994; Neven and Rehfield, 1995; Mangan et al., 1998; Tang et al., 2000; Wang et al., 2001a, 2002; Birla et al., 2004; Feng et al., 2004; Hansen et al., 2004). Insect pests in those commodities can be controlled by heat, but different heat treatment methods result in different heating rates and final temperature distributions in commodities, which further result in varying degrees of efficacy. While heat treatments can be used for a large range of agricultural commodities, for simplicity, we will use fruit as the subject of choice in presentation of heat transfer theory.

Conventional heating consists of convective heat transfer from the heating medium to the fruit surface and then conductive heat transfer from the surface to the fruit centre. A common difficulty with hot air or water heating methods is the slow rate of heat transfer, resulting in hours of treatment time (Hansen, 1992; Opoku et al., 2002). Generally, thermal energy delivered to the interior of a fruit is significantly influenced by its size, heating medium temperature and heating methods. Reported heating times for fruit centre temperatures to reach the required maximum temperatures for disinfestation range from 23 min for cherries to 6 h for apples.

As a result, external and internal damage caused by heat over long exposure times includes peel browning, pitting, poor colour development and abnormal softening (Lurie, 1998). In contrast, MW or RF energy (collectively known as dielectric heating) interacts directly with the entire volume of fruit and thus provides fast and volumetric heating. The common difficulties in dielectric heating are the unpredictable final temperatures and the heating non-uniformity in samples. There is a need to fully understand the influence of various factors on heat transfer in fruits to minimize adverse effects on fruit quality.

Heat transfer theory has been well established (Campbell, 1977; Incropera and DeWitt, 1996; Dincer, 1997; Holdsworth, 1997). However, little has been reported in the horticultural literature on the application of heat transfer theory to the study of heating rates for fruit disinfestation. Past research efforts on thermal quarantine treatments were mostly empirical and pest- and commodity-specific. Such tests are very labour-intensive and costly, and the results are only useful for the fruits tested under the specific conditions investigated.

Investigations of the influence of physical parameters on heat transfer rates can be based on fundamental heat transfer theory via computer simulation models. A major advantage of the computer simulation model is its ability to assess the effect of various physical parameters on heating profiles in fruits. In this chapter, a fundamental heat transfer mechanism in fruit will be discussed based on basic heat transfer methods such as conduction, convection and radiation. A common simulation model will be provided to predict the temperature profiles in fruits as affected by thermal properties and boundary conditions. With brief discussions of the heating mechanisms of dielectric heating, some typical cases of thermal treatments on fruit will provide detailed comparisons between dielectric and conventional heating.

2.2 Conventional Heat Transfer Theory

An understanding of heat transfer in fruit is necessary to determine the advantages and limitations of conventional heating methods and to control the target temperature in fruit within a specific range. The first law of thermodynamics, which is the energy conservation principle applied to a closed system, can be applied to the heat transfer in fruits. Based on the energy balance, the change in internal energy within the fruit results from the heat transferred into the fruit (Q_t: W) and heat generated within the fruit (Q_g, W), as shown in Fig. 2.1.

Since the solid system has been taken to be incompressible, the special form of the first law of thermodynamics can be written as:

$$\rho v C_p \frac{\partial T}{\partial t} = Q_t + Q_g \qquad (2.1)$$

where C_p is the specific heat of the fruit (J/kg/K), ρ is the fruit density (kg/m^3), v is a small control fruit volume (m^3), T is the fruit temperature (K) and t is the time (s). For conventional heating, the heat transfer into the fruit (Q_t) consists of three important heating modes: heat conduction, heat convection and thermal radiation.

Conduction

Conduction of relevance to this application takes place in solid materials and contacting points between two solid materials, which is defined as molecular

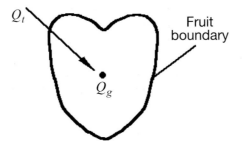

Fig. 2.1. Application of energy conservation to a fruit.

interchange; the vibration motion of one molecule causes neighbouring molecules to vibrate. Conduction can also be the result of electron drift, where heat transfer is analogous to electron flow in electricity over a very short distance with energy exchange (Mohsenin, 1980).

The phenomenological law governing this heat transfer is Fourier's law of heat conduction. For example, the heat across a surface area (A, m^2) of a plane wall or plate is proportional to the local temperature gradient dT (°C), as shown in Fig. 2.2:

$$Q_t = -kA \frac{dT}{dx} \tag{2.2}$$

where k is the thermal conductivity of the material (W/m/K) and x is the coordinate in the flow direction (m). The magnitude of the thermal conductivity for a given material depends on its microscopic structure and the applied temperatures. Under the same heat flux, small temperature gradients appear in larger heat conductivity materials such as metals, but larger temperature gradients exist in smaller heat conductivity materials such as fibreglass for insulation purposes.

Equation (2.2) is known as the one-dimensional Fourier's law of heat conduction expressed in Cartesian coordinates. Since different shapes of fruits are found, Fourier's law of heat conduction (Eq. 2.2) can be expressed in a variety of forms depending on the coordinate system being used. For an elementary volume (v, m^3) three-dimensional rectangular-shaped objects, Eq. 2.2 becomes:

$$Q_t = -kv \left(\frac{\partial^2 T}{\partial x^2} + \frac{\partial^2 T}{\partial y^2} + \frac{\partial^2 T}{\partial z^2} \right) \tag{2.3}$$

For the radial flow of heat in cylindrical-shaped objects with a radius of r (m), Eq. 2.2 becomes:

$$Q_t = -kv \left(\frac{\partial^2 T}{\partial r^2} + \frac{1}{r} \frac{\partial T}{\partial r} \right) \tag{2.4}$$

For spherical-shaped objects, Eq. 2.2 becomes:

$$Q_t = -kv \left(\frac{\partial^2 T}{\partial r^2} + \frac{2}{r} \frac{\partial T}{\partial r} \right) \tag{2.5}$$

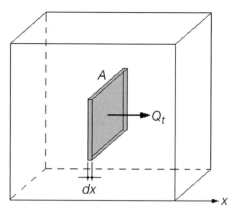

Fig. 2.2. Steady one-dimensional conduction across an elemental plane with a thickness of *dx* and surface area *A*.

Convection

Convection describes heat transfer from a solid surface to a moving fluid, for example hot air or water flowing over fruits. Whenever there is a temperature difference between the two, heat exchange occurs due to convection. Depending on whether the flow of the fluid is artificially or naturally induced, convective heat transfer is classified as either forced or free (or natural) convection. Forced convection includes use of mechanical means, such as a pump or a fan, to induce the movement of the fluid. In contrast, free convection occurs due to density differences in fluid caused by temperature gradients. Both these convective heat transfers include either a laminar or turbulent flow of fluid.

A fundamental analysis of convective heat transfer can be extremely complicated and may not be possible for many surface shapes because the rate of heat transfer by convection is usually a function of surface geometry, temperature, velocity and thermal properties. For this reason, an empirical approach is used to link the convective heat transfer coefficients to suitable physically significant non-dimensional groups.

The convective heat transfer between a fluid and fruit surface can be simplified to a function of the temperature difference between the fluid T_i and the fruit surface T_f, as shown below:

$$Q_t = h \cdot A(T_i - T_f) \tag{2.6}$$

where h is the convective heat transfer coefficient (W/m²/K). To use Eq. 2.6 effectively for describing convective heat transfer, a convenient method is required to determine h for common geometries encountered in heat transfer analysis.

With the further work established by Monteith (1973) and Campbell (1977), convective heat transfers through a laminar boundary layer of uniform thickness (δ [m]) between a solid surface and a fluid can be written as:

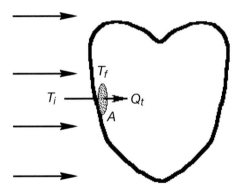

Fig. 2.3. Schematic view of convective heat transfer to a fluid at temperature T_i flowing past a surface area of A at temperature T_f.

$$Q_t = k \cdot A(T_i - T_f) / \delta \tag{2.7}$$

Since the boundary layer thickness δ cannot be measured directly, the above equation is rewritten by introducing the Nusselt (Nu) number for comparing the convective heat loss from similar solid surfaces at different scales:

$$Q_t = Nu \cdot k \cdot A(T_i - T_f) / d \tag{2.8}$$

with

$$Nu = d/\delta \tag{2.9}$$

Comparing Eq. 2.6 with 2.8 gives:

$$h = kNu / d \tag{2.10}$$

where d is defined as a characteristic length of the solid surface (m). This parameter is related to the shape of the studied objects and generally measures the length along which the laminar flow laps against the solid surface. The Nu number can be expressed as a function of other non-dimensional groups. In natural convection, heat transfer takes place through the motion of fluid induced by temperature gradients. In this case, the general expression for Nu can be written as a function of Grashof (Gr) and Prandtl (Pr) numbers (Monteith, 1973):

$$Nu = A(Gr \cdot Pr)^n \tag{2.11}$$

where A and n are constants depending on geometry and flow type and are determined by experiments. The non-dimensional Grashof number depends on the buoyancy and inertial and viscous forces and is calculated as follows:

$$Gr = \frac{a \cdot g \cdot d^3 \cdot (T_f - T_i)}{v^2} \tag{2.12}$$

where a is the coefficient of thermal expansion of the fluid (K^{-1}), and v is the kinematic viscosity of the fluid (m^2/s).

The non-dimensional Prandtl number, which is related to the momentum and heat transfer, is defined as:

$$Pr = \frac{\nu}{\kappa} \tag{2.13}$$

where κ is the thermal diffusivity of the fluid (m^2./s), and ν is the kinematic viscosity of the fluid (m^2/s). In forced convection, the Nu number is usually expressed as:

$$Nu = B \cdot Re^p \cdot Pr^q \tag{2.14}$$

where B, p and q are constants depending on geometry and flow type. The Reynolds number, Re, which is the ratio of inertia forces in a fluid to viscous forces, is written as:

$$Re = \frac{u \cdot d}{\nu} \tag{2.15}$$

The Nu for laminar and turbulent flows in natural and forced convection modes along a flat plate can be found in Table 2.1, as in Monteith (1973) and Campbell (1977).

In order to introduce these expressions (from Table 2.1) in our model, a criterion has to be defined for distinguishing the convective mode (forced and natural) and the type of flow (laminar and turbulent). Distinction between laminar and turbulent flow is based on Gr for natural convection and Re for forced convection. The ratio Gr/Re^2 introduces a criterion to distinguish natural from forced convection. When Re^2 is much larger then Gr, the buoyancy forces are negligible, forced convection is dominant while the inverse condition results in natural convection. The critical values of Gr/Re^2 are given in Table 2.2 for horizontal flat plates. This table can easily be used to set the criterion corresponding to a given fluid, such as air or water at 20°C.

With forced convection and turbulent flow, the convective heat transfer coefficient can be estimated based on boundary layer similarity for a sphere (Campbell, 1977; Dincer, 1997):

$$h = 0.34 \frac{k_f}{d} \left(\frac{ud}{\nu_f} \right)^{0.6} \tag{2.16}$$

The value of h may vary from 6 W/m^2/K in still air up to 80 W/m^2/K in moving air, and to 1200–6000 W/m^2/K in circulating water (Earle, 1983). Therefore, increasing air velocity, or using water instead of air, can increase heat transfer at the surface.

Table 2.1. Nusselt number calculation for convective heat exchange (from Monteith, 1973; Campbell, 1977).

	Laminar flow	Turbulent flow
Natural convection	$Nu = 0.54(Gr\,Pr)^{\frac{1}{4}}$	$Nu = 0.14(Gr\,Pr)^{\frac{1}{3}}$
Forced convection	$Nu = 0.67Re^{\frac{1}{2}}\,Pr^{\frac{1}{3}}$	$Nu = 0.036Re^{\frac{4}{5}}\,Pr^{\frac{1}{3}}$

Table 2.2. Criterion of convective mode and flow type (from Monteith, 1973; Campbell, 1977).

Choice criterion of convective mode			Laminar flow	Turbulent flow
General criterion	$\dfrac{Gr}{Re^2} < 0.1$	Forced convection	$Re < 5 \times 10^4$	$Re > 5 \times 10^4$
General criterion	$\dfrac{Gr}{Re^2} > 16$	Natural convection	$Gr < 10^8$	$Gr > 10^8$

Thermal Radiation

Radiation heat transfer takes place every day, such as in heat from a fire and toaster coils, etc. All materials continuously emit energy because of their temperature, and the emitted energy is defined as thermal radiation, which can be absorbed, reflected and transmitted depending on the radiation incident materials. The radiation emitted by a real body at an absolute temperature T (K) is given as:

$$Q_t = A\varepsilon\sigma T^4 \tag{2.17}$$

where ε is the surface emissivity and σ is the Stefan-Boltzmann constant (5.73×10^{-8}/W/m^2/K^4). For a highly polished surface, ε can be as low as 0.03, and for a black body ε is 1. When two materials at different temperatures 'see' each other through air, heat is exchanged between them by thermal radiation. The hot material experiences a net heat loss and the cold one a net heat gain as a result of the thermal radiation heat exchange. The net radiation loss ($Q_{t1,2}$, W) from materials 1 to 2 facing in parallel is given as:

$$Q_{t1,2} = \frac{A\sigma(T_1^4 - T_2^4)}{\dfrac{1}{\varepsilon_1} + \dfrac{1}{\varepsilon_2} - 1} \tag{2.18}$$

where T_1 and T_2 are the absolute temperature (K) of materials 1 and 2, respectively, and ε_1 and ε_2 are the surface emissivity of materials 1 and 2, respectively.

We will not focus on radiation in this chapter because it is negligible in the heating of fruit by hot air and water. Our focus is on the simulation of conventional heating, including convection from the fluid medium to the fruit surface and conduction from the surface to the fruit core, which should help to explain the temperature distribution and heating rate within fruit under different initial, boundary and operational conditions.

Modelling for Conventional Heating

For simplicity, we can consider here only spherically shaped fruits, such as apple, cherry, longan, orange and passion fruit. Heat transfer in these fruits can be considered as symmetric about the centre when they are exposed to uniform ambient air or water. Temperature distribution in these fruits is thus a function of the radial position and treatment time. In conventional heating, air and/or water is the source of thermal energy. Thermal energy is transferred from the heating medium to the fruit surface ($r = r_o$, radius of a fruit) by convection (Eq. 2.6), as described by the following boundary heat flux equation (Wang et al., 2001b):

$$-k\frac{\partial T}{\partial r}\bigg|_{r=r_o} = h\Big[T(r_o,t) - T_e\Big]$$

Heat flow   Heat flow from heating medium   (2.19)
into fruit   to fruit surface

where r is the radial co-ordinate (m), T is the fruit temperature (°C) and T_e is the ambient temperature (°C).

Once the thermal energy is transferred to the fruit surface, a large temperature gradient is established at the surface. Thermal energy is then transferred into the fruit interior by heat conduction. Temperature as a function of time at any location within a homogeneous and isotropic sphere is governed by a general energy balance equation (Wang et al., 2001b):

$$\rho C_p \frac{\partial T}{\partial t} = k\left(\frac{\partial^2 T}{\partial r^2} + \frac{2}{r}\frac{\partial T}{\partial r}\right) + \quad Q$$

(2.20)

Product   Heat conduction   Heat
heating rate   within fruit   generation

where Q is the heat generation within the fruit (W/m³). In conventional heating, the only likely source of heat generation within the fruit is respiration. However, relative to externally applied energy, the heat of respiration is very small. Thus, by setting $Q = 0$, dividing all items in Eq. 2.20 by ρC_p and substituting α for $k/\rho C_p$, Eq. 2.20 becomes:

$$\frac{\partial T}{\partial t} = \alpha\left(\frac{\partial^2 T}{\partial r^2} + \frac{2}{r}\frac{\partial T}{\partial r}\right)$$

(2.21)

where α is thermal diffusivity (m²/s). The right-hand side of Eq. 2.21 represents the rate of change of internal energy. Conductive heat transfer within fruit is slow due to the relatively small value of thermal diffusivity. As a result, the heating rate at the fruit centre can be very slow, especially for large fruits such as mango.

Thermal conductivity, specific heat capacity and density are three important engineering properties of a fruit related to heat transfer characteristics, together with the convective heat transfer coefficient. These parameters are essential in studying and analyzing heating and cooling processes for postharvest pest control. We will summarize the typical values of these parameters reported in the literature in the following section.

Physical Properties Related to Heat Transfer in Agricultural Commodities

The rate of heating within a fruit is a function of three major factors, including the thermal conductivity (k), density (ρ) and specific heat (C_p) of the fruit. These thermal properties of agricultural commodities make up thermal diffusivity ($a=k/\rho C_p$), which is the ratio of the heat conducted to the heat stored. Table 2.3 shows the typical thermal properties of selected materials at room temperature.

Generally, larger diffusivity results in faster heat conduction and smaller heat storage in materials of similar dimensions. The heat conduction in fruits is much slower than that in metals because of smaller α (1.6×10^{-7} m²/s) for fruits compared to $1.5–17 \times 10^{-5}$ m²/s for metals. The thermal diffusivity of fruits is close to that of water. Though thermal properties vary with temperature, these values for insect pest controls have a limited operational temperature range between 20 and 60°C. The thermal properties for other materials can be found in Incropera and DeWitt (1981).

Thermal properties of fruit may vary in different cultivars. Precise simulation depends on accurate values for the thermal properties of the targeted materials. For those thermal properties unavailable in the literature, direct measurement may be necessary. The average bulk density of a material can be obtained by dividing the sample weight recorded each time upon completion of sample loading using the known effective volume.

Thermal conductivity and specific heat can be measured by a line source method using either a bare wire or a thermal conductivity probe with a handheld Decagon thermal properties meter (Mohsenin, 1980; Yang *et al.*, 2002) and differential scanning calorimetry (DSC) (Tang *et al.*, 1991; Yang

Table 2.3. Thermal properties of selected materials at room temperature.

Material	Density ρ (kg/m³)	Specific heat C_p (J/kg/K)	Thermal conductivity k (W/m/K)	Thermal diffusivity α ($\times 10^{-7}$ m²/s)	Source
Metals					
Copper	8933	385	401	1166	Incropera and DeWitt (1981)
Aluminum	2702	903	237	971	Incropera and DeWitt (1981)
Fluids					
Air	1.16	1007	0.263	225	Incropera and DeWitt (1981)
Water	1000	4179	0.613	1.47	Incropera and DeWitt (1981)
Fruits/vegetables					
Apple (green)	790	3700	0.422	1.44	Rahman (1995)
Apple (red)	840	3600	0.513	1.70	Rahman (1995)
Cherry	1010	3643	0.511	1.39	Mohsenin (1980)
Cherry tomato	1010	3300	0.527	1.58	Rahman (1995)
Orange	1030	3661	0.580	1.54	Rahman (1995)
Papaya	N/A	3433	N/A	1.52	Hayes and Young (1989)
Pear	1000	3700	0.595	1.61	Rahman (1995)
Potato	1100	3515	0.560	1.45	Rahman (1995)

N/A, not available.

et al., 2002), respectively. Thermal diffusivity is thus calculated based on the measured thermal conductivity, density and specific heat.

Table 2.4 presents typical values of convective heat transfer coefficient under different conditions. Such values may vary for different geometry and conditions, but are the references for basic calculations. The *h* values are smaller in free convection than in forced convection. The fluid medium of water results in the largest *h* values compared with engine oil and air. Increasing the velocity of a medium increases *h* values.

The influence of surface and internal thermal resistances on heating rates can be described by the non-dimensional Biot number (Bi), defined as (Incropera and DeWitt, 1981):

$$Bi = \frac{\text{Internal resistance } (r_0/kA)}{\text{Surface resistance } (1/hA)} = \frac{hr_0}{k} \qquad (2.22)$$

When Bi < 0.1, the surface thermal resistance dominates the rate of heat transfer and the internal temperature gradients are small. When the Bi number is large (e.g. 10), internal thermal resistance plays the most important role and the internal temperature gradients are significant.

Effect of Typical Parameters on Heat Transfer

Influence of fruit thermal diffusivity

Thermal diffusivity is a general factor that thermal conductivity, specific heat and density in thermal processes. It is important to note from Eq. 2.21 that

Table 2.4. Typical values of the convective heat transfer coefficient.

Parameter	h (W/m²/K)	Source
Free convection		
0.25 m vertical plate in		Ozisik (1985)
Air	5	
Engine oil	37	
Water	440	
0.02 m diameter sphere		Ozisik (1985)
Air	8	
Engine oil	62	
Water	741	
Forced convection		Ozisik (1985)
Air at 25°C with speed of 10 m/s over a flat plate		
$L = 0.1$ m	39	
$L = 0.5$ m	17	
Flow at 5 m/s across 1 cm-diameter cylinder		
Air	85	
Engine oil	1800	
8 cm-diameter apples in		Tang *et al.* (2000)
Air heating at air velocity 1 m/s	17	
Air heating at air velocity 4 m/s	39	
Water heating at water velocity 1 m/s	1911	

thermal diffusivity is the only internal thermal property of fruit that governs the temperature variation within fruits. Although large variations may exist in the density (ρ), specific heat (C_p) and thermal conductivity (k) of different fruits and varieties (Table 2.3), the values of thermal diffusivity for fruits fall between 1.4×10^{-7} m^2/s and 1.7×10^{-7} m^2/s (Mohsenin, 1980).

Figure 2.4a shows the effect of thermal diffusivities on the heating rate at a fruit centre when subjected to hot air (55°C) at a velocity of 1 m/s. The higher the thermal diffusivity, the faster the heating rate. The heating periods are 84, 81 and 79 min to reach 50°C for thermal diffusivities of 1.4, 1.6 and 1.8 $\times 10^{-7}$ m^2/s, respectively, which may simulate different fruit type (from apple to pear). That is, regardless of the fruit variety or type, the heating rates are close for a given size of fruit when heated with hot air.

Similar simulation results are obtained for the diffusivity effect on the heating rate when subjected to hot water (55°C) circulating at 1 ms (Fig. 2.4b). It takes around 26, 23 and 21 min for a fruit of 6 cm diameter with the above three diffusivities, respectively, to reach 50°C. The difference in thermal diffusivity from its middle value (1.6 $\times 10^{-7}$ m^2/s) results in differences of about 4 and 13% in heating times for hot air and water treatments, respectively.

The different effects of thermal diffusivity during hot air or water heating result from the relative magnitude between the internal and surface heat resistance. The Bi number (Eq. 2.22) for hot air heating of a 6 cm-diameter fruit is between 1 and 3, but between 82 and 130 for hot water. This suggests that the internal heat resistance in fruit during water heating is a more dominant factor in controlling the heat transfer rate than with hot air.

As a result, variations in the thermal diffusivity cause a relatively large variation in heating time during hot water heating compared with hot air heating. However, this difference in heating time is small compared to the effect of fruit size or heating methods (see later sections). Overall, the variations in thermal diffusivity among fruit varieties and types may not influence heating time to cause any practical concern in real treatment systems.

Influence of fruit size

The effect of fruit sizes (3, 6 and 9 cm diameter) on heat transfer to a fruit centre is reported in detail in Wang *et al.* (2001b) using a simulation model. Figure 2.5a shows a fruit centre temperature as influenced by different fruit diameters when subjected to forced hot air (air temperature, 55°C; air velocity, 1 m/s). The centre temperatures of the fruit take about 28, 81, and 153 min to reach 50°C for diameters of 3, 6 and 9 cm, respectively. The heating process for small fruits is much faster than for large fruits, making it clear that the effect of fruit size is very important compared with variations in thermal diffusivity among fruit varieties.

When fruits of different sizes are subjected to hot water treatments (55°C; 1 m/s), it takes about 6, 23 and 52 min to reach 50°C for fruit diameters of 3, 6 and 9 cm, respectively (Fig. 2.5b). The heating time in water immersion is

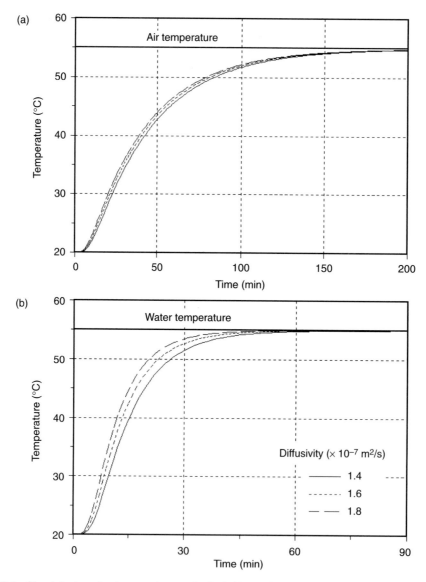

Fig. 2.4. Simulated centre temperatures of a fruit (6 cm diameter) influenced by thermal diffusivity when subjected to hot air (a) and hot water (b) at 55°C and 1 m/s (from Wang *et al.*, 2001b).

much shorter than with hot air heating. A 1 cm-diameter difference (17%) among medium-sized fruits results in a 27% and 35% difference in time for fruits to reach 50°C for hot air and hot water treatments, respectively. Little can be done to increase internal heat conduction. Therefore, sorting according to size is very important to help achieve uniform heating among fruits when using hot air or hot water treatments.

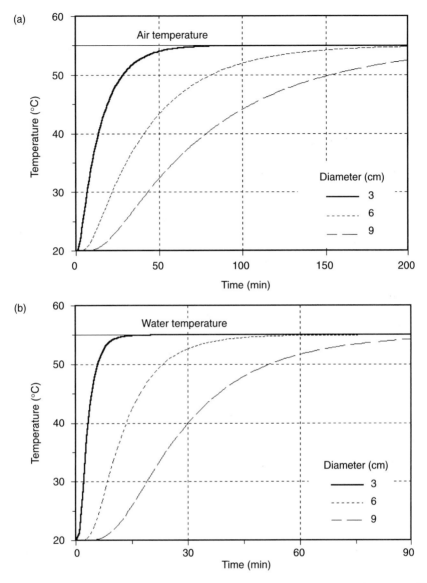

Fig. 2.5. Simulated centre temperatures of a fruit with a thermal diffusivity of 1.6×10^{-7} m²/s influenced by diameter when subjected to hot air (a) and hot water (b) at 55°C and 1 m/s (from Wang *et al.*, 2001b).

Influence of heating medium velocity

Computer simulation results are presented in Fig. 2.6 for a fruit of 6 cm diameter in hot air under different air circulating velocities (Wang *et al.*, 2001b). When heated in circulating air at 0.5 m/s and 55°C, the core temperatures of the fruit increase very slowly to approach the heating medium temperature. For example, it takes about 113 min for the centre temperature to

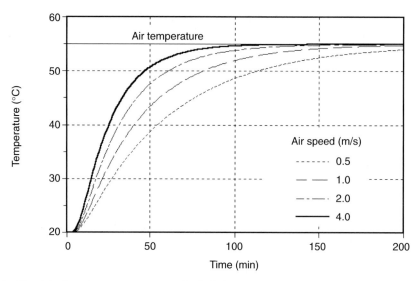

Fig. 2.6. Simulated centre temperatures of a fruit (6 cm diameter) with a thermal diffusivity of 1.6×10^{-7} m²/s influenced by air velocity when heated by hot air at 55°C (from Wang *et al.*, 2001b).

increase from 20 to 50°C (see Fig. 2.6). By increasing circulating air velocity to 1, 2 and 4 m/s, the time for the same centre temperature rise is reduced to about 81, 61 and 47 min, respectively.

The value of the Bi number for hot air heating is a little larger than 1, so the thermal resistance at the surface boundary layer is comparable to that within fruits. The effect of increasing air velocity on the heat transfer is the result of an increased heat transfer coefficient and increased Bi number as shown in Eqs 2.16 and 2.22. Attention should be paid to uniform air distribution to ensure uniform heating among fruit in bins and chambers. For example, if air velocities at the centre and corner of a chamber vary from 0.5 to 2 m/s, the heating time of the fruits at those positions will vary by about 46%.

When heated in a circulating water bath at 55°C, the heating rate of a fruit centre is significantly increased compared with hot air heating. Nevertheless, it still takes about 23 min for the centre temperature to rise to 50°C. Water circulating velocity has little effect on heating time because, in all cases, the Bi number is greater than 50, indicating that the surface thermal resistance is very small compared with the internal thermal resistance (Wang *et al.*, 2001b).

When using water as the heating medium, further reduction in the surface thermal resistance by increasing water circulation velocity will not increase the heating rate noticeably. In this case, the only benefit of water circulation is to ensure temperature uniformity among the individual fruit. Circulating water as a heating medium for fruits represents a practical best-case scenario in terms of heat transfer to deliver thermal energy to a fruit surface with conventional heating methods, but little can be done with heat conduction within fruits.

In practice, only a low circulating water velocity is needed to provide uniform water temperature in a conveying flume. Based on the criterion of delivering adequate thermal energy to a fruit's centre to kill infested insects, fruit surfaces are generally exposed to high temperatures for an extended period, which may cause severe and visible thermal damage (Wang *et al.*, 2001b).

2.3 Dielectric Heating

Heating Mechanisms

Dielectric heating is a term that covers radio frequency (RF) and microwave (MW) heating. RF and MW are high-frequency electromagnetic waves generated by vacuum tubes, magnetrons or klystrons. The rotational responses of polarized molecules to an alternating electric field (electronic polarization) and migration of charged ions (ionic polarization) are the main contributing mechanisms for heat generation in a material (Barber, 1983).

Generally, a distinction is made between RF dielectric heating (1–300 MHz) and MW heating (300–30,000 MHz) (Orfeuil, 1987), not only in the frequency ranges, but also in generators, applicators and related material dielectric properties. Because of the congested bands of RF and MW already being used for communication purposes, the US Federal Communications Commission (FCC) allocated several frequencies for industrial, scientific and medical (ISM) applications. The ISM band assignments are summarized in Table 2.5.

Although it is possible to use other frequencies for dielectric heating as long as the system provides adequate shielding to prevent leakage of electromagnetic waves, the added costs in design (especially for continuous systems), manufacturing and the maintenance of effective shielding can be prohibitive. It is common practice, therefore, to design dielectric heating equipment within the ISM bands. The wavelength of electromagnetic waves in free space decreases with increasing frequency (see Table 2.5). That is, the RF wavelength is longer than that of MW.

RF heating is used in many industries, such as plastic welding, curing of glue in plywood processing, textile drying and finish drying of bakery products (Orfeuil, 1987). There is a definite advantage in using RF heat in drying applications over conventional heating methods. Since electromagnetic energy tends to act on the water and aqueous ions in a material, it is dissipated to a greater extent in more damp regions. Once adequate thermal energy has been converted from electromagnetic energy, the water is evaporated or driven into dryer areas, and less energy is absorbed thereafter. This leads to a frequently desirable phenomenon known as the 'temperature levelling effect'; that is, the high-temperature portion of a material absorbs less electromagnetic energy than the low-temperature part.

A less desirable phenomenon in dielectric heating is thermal runaway. Many materials have dielectric loss factors that increase with temperature in certain frequency ranges, which causes those materials to absorb more energy

Table 2.5. Federal Communications Commission allocated frequency bands designated for ISM applications (from FCC, 1998).

Dielectric heating range	Centre frequency (MHz)	Deviation (MHz)	Deviation (%)	Wavelength in free space (m)
Radio frequency (RF)	6.78	0.015	0.221	44.2
	13.56	0.007	0.052	22.1
	27.12	0.163	0.60	11.1
	40.68	0.020	0.049	7.4
Microwave (MW)	915	13	1.421	0.328
	2,450	50	2.041	0.122
	5,800	75	1.293	0.052
	24,125	125	0.518	0.0124
	61,250	250	0.408	0.0049
	122,500	500	0.408	0.0024
	245,000	1,000	0.408	0.0012

as their temperatures increase. This serves as positive feedback during heating, by which the portion of the material at a higher temperature tends to absorb more energy, leading to even greater temperature differences. It is therefore essential that the electric field for those materials be uniform to ensure even heating. Since the temperature distribution within a material in a dielectric heating process is seldom as predictable as that for a conventional process, heat distribution and heat penetration studies are not as straightforward.

A unique feature of RF treatments is that RF energy is directly coupled to a dielectric material, such as most agricultural produce, to generate heat (Q in Eq. 2.20) within the material, thus eliminating or reducing the need for internal heat transfer. This can significantly increase heating rates and reduce heating time. The magnitude of heat generation is proportional to the relative loss factor, ε'', at a given frequency f (Hz), and the electric field intensity E (V/m):

$$Q = 5.563 \times 10^{-11} f\varepsilon'' E^2 \tag{2.23}$$

For a rapid and uniform heating process in which heat conduction is relatively small, Eq. 2.20 reduces to:

$$\rho C_p \frac{\partial T}{\partial t} = Q \text{ or simply, } \Delta T = \frac{Q}{\rho C_p} \Delta t \tag{2.24}$$

That is, the temperature increases linearly with treatment time. Adjusting RF power Q in Eq. 2.24 can control the rate of the temperature increase.

Dielectric Properties

Knowledge of dielectric properties, among other parameters, is helpful in gaining insight into the interaction between materials and MW or RF energy. Dielectric properties can be used in estimating the penetration depth of MW or RF waves in selected materials and predicting the uniformity of MW and RF heating. Permittivity is a quantity commonly used to describe dielectric properties that influence the reflection of electromagnetic waves at interfaces and the attenuation of wave energy within materials. The complex relative permittivity, ε^*, of a pure liquid material is given by the following equation, expressed in rectangular form (Murphy and Morgan, 1938):

$$\varepsilon^* = \varepsilon_\infty + \frac{\varepsilon_s - \varepsilon_\infty}{1 + \omega^2 \tau^2} - j\frac{(\varepsilon_s - \varepsilon_\infty)\omega\tau}{1 + \omega^2 \tau^2} \tag{2.25}$$

where ε_∞ is the infinite or high-frequency relative permittivity, ε_s is the static or zero frequency relative permittivity, ω is the angular frequency and τ is the time in s for relaxation of the dielectric material. Equation 2.25 can be expressed in the following complex form (Murphy and Morgan, 1938):

$$\varepsilon^* = \varepsilon' - j\varepsilon'' \tag{2.26}$$

The real part ε' is referred to as the dielectric constant (energy stored), which influences the electric field distribution and phase of waves travelling through a material. The imaginary part ε'', referred to as the dielectric loss factor, mainly influences energy absorption and attenuation. The mechanisms that contribute to the dielectric loss in heterogeneous mixtures include polar, electronic, atomic and Maxwell-Wagner responses (Metaxas and Meredith, 1993).

At the RF and MW frequencies of practical importance and current application in agricultural processing (RF: 1–50 MHz; MW: 915 and 2450 MHz), ionic conduction and dipole rotation dominate loss mechanisms (Ryynänen, 1995):

$$\varepsilon'' = \varepsilon_d'' + \varepsilon_\sigma'' = \varepsilon_d'' + \frac{\sigma}{\varepsilon_0 \omega} \tag{2.27}$$

where subscripts d and σ stand for contributions due to dipole rotation and ionic conduction, respectively ε_0 is the permittivity of free space or a vacuum (8.854×10^{-12} F·/m). The ionic conductivity σ (S·/m) of a material is usually considered as the property that depends on the ease with which electric charges migrate in the material upon the application of an electric field. An effective ionic conductivity can be expressed as (Nelson, 1991):

$$\sigma = \omega\varepsilon_0\varepsilon_\sigma'' \tag{2.28}$$

Penetration Depth

The penetration depth of MW and RF power is defined as the depth where the power is reduced to $1/e$ ($e = 2.718$) of the power entering the surface. The penetration depth (dp, m) of RF and MW in a lossy material can be calculated from dielectric properties, as per von Hippel (1954):

$$dp = \frac{c}{2\pi f \sqrt{2\varepsilon' \left[\sqrt{1 + \left(\frac{\varepsilon''}{\varepsilon'} \right)^2} - 1 \right]}}$$

(2.29)

where c is the speed of light in free space (3×10^8 m·/s). The calculated penetration depth of RF and MW in several fruits is listed in Table 2.6.

As shown in Eq. 2.29, the penetration depths of electromagnetic waves are inversely proportional to the frequency. For example, penetration depths of 27 and 915 MHz waves in Red Delicious apples are 18.9 and 4.6 cm, respectively (see Table 2.6). Penetration depths of electromagnetic waves in commodities are deeper in dry commodities. For example, RF at 27 MHz has an estimated penetration depth of 538 cm and 654 cm in almonds and walnuts, respectively.

Deep penetration depths in nuts at 27 MHz make it possible to design a continuous operation, with conveyor belts transporting multilayered products through two RF plate electrodes. Microwaves at 915 and 2450 MHz, however, may not be able to treat more than two layers of nuts because of smaller penetration depths (2–3 cm). Limited penetration depths lead to non-uniform heating in large agricultural commodities. This may in part explain the disappointing results reported by Seo *et al.* (1979) and Hayes *et al.* (1984) when using 2450 MHz MW to treat mangoes and papaya. Feng *et al.* (2002) studied the penetration depth of MW in Red Delicious apple slices as a function of moisture content at two temperatures (22 and 60°C) and found that the penetration depth increased significantly as moisture was removed from the sample.

RF energy has larger penetration depth in fresh fruits and nuts than MW because of its longer wavelength. In fact, there is generally no penetration depth limitation associated with RF heating of fruits in anticipated commercial applications in which single or multilayered produce would be treated in a continuous RF system.

Table 2.6. Penetration depths (cm) of fruits calculated from the measured dielectric properties at 20°C (from Wang *et al.*, 2003, 2005).

Fruit	27 MHz	915 MHz	1800 MHz
GD apple	15.2	5.3	2.2
RD apple	18.9	4.6	2.1
Almond	538.2	1.9	2.3
Cherry	8.5	2.8	1.4
Grapefruit	10.9	3.7	1.8
Orange	10.1	2.7	1.5
Walnut	653.6	3.1	2.3
Avocado	5.1	1.5	1.1
Cherimoya	9.4	1.5	1.1
Longan	9.7	3.3	1.5
Passion fruit	9.0	2.7	1.5
Persimmon	10.5	2.1	1.5
White sapote	9.0	1.8	1.3

GD, Golden delicious; RD, Red delicious.

Typical RF Systems

A typical RF heating system consists of two principal parts: an applicator and a RF power generator (Roussy and Pearce, 1995). Systems used for RF heating can generally be categorized into conventional and 50 Ω RF heating equipment, depending upon how the RF power is coupled into the material being heated. A conventional RF system consists of a transformer, rectifier, oscillator, inductance-capacitance pair (commonly referred to as the 'tank circuit') and the work circuit. The transformer raises the voltage to about 9 kV and the rectifier changes the alternating current to a direct current, which is then converted by the oscillator into RF energy.

The frequency of the RF energy is determined by the values of the inductance and capacitance in the tank circuit. In a conventional system, the RF applicator functions as a capacitor in the working circuit. The applicator can be one of three basic types: (i) parallel plates that are the most suitable for bulky materials (Fig. 2.7a); (ii) rod structures producing a stray-field for planar materials (Fig. 2.7b); and (iii) a stagger-through type for very thinly layered materials (Fig. 2.7c). In all those applicators, it is the distance of the electrodes, the geometry and the dielectric properties that determine the amount of RF power coupled to the sample.

Currently, conventional RF generators are available in power ranging up to 250 kW. It is possible to obtain even higher powers by placing several generators together using RF combiners. One manufacturer has plans to produce a RF heater using combiners to develop an output power of 2.5 megawatts for use in wood drying (HeatWave Technologies Inc., Vancouver, Canada).

Power amplifier generators, also known as 50 Ω generators, use a different approach for the coupling of RF power. In these systems, a stable, fixed-frequency oscillator supplies a RF signal to a power amplifier, which supplies power to the load. The output of the generator has a characteristic impedance of 50 Ω, and will transfer all of its power into a load that also has an impedance of 50 Ω. If the generator is presented with an impedance that is higher or lower than 50 Ω, a portion of the power will be reflected back toward the generator. Rather than design the applicator to have a constant impedance of 50 Ω in operation, which would be difficult, a matching network of other electronic components is used to tune the generator impedance . The matching is often performed automatically, using motorized, high-voltage vacuum capacitors.

While, theoretically, almost any characteristic impedance may be chosen for power amplifier generators, 50 Ω has become a standard for fixed-frequency generators due to the wide availability of compatible cable, instrumentation and other hardware. Generators of 50 Ω may be separated from the applicator, connected by a section of coaxial cable (Jones and Rowley, 1996), but high cost and lower power output may limit the use of 50 Ω technology in large-scale industrial applications.

At RF frequencies, the wavelength of the applied electromagnetic waves is usually much larger than the dimensions of the applicator. Thus, RF applicators

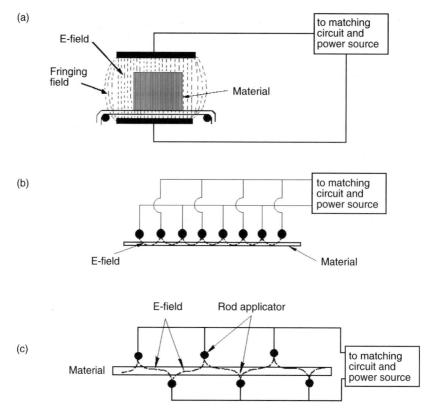

Fig. 2.7. Electrode configurations for an RF: (a) parallel plates applicator; (b) stray-field rod applicator; and (c) stagger-through applicator (from Metaxas and Clee, 1993).

typically do not have 'hot-spots' caused by the presence of standing waves, a common phenomenon in MW applicators.

There are, however, other considerations that may affect heating uniformity. Abrupt transitions in material properties, particularly between the heated materials and the surrounding air, can cause edge heating due to concentration of the electric field. The high relative permittivities of commodities with high moisture content make them especially susceptible to this phenomenon. Another phenomenon is fringing, where the electric field near the edges of the applicator is distorted, curving into the surrounding space. Fringing effects can be mitigated by extending the plates beyond the edge of the product.

2.4 Case Studies to Demonstrate the Differences between Conventional and Dielectric Heating

Comparison between Hot Air and RF Heating for Walnuts

RF and MW treatments have particular advantages over conventional hot air heating in treating in-shell walnuts. The shell and the air spaces within the shell act as layers of insulation and slow down heat transfer with conventional heating methods, while electromagnetic energy directly interacts with the kernel inside the shell to generate heat. Figure 2.8 shows typical temperature/time profiles for in-shell walnut kernels during hot air treatment (53°C at 1 m/s air velocity) and a 27.12 MHz RF treatment.

With hot air treatment, the heating rates decrease as the product temperature approaches a medium temperature, resulting in a prolonged heating time of more than 40 min (see Fig. 2.8). The kernel temperature lags significantly behind that of the hot air temperature, increasing very slowly with treatment time once it reaches within ~10°C of the air temperature. The small thermal conductivity of the porous walnut shell and the in-shell void hinders the transfer of thermal energy from the hot air outside the walnut shell. For a large quantity of walnuts in containers treated with hot air, a large temperature gradient exists between entering and outgoing places where the air passes through the containers.

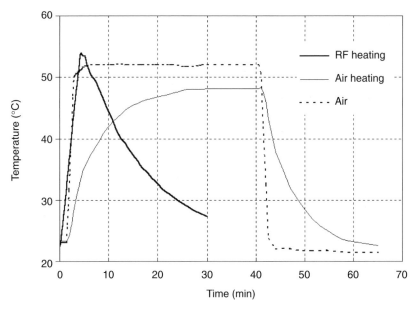

Fig. 2.8. Typical heating and cooling curves for an in-shell walnut kernel when subjected to forced hot air (air temperature 53°C; air velocity 1 m/s) and radio frequency treatments (from Tang *et al.*, 2000).

With RF treatment, walnut kernel temperature increases linearly with process time (see Fig. 2.8), as predicted by Eq. 2.24. Increasing RF power input can further increase the heating rate (Tang *et al.*, 2000; Wang *et al.*, 2001a, b, 2002). Heating non-uniformity becomes a problem in RF systems when scaling-up from small-scale laboratory tests to large-scale industrial applications because of different properties and shapes of individual walnuts and the different locations of walnuts in a non-uniform electromagnetic field. However, it is possible to achieve desired temperature distributions for pest control and walnut quality because industrial systems can be designed as a continuous process using several RF units with stirrings in between (Wang *et al.*, 2005).

Comparison between Hot Water and RF Heating for Apples

Figure 2.9 shows experimental time/temperature profiles for an apple (7.2 cm diameter) centre when subjected to hot air treatment (55°C at 4 m/s air velocity) and 27.12 MHz RF treatment. With hot air treatment, the heating rates are very low at the beginning, and decrease as the apple temperature approaches the heating medium temperature due to limited conductive heat transfer within the apple. In the case shown in Fig. 2.9, it took about 56 min for the apple centre to reach 50°C.

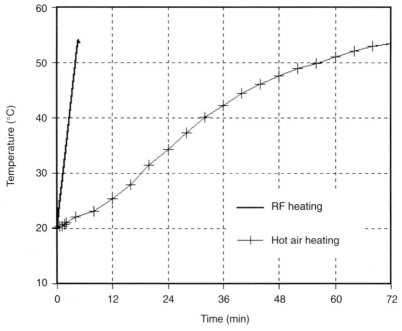

Fig. 2.9. Experimental heating curves for an apple (7.2 cm diameter) centre when subjected to forced hot air (air temperature 55°C; air velocity 4 m/s) and RF: 27.12 MHz treatments (from Wang *et al.*, 2001b).

With RF treatment, the apple centre temperature increased linearly with process time, as indicated by Eq. 2.24. The heating rate can be further increased by increasing RF power – as used for walnuts, above. After only 4.9 min of RF treatment, the centre temperature reached 54°C in air. Radio frequency treatments have particular advantages over conventional heating in treating large fruits. To avoid overheating at the contact points between fresh fruits, water immersion is used in RF heating (Ikediala *et al.*, 2002; Birla *et al.*, 2004). That is, fresh fruits are immersed in tap water whilst heated in RF systems. Since dielectric loss factors of fresh fruit are higher than those of tap water at the RF range (Wang *et al.*, 2003), the fruit may absorb much more RF energy than the tap water, as predicted by Eq. 2.24, thus resulting in different heating rates.

Matching the electrical conductivity of the water with that of fruit pulp or combining water surface heating and RF core heating may help to improve the heating non-uniformity between the surface and the core (Birla *et al.*, 2004, 2005). The simulation of temperature/time history in fruit based on the heat transfer theory can be used to determine heating parameters, which is a potential way of controlling insect pests in fruits without causing significant damage to fruit quality.

Comparison between Hot Water and RF Heating for Cherries

Figure 2.10a shows the temperature/time history at the core of a cherry with a 2.4 cm diameter when immersed in hot water (47°C at 1 m/s). It takes 8 min for the cherry core to reach the medium temperature. RF heating allows the cherry core temperature to reach a selected treatment temperature of 52°C in 2 min in a 0.15% saline solution (*see* Fig. 2.10b). The RF heating rate is linear and rapid (12 ± 1°C/min) (Ikediala *et al.*, 2002). The difference between the heating time for cherries in hot water *vs* RF heating is small compared with that for large fruits such as apples. This short heating time in RF systems may help a little to reduce thermal damage to fruit based on the effect of fruit size on the heat transfer, as discussed above.

Comparison between Hot Air or Water and RF Heating for Oranges

Figure 2.11 shows experimental time/temperature histories for oranges (fruit diameter 9 cm) when subjected to hot air, hot water and RF heating. It takes about 66 and 173 min for core temperatures to reach 50°C in hot water and hot air heating, respectively. With RF heating, core temperatures increase linearly with the process time and a heating rate varying from 3–5°C/min from fruit to fruit.

There appears to be no marked temperature difference between fruit core and subsurface with respect to heating time during RF heating, while the temperature difference between subsurface (2 cm deep) and core was as high as 8°C with hot water and hot air (*see* Fig. 2.11). With hot air treatment, the temperature difference persists even after 3 h of heating. RF heating is

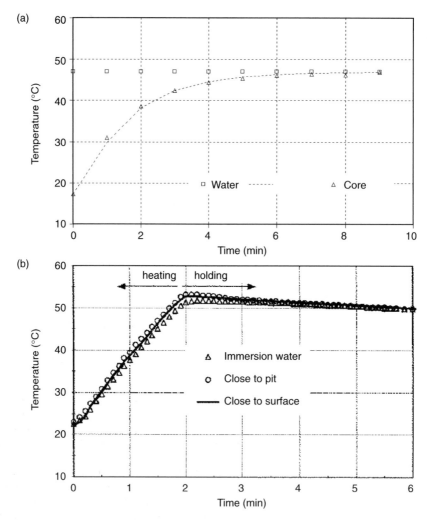

Fig. 2.10. Experimental heating curves for a cherry centre when subjected to hot water (47°C; 1 m/s) (a) and RF (27.12 MHz) treatments (b) during saline water immersion followed by holding (from Wang *et al.*, 2001b; Ikediala *et al.*, 2002).

approximately 8.8 and 23 times faster compared with hot water and air, respectively, as it takes only 7.5 min to heat from 20 to 50°C.

Figure 2.12 illustrates the temperature distributions inside an orange measured with infrared thermal imaging when subjected to RF heating for 5 and 10 min, and to hot water and hot air heating at 53°C for 10 and 20 min, respectively, from an initial fruit temperature of 20°C. RF heating results in fairly uniform temperatures over the entire orange and achieves the target temperature in a short time. With hot water and hot air treatments, a large temperature gradient is observed from the surface to the core. The core temperatures are 19, 25 and 56°C after hot air, hot water and RF heating for 10 min, respectively (Birla *et al.*, 2004).

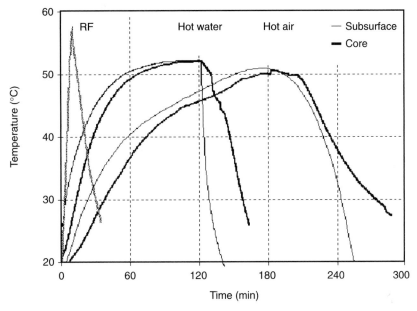

Fig. 2.11. Experimental heating and cooling curves for an orange (9 cm diameter) centre and subsurface (2 cm deep) when subjected to forced hot air (air temperature 53°C; air velocity 1 m/s), hot water (53°C) and RF (27.12 MHz) treatments (from Birla *et al.*, 2004).

In summary, for conventional heating, heating rates in the interior of fruits are small at the initial period due to thermal inertia over a large volume, and become small again as the core temperature approaches that of the medium temperature due to the reduced temperature gradient between the core and fruit surface. The heating rate in RF heating is much faster than in conventional heating, and is not dependent on heat transfer through convection at the surface and conduction within the fruit.

2.5 Closing Remarks

Delivery of thermal energy to commodities can be achieved by two different general methods: (i) conventional surface dielectric heating; and (ii) volumetric heating, each governed by a different set of physical principles. All current industrial thermal treatments rely on the first method. Understanding the fundamental principles and factors that influence the heating rate and uniformity with this method can help in improving the effectiveness and efficiency of conventional heat treatments.

Treatments based on volumetric heating using electromagnetic energy are still in research and development stages, but intensive research has demonstrated its potential for a selected group of commodities, especially dried nuts. For future applications, it is very likely that a combination of surface heating and volumetric heating will be used, to take advantage of both methods.

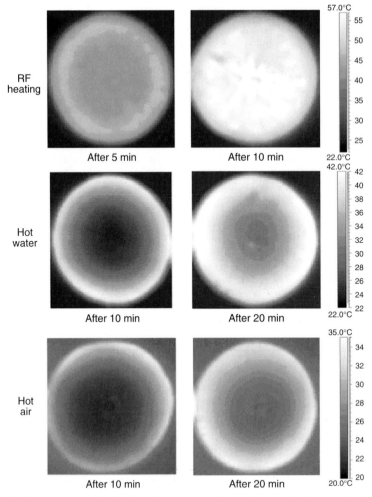

Fig. 2.12. Temperature distributions in oranges (9 cm diameter) when subjected to: (i) RF heating for 5 and 10 min; (ii) hot water and (iii) hot air heating at 53°C for 10 and 20 min from a 20°C initial fruit temperature (from Birla *et al.*, 2004).

Computer models based on well-established numeric schemes, such as the finite element method or finite difference methods, are commonly used in other engineering fields to study heat transfer phenomena involving heat conduction and convection. Sharply increased computation power of desktop computers now allows reliable simulation of detailed object geometries in complicated heating environments. The same approach can be used in optimizing heat treatment pest control treatments.

A major advantage of computer simulation is its ability to incorporate an intrinsic thermal death kinetics model of the target insects with predicted temperature histories at different locations of treated commodities to study the influence of treatment parameters on insect mortality. Product quality kinetic models can also be included in the simulation to optimize the treatments that can completely control insects while causing minimum thermal damage to the treated commodities. Modelling volumetric heating involving electromagnetic energy is more difficult compared with the surface heating methods because of the various complications associated with interaction between commodities and electromagnetic waves in confined application systems.

2.6 References

Barber, H. (1983) *Electroheating*. Granada Publishing Limited, London.

Birla, S.L., Wang, S. and Tang, J. (2004) Improving heating uniformity of fresh fruit in radio frequency treatments for pest control. *Postharvest Biology Technology* 33, 205–217.

Birla, S.L., Wang, S., Tang, J., Fellman, J., Mattinson, D. and Lurie, S. (2005) Quality of oranges as influenced by potential radio frequency heat treatments against Mediterranean fruit flies. *Postharvest Biology and Technology* 38, 66–97.

Campbell, G.S. (1977) *An Introduction to Environmental Biophysics*. Springer-Verlag, New York.

Dincer, I. (1997) *Heat Transfer in Food Cooling Applications*. Taylor & Francis, Washington, DC.

Earle, R.L. (1983) *Unit Operations in Food Processing*. Pergamon Press, New York.

FCC (1998) *Title 47 CFR 18.301*. Federal Communications Commission. Washington, DC.

Feng, H., Tang, J. and Cavalieri, R.P. (2002) Dielectric properties of dehydrated apples as affected by moisture and temperature. *Transactions of the ASAE* 45, 129–135.

Feng, X., Hansen, J.D., Biasi, B. and Mitcham, E.J. (2004) Use of hot water treatment to control codling moths in harvested California 'Bing' sweet cherries. *Postharvest Biology and Technology* 31, 41–49.

Hansen, J.D. (1992) Heating curve models of quarantine treatments against insect pests. *Journal of Economic Entomology* 85, 1846–1854.

Hansen, J., Wang, S. and Tang, J. (2004) A cumulated lethal time model to evaluate the efficacy of heat treatments for codling moth *Cydia pomonella* (L.) (Lepidoptera: Tortricidae) in cherries. *Postharvest Biology and Technology* 33, 309–317.

Hayes, C.F. and Young, H. (1989) Extension of model to predict survival from heat transfer of papaya infested with oriental fruit flies (Diptera: Tephritidae). *Journal of Economic Entomology* 82, 1157–1160.

Hayes, C.F., Chingon, H.T.G., Nitta, F.A. and Wang, W.J. (1984) Temperature control as an alternative to Ethylene Dibromide fumigation for the control of fruit flies (Diptera: Tephritidae) in papaya. *Journal of Economic Entomology* 77, 683–686.

Holdsworth, S.D. (1997) *Thermal Processing of Packaged Foods*. Blackie Academic & Professional, London.

Ikediala, J.N., Hansen, J.D., Tang, J., Drake, S.R. and Wang, S. (2002) Development of a saline water immersion technique with RF energy as a postharvest treatment against codling moth in cherries. *Postharvest Biology and Technology* 24, 209–221.

Incropera, F.P. and DeWitt, D.P. (1981) *Fundamentals of Heat Transfer*. John Wiley & Sons, New York.

Jones, P.L. and Rowley, A.T. (1996) Dielectric drying. *Drying Technology* 14, 1063–1098.

Lurie, S. (1998) Postharvest heat treatments. *Postharvest Biology and Technology* 14, 257–269.

Mangan, R.L., Shellie, K.C., Ingle, S.J. and Firko, M.J. (1998) High temperature forced-air treatments with fixed time and temperature for 'Dancy' tangerines, 'Valencia' oranges, and 'Rio Star' grapefruit. *Journal of Economic Entomology* 91, 933–939.

Metaxas, A.C. and Clee, M. (1993) Coupling and matching of radio frequency industrial applications. *Power Engineering Journal* 7, 85–93.

Metaxas, A.C. and Meredith, R.J. (1993) *Industrial Microwave Heating.* Peter Peregrinus Ltd., London.

Moffitt, H.R., Drake, S.R., Toba, H.H. and Hartsell, P.L. (1992) Comparative efficacy of methyl bromide against codling moth (Lepidoptera: Tortricidae) larvae in Bing and Rainier cherries and confirmation of efficacy of a quarantine treatment for Rainier cherries. *Journal of Economic Entomology* 85 (5), 1855–1858.

Mohsenin, N.N. (1980) *Thermal Properties of Foods and Other Agricultural Materials.* Gordon and Breach Science Publishers, New York.

Monteith, J.L. (1973) *Principles of Environmental Physics.* Edwin Arnold, New York, 241 pp.

Murphy, E.J. and Morgan, S.O. (1938) The dielectric properties of insulating materials. Part II: dielectric polarizability and anomalous dispersion. *Bell System Technical Journal* 18, 502–537.

Nelson, S.O. (1991) Dielectric properties of agricultural products: measurements and applications. *IEEE Transactions of Electrical Insulation* 26, 845–869.

Neven, L.G. (1994) Combined heat treatments and cold storage effects on mortality of fifth-instar codling moth (Lepidoptera: Tortricidae). *Journal of Economic Entomology* 87, 1262–1265.

Neven, L.G. and Rehfield, L.M. (1995) Comparison of pre-storage heat treatments on fifth-instar codling moth (Lepidoptera: Tortricidae) mortality. *Journal of Economic Entomology* 88, 1371–1375.

Opoku, A., Sokhansanj, S., Crerar, W.J., Tabil, L.G. and Whistlecraft, J.W. (2002) Disinfestation of Hessian fly pupae in small rectangular baled hay using experimental thermal treatment unit. *Canadian Biosystems Engineering* 44, 327–333.

Orfeuil, M. (1987) *Electric Process Heating – Technologies/Equipment/Applications.* Battelle Press, Columbus, Ohio.

Ozisik, M.N. (1985) *Heat Transfer: a Basic Approach.* McGraw-Hill, New York.

Rahman, S. (1995) *Food Properties Handbook.* CRC Press, Boca Raton, Florida.

Roussy, G. and Pearce, J.A. (1995) *Foundations and Industrial Applications of Microwave and Radio Frequency Fields: Physical and Chemical Processes.* John Wiley & Sons, Chichester, UK.

Ryynänen, S. (1995) The electromagnetic properties of food materials: a review of the basic principles. *Journal of Food Engineering* 29, 409–429.

Seo, S.T., Akamine, E.K., Goo, T.T.S., Harris, E.J. and Lee, C.Y.L. (1979) Oriental and Mediterranean fruit flies: fumigation of papaya, avocado, tomato, bell pepper, eggplant, and banana with phosphine. *Journal of Economic Entomology* 72, 354–359.

Sharp, J.L., Gaffney, J.J., Moss, J.I. and Gould, W.P. (1991) Hot-air treatment device for quarantine research. *Journal of Economic Entomology* 84, 520–527.

Sokhansanj, S., Venkatesan, V.S., Wood, H.C., Doane, J.F. and Spurr, D.T. (1992) Thermal kill of wheat midge and Hessian fly (Diptera: Cecidomyiidae). *Postharvest Biology and Technology* 2, 65–71.

Sokhansanj, S., Wood, H.C., Whistlecraft, J.W. and Koivisto, G.A. (1993) Thermal disinfestation of hay to eliminate possible contamination with Hessian fly (Mayetiola Destructor (Say)). *Postharvest Biology and Technology* 3, 165–172.

Tang, J., Ikediala, J.N., Wang, S., Hansen, J. and Cavalieri, R. (2000) High-temperature, short-time thermal quarantine methods. *Postharvest Biology and Technology* 21, 129–145.

Tang, J., Sokhansanj, S., Yannacopoulos, S. and Kasap, S.O. (1991) Specific heat of lentils by differential scanning calorimetry. *Transactions of the ASAE* 34 (2), 517–522.

Von Hippel, A.R. (1954) *Dielectric Properties and Waves.* John Wiley, New York.

Wang, S., Ikediala, J.N., Tang, J., Hansen, J.D., Mitcham, E., Mao, R. and Swanson, B. (2001a) Radio frequency treatments to control codling moth in in-shell walnuts. *Postharvest Biology and Technology* 22, 29–38.

Wang, S., Tang, J. and Cavalieri, R.P. (2001b) Modeling fruit internal heating rates for hot air and hot water treatments. *Postharvest Biology and Technology* 22, 257–270.

Wang, S., Tang, J., Johnson, J.A., Mitcham, E., Hansen, J.D., Cavalieri, R.P., Bower, J. and Biasi, B. (2002) Process protocols based on radio frequency energy to control field and storage pests in in-shell walnuts. *Postharvest Biology and Technology* 26, 265–273.

Wang, S., Tang, J., Johnson, J.A., Mitcham, E., Hansen, J.D., Hallman, G., Drake, S.R. and Wang, Y. (2003) Dielectric properties of fruits and insects as related to radio frequency and microwave treatments. *Biosystems Engineering* 85, 201–212.

Wang, S., Yue, J., Tang, J. and Chen, B. (2005) Mathematical modelling of heating uniformity of in-shell walnuts in radio frequency units with intermittent stirrings. *Postharvest Biology and Technology* 35, 94–104.

Yang, W., Sokhansanj, S., Tang, J. and Winder, P. (2002) Estimation of thermal conductivity of oilseeds using maximum slope method. *Biosystems Engineering* 82 (2), 169–176.

Yokoyama, V.Y., Miller, G.T. and Dowell, R.V. (1991) Response of codling moth (Lepidoptera: Tortricidae) to high temperature, a potential quarantine treatment for exported commodities. *Journal of Economic Entomology* 84, 528–531.

3   Temperature Measurement

J. Tang[1] and S. Wang[2]

Department of Biological Systems Engineering, Washington State University, Pullman, Washington, USA; e-mails: [1]jtang@wsu.edu; [2]shaojin_wang@wsu.edu

3.1 Introduction

Thermal treatments for postharvest insect control in agricultural commodities have been extensively investigated as potential alternatives to chemical fumigation, including hot air (Heather *et al.*, 1997), hot water (Follett and Sanxter, 2001; Hansen *et al.*, 2004), vapour (Neven *et al.*, 1996), microwave (MW) (Ikediala *et al.*, 1999) and radio frequency (RF) energy (Wang *et al.*, 2001a, 2002). Since insect pests need to be killed thermally, accurate measurement of the temperature sensed by insects in commodities is therefore essential in the development and control of thermal treatments.

Temperature sensors used in research on thermal treatments for postharvest insect control can be based on different temperature-dependent physical phenomena, including thermal expansion, thermoelectricity, electrical resistance and thermal radiation. Such sensors vary from simple liquid-in-glass thermometers to sophisticated state-of-the-art thermal imaging (Doebelin, 1983; Magison, 1990). This chapter focuses only on temperature sensors that generate digital signals readily used for online monitoring and automatic temperature control. We will discuss mainly commonly used temperature sensors, including: (i) their principles and properties; (ii) advantages and limitations; (iii) calibration and precision; (iv) response times; and (v) practical applications for postharvest insect control research.

3.2 Principles and Properties

Temperature is one of the seven basic SI units and is defined qualitatively as a measure of hotness. A primary measuring scale is the thermodynamic temperature scale (Kelvin temperature scale), which is based on the equal increment between absolute zero and the triple point of pure water (273.15 K).

Another scale commonly used in science and engineering is the Celsius scale, in which the magnitude of one degree Celsius is numerically equal to one Kelvin, as follows:

$$T\ (°C) = T(K) - 273.15 \tag{3.1}$$

where 0 and 100 degrees Celsius correspond to the freezing and boiling points of water, respectively.

The Fahrenheit scale is widely used in US industry and is converted to the Celsius scale by the following relationship:

$$T\ (°C) = (T\ (°F) - 32)/1.8 \tag{3.2A}$$

and conversely:

$$T\ (°F) = 1.8\ T\ (°C) + 32 \tag{3.2B}$$

Temperature measurement techniques can be divided into three categories depending on the nature of contact between the measuring device and the medium of interest: invasive, semi-invasive and non-invasive (Childs *et al.*, 2000). Based on the foci of electrical sensors, resistance thermometers, thermocouples and fibre-optic thermometers belong to the category of invasive sensors, sonic thermometers belong to the semi-invasive sensors and infrared thermometers and magnetic resonance imaging (MRI) belong to non-invasive sensors. The following discusses each of these electrical sensors in detail, all of which are widely used in thermal treatments for postharvest pest control.

Resistance Thermometers

Resistance thermometers are based on well-defined, temperature-dependent electrically resistant metals. Resistance temperature devices can be highly accurate, and are widely used in industrial and laboratory applications. The main types of resistance thermometers include resistance temperature devices and semiconductor thermometers (thermistors).

Resistance temperature devices

Resistance temperature devices (RTD) use a metal material (e.g. platinum) with a highly resistant temperature coefficient and stable physical properties as the sensing element. In general, the electrical resistance in a metal changes in a predictable manner with temperatures that can be expressed by Eq. 3.3. This relationship can be used to convert electrical resistance R to temperature T:

$$R_t = R_0(1 + aT + bT^2 + cT^3 + \cdots) \tag{3.3}$$

where R_t is the electrical resistance at temperature T, R_0 is the electrical resistance at a reference temperature, usually 0°C, and a, b and c ... are material constants. The number of terms used in Eq. 3.3 depends on the material used in the sensor, the temperature range to be measured and the accuracy required. In many cases, only constant a is used to provide satisfactory accuracy over limited temperature ranges.

Figure 3.1 shows the temperature range and output resistance ratio of the three most commonly used metals for temperature measurement: platinum, nickel and copper. Copper RTD is commonly used in the temperature range −100–100°C and its cost is relatively low. Nickel RTD has high values of temperature coefficient of resistance, but its variation in electrical resistance with temperature is nonlinear. Platinum RTD (0.10 μΩm at 0°C) has a well-established temperature coefficient of resistance and its resistivity is six times that of copper RTD (0.017 μΩm) (Michalski *et al.*, 2001).

The most widely used RTD is a platinum resistance sensor measured to be 100 Ω and 139 Ω at 0 and 100°C, respectively. Lead wire resistance and changes in lead wire resistance with temperature may have significant effects on resistance readings. In application, platinum RTD resistance is transformed into voltage and measured in a bridge circuit connected by two, three or four wires according to application objectives and the need to minimize the influence of lead wire resistance. Platinum RTD can be applied to measure solid, liquid and gaseous substances at temperatures between −200 and 800°C. The popular structures for RTD are wire-wound and thin film elements. A fine wire winding of about 0.01–0.10 mm diameter is commonly used for RTD sensors made from cylindrical forms of glass, quartz and ceramics.

A major problem with RTD sensors is their relatively slow response time due to a protective sheath (ceramic, glass or stainless steel). A thin film as small as 10 × 3 × 1 mm can be made to measure the surface temperature with a fast response time (Michalski *et al.*, 2001). The smallest thin-film RTDs, which can be used from −50°C to +200°C, are 2.0 × 2.3 × 1.0 mm (Omega

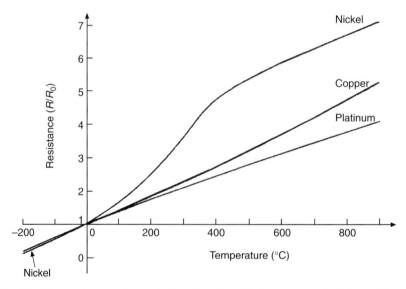

Fig. 3.1. Resistance–temperature relationships for nickel, copper and platinum (from Simpson *et al.*, 1991).

Engineering Inc., Stamford, Connecticut, USA). Sensor accuracy on the order of 0.2°C can be attained with careful calibration in specific temperature ranges (Wang *et al.*, 2003).

Thermistors

A thermistor uses a semiconductor as the sensing element. The electrical resistance of the semiconductor, made mostly from metallic oxides (e.g. N_iO), decreases sharply as the temperature increases, following the general relationship (Simpson *et al.*, 1991):

$$R = R_0 e^{\beta(1/T - 1/T_0)} \tag{3.4}$$

where R is the electrical resistance (Ω) at temperature T (K), and R_0 is the electrical resistance at a reference temperature, T_0, which is generally taken as 298 K (25°C). β is a constant for specific materials (K) and is on the order of 4000. The sensitivity of the electric resistance of a thermistor to temperature changes is more than a hundred times that of RTDs and a thousand times that of thermocouples.

The sensing elements of commercially available thermistors are protected by non-corrosive materials (e.g. epoxy, glass or stainless steel) and are manufactured in the forms of beads, rods, probes and disks. Sensors can be made as small as 0.1 mm in diameter to give a short response time. Glass- and stainless steel-sheathed probes 2–3 mm are widely used.

During measurement, a thermistor is connected to one leg of a Wheatstone bridge to yield a resolution as precise as 0.0005°C (Simpson *et al.*, 1991). Thermistor sensor accuracy is often limited by read-out devices with only ±0.1°C resolutions. Nonlinearity in electrical resistance output limits thermistor applications to relatively narrow temperature ranges (e.g. within 100°C). Long-term output stability is a common problem with thermistor probes because of slow drifts in the material properties of semiconductors.

Thermocouples

Thermocouples are made with two dissimilar metals (see Fig. 3.2). When these two metals are joined at two junctions, each at a different temperature, an electromotive force (emf) develops, causing an electric current to flow through the circuit. The magnitude of this emf is proportional to the temperature difference between the two junctions. This physical phenomenon, referred to as thermoelectricity, was discovered by T.J. Seebeck in 1821.

Thermoelectricity is used for temperature measurement in thermocouple systems in which one junction is maintained at a known reference temperature (T_{ref}) while the other is positioned for temperature measurement. The temperature to be measured (T) is converted from thermoelectric voltage (V) according to:

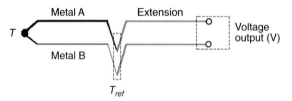

Fig. 3.2. Thermocouple circuit for measuring temperature (T) with a cold junction (T_{ref}) and an extension from the output voltage (V).

$$T = T_{ref} + \frac{V}{\alpha}$$
(3.5)

where V is the measured voltage in the thermocouple circuit and α is the voltage/temperature proportionality constant (V/°C). Voltage can be measured with a potentiometer or a high-impedance, solid-state digital voltmeter. The value of constant α depends on the type of wires used to make the thermocouples. Theoretically, for precision work, the reference junctions should be kept in a triple-point-of water apparatus that maintains a fixed temperature of 0.0100 ± 0.0005°C. In reality, the reference temperature, T_{ref}, can be room temperature, measured with a thermistor in most computer-based data acquisition systems.

Different pairs of metals are used for thermocouple types, commercially labelled as E (chrome/constantan), K (chrome/aluminium), T (copper/constantan), J (iron/constantan), R (platinum/platinum + 13% rhodium), S (platinum/platinum + 10% rhodium), and B (platinum + 30% rhodium/platinum + 6% rhodium). Typical characteristics of these thermocouples are shown in Fig. 3.3 (Simpson *et al.*, 1991).

Thermocouples give relatively low voltage outputs (e.g. 0.04 mV/°C for type T). Type E thermocouples have the highest voltage output per degree among all commonly used thermocouple types, with the lowest voltage output being for Type B. Type E thermocouples are ideally suited to temperature measurement around ambient because of low thermal conductivity and corrosion resistance. Type E thermocouples, however, are not suited for reducing, sulphurous, vacuum or low-oxygen environments. Type K thermocouples have a nearly constant voltage versus temperature coefficient over a temperature range 0–1000°C and thus can be used to obtain a moderately accurate direct read-out.

Types T and J are commonly used in industrial applications because of their resistance to corrosion and harsh environments. For example, Type J can be used between –150°C and 760°C in both oxidizing and reducing environments, while Type T can be used between –200°C and 350°C. The upper temperature limit for Type T is due to oxidation of copper at temperatures > 350°C.

The temperature-sensitive element in thermocouples can be made very small to provide a response time of only a few seconds (Baker *et al.*, 1953). Thermocouples can be used to detect the transient temperature in a solid, liquid or gas with an accuracy of $\pm$ 0.25% for Types R and S, and $\pm$ 0.5% for Type T over a measurement range of 100°C. Generally, thermocouples cannot

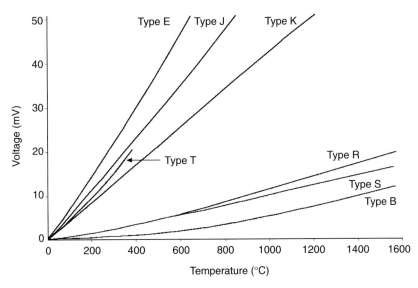

Fig. 3.3. Typical voltage–temperature relationship for common thermocouple materials when the reference temperature is 0°C (from Simpson *et al.*, 1991).

be used in electromagnetic fields because of their interaction with metallic elements (Wang *et al.*, 2003).

Because they consist only of thermoelectric wires, thermocouples require insulating and encapsulating in a protective sheath. To measure solid surface temperatures, thermocouples are generally attached on the surface, which may disturb and thus underestimate the surface temperature due to partial sensing of the medium temperature near the surface. A thermocouple placed in a gaseous environment will experience heat transfer by conduction along its wire and support, convection with the sounding gas both at the tip and along the wire, and thermal radiation with soundings. Therefore, a gas temperature measurement may require the use of a stagnation probe and radiation shield (Childs, 2001).

When the measurement location and read-out instrumentation are separated by a considerable distance, extension wires (leads) are needed for thermocouple measurements. There are two types of extension wires: those with similar physical composition to those with thermocouple wire and the compensation leads manufactured using different materials. Normally, the effect of extension wires on the overall emf is small; otherwise a re-calibration is needed to reduce the measurement error.

Fibre-optic Thermometers

In RF and MW heating, temperature sensors with metal components cannot be used directly during treatments because of electromagnetic noise and interaction between the metal parts and electromagnetic fields. For this reason, fibre-optic thermometers are often used (Kyuma *et al.*, 1982; Wang *et al.*, 2001a, 2002). Generally, typical fibre-optic sensors include those based on optical reflection, scattering, interference, absorption, fluorescence and thermally generated radiation. Major fibre-optic temperature sensors are practically developed from one of three methods: fluoroptic thermometry, Fabry-Perot interferometry and absorption shift of semiconductor crystals (Wang *et al.*, 2003).

The fundamental mechanisms behind fluoroptic thermometry rely on the use of phosphor (magnesium fluorogermanate activated with tetravalent manganese) as the sensing element. This material fluoresces in the deep red region when excited by ultraviolet or blueviolet radiation. The rate of decay of the afterglow of this material varies with temperature (e.g. 5 ms at –200°C, 0.5 ms at 450°C). Measuring the rate of afterglow decay allows an indirect determination of the temperature (Grattan and Zhang, 1995). The typical fluorescent thermometer (Model 3000, Luxtron Corporation, Santa Clara, California, USA) provides high measurement precision without periodical calibration.

A Fabry-Perot interferometer (FPI) consists of two parallel reflective surfaces that form a cavity resonator. The space separating these two surfaces, also called the FPI cavity (1–2 wavelengths deep), varies with temperature. Changes in temperature result in the change of optical length (refractive index multiplied by cavity depth) of this resonator, even though the actual physical thickness of the film exhibits no measurable changes. Changes in the FPI cavity path, or in the optical path length of the resonators for light from a white-light source, are measured to determine the temperature of the sensing element. A typical representative of the FPI fibre-optic systems has been developed by FISO Technologies Inc., Quebec, Canada, which includes 3–4 m-long sensor cables, one to eight channel data acquisition and software (see Fig. 3.4).

A fibre-optic sensor based on absorption shift of semiconductor crystals relies on the temperature-dependent light absorption/transmission characteristics of gallium arsenide (GaAs). A narrow band light source emits the light along an input light guide to the GaAs prism sensor. After passing the prism, the radiation returns by an output light guide to the diode photo detector. The unique feature of this crystal is that, when temperature increases, the crystal transmission spectrum shifts to higher wavelengths (see Fig. 3.5). The position of the absorption shift indicates the temperature measurement of the sensing element (Belleville and Duplain, 1993).

Fibre-optic sensors generally have small probe sizes (only 0.8 mm in diameter, FISO Technologies Inc., Quebec, Canada) with low thermal conductivity and can provide accuracy comparable to that of thermocouples. They generally have short response times (0.05–2.00 s in liquid foods) and are well suited to relatively fast MW or RF heating because they neither deform the existing temperature field nor influence electromagnetic fields. However, fibre-optic sensors are expensive, unstable and fragile.

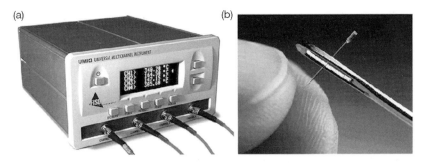

Fig. 3.4. FISO optical thermometer with four-channel data acquisition (a); FISO Technologies UMI-4 model signal conditione with FOR-MEM model fibre-optic pressure sensor (b) (both from FISO Technologies, Quebec, Canada).

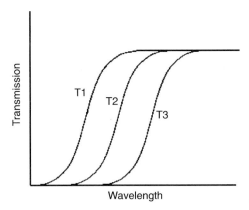

Fig. 3.5. Temperature-dependent light transmission of gallium arsenide crystal as the sensing element of a fibre-optic thermometer. T1, −40°C; T2, 40°C; T3, 120°C (from Wang *et al.*, 2003).

Sonic Thermometers

The basic principle of a sonic thermometer is the measurement of the time taken by an ultrasound pulse to travel between two transducers (see Fig. 3.6). It is well known that the speed of sound increases when air moves in the same direction as sound. A pair of transducers (a, b) can act alternately as transmitters and receivers, exchanging high-frequency ultrasound pulses. The speed of sound in air is useful, for example, for rapid calculation of air temperature when the effect of humidity and other gas concentration is negligible (Coppin and Taylor, 1983).

The output of the sonic temperature (T_s, °C) according to Kaimal and Gaynor (1991) is:

$$T_s = \frac{L^2}{1612}\left(\frac{1}{t_1} + \frac{1}{t_2}\right)^2 - 273.15 \qquad (3.6)$$

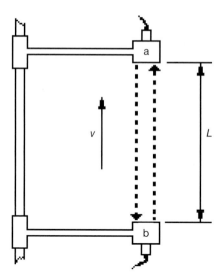

Fig. 3.6. Sonic anemometer. v, air speed; L, distance between transducers a and b (from Campbell and Unsworth, 1979).

where L is the distance between the transducers (m) and t_1 and t_2 are the travel times in each direction. This sensor has a high sampling rate of up to 50 times per second. Measurement precision may reach 0.4% after careful calibration. An ultrasonic system usually measures both the air temperature and air velocity simultaneously and thus can be used for studying air thermal dynamic characteristics (Wang *et al.*, 1999; Boulard *et al.*, 2000). This system can also be used to determine the air temperature and velocity in a large space, but its measurement is costly and sensitive to the environment.

Infrared Thermometers

Infrared is an invisible portion of the electromagnetic spectrum. All substances, regardless of temperature, emit energy in the form of electromagnetic radiation. Planck's distribution demonstrates the thermal radiative power from a black body as a function of temperature and wavelength (see Fig. 3.7).

The wavelength at the maximum emissive power for each temperature increases with decreased temperature. Radiation thermometers relate measured radiation to temperature based on one of the following relationships (Childs, 2001):

$$Q_{a,b} = C_1 \int_a^b \frac{\varepsilon_{\lambda,T}}{\lambda^5 (e^{C_2/\lambda T} - 1)} d\lambda \tag{3.7}$$

$$Q_{a,b} = \sigma F_{sh} (\varepsilon_{surf} T_{surf}{}^4 - \varepsilon_{sensor} T_{sensor}{}^4) \tag{3.8}$$

where C_1 and C_2 are constants, λ is the wavelength of radiation (m), $\varepsilon_{\lambda,T}$ is the hemispherical emissivity, σ is the Stefan-Boltzmann constant (W/m^2/K^4), F_{sh} is a

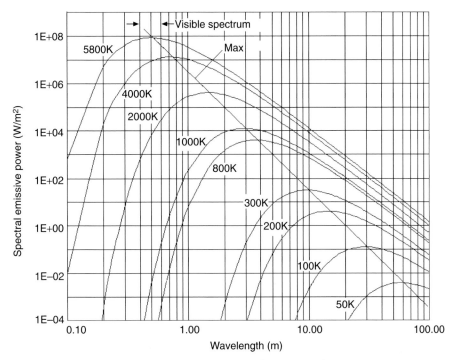

Fig. 3.7. Spectral black body emissive power (from Childs, 2001).

shape view factor between two surfaces (m^2), ϵ is the emissivity, $Q_{a,b}$ is the total power of an actual body between a and b of the wavebands (W/m^2), Q_{rad} is the radiative heat transfer flux (W) and T is the absolute temperature (K).

Infrared thermometers are most widely used in non-invasive temperature measurements. Infrared devices often refer to radiometers, radiation pyrometers, optic pyrometers or thermal imagers and operate using electromagnetic radiation in the visible spectrum (0.3–0.72 μm) and a portion of the infrared region (0.72–40 μm) (see Fig. 3.8).

A typical infrared measurement system consists of the following components: (i) a source of light; (ii) the medium through which the radiant energy is transmitted; (iii) an optical system to gather the thermal radiation; (iv) a transducer to convert the radiation into a signal related to temperature, amplification and interface circuit to control; and (v) a display and record of the measurement (Childs, 2001). A radiation thermometer may use either a thermal detector or photon detectors as its sensing element. Thermal detectors are blackened elements designed to absorb incoming radiation at all wavelengths radiating from the measured object. Temperature rise in a detector caused by absorbed radiation is measured by specially made RTDs, thermistors or thermopiles (consisting of up to 30 thermocouples).

Another type of thermal detector is based on pyroelectric crystals (e.g. lithium tantalate crystals). Pyroelectric thermal detectors have shorter response times than thermal detectors using a RTD, thermistors or thermopiles and are

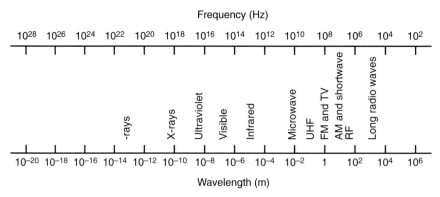

Fig. 3.8. The electromagnetic spectrum.

responsive to a wide radiation range between X-ray (0.001 μm) and far-infrared (1000 μm) (*see* Fig. 3.8).

Radiation thermometers can measure the targeted surface temperature from a distance, so there is no contact or direct interaction between the thermometer and the object. Since infrared sensors measure an object temperature at a distance, there is no limitation to the thermal tolerance of materials used in manufacture of the sensor. The time constant of radiation thermometers is short, often in the range of several microseconds to a few seconds. Radiation thermometers, however, only measure the surface temperature of a relatively opaque material. They cannot be used to measure transparent materials because the sensor will detect surface temperatures behind the object.

Radiation thermometers require information on emissivity of the measured body surface and an internal or external reference temperature to provide accurate temperature information. The emissivity of a material is wavelength dependent and a function of its dielectric constant and, subsequently, its refractive index. Table 3.1 lists typical total emissivities for a number of materials. This information is essential to accurately measure temperature because infrared thermometers need adjustment depending on the emissivity of their target.

Thermal imaging of temperature fields has now become a popular method of real-time temperature measurement in thermal treatments. It involves determining the spatial temperature distribution of a target and usually consists of an optical system, a detector, processing electronics and a display. The principle of thermal imaging can be divided into surfacial systems using scanning, matrix methods and linear systems. Many commercial thermal imaging systems are available, such as FLIR Systems ThermaCAM, Mitsubishi IR and Indigo Systems, with different resolutions, detectors, temperature ranges and precision.

The application range of thermal imaging is extremely large, including the analysis of thermal problems, monitoring of operations of industrial thermal treatments such as hot and cold spots, and investigations of

Table 3.1. Typical total emissivities for a variety of materials at room temperature (from Michalski *et al.*, 2001).

Material	Emissivity
Aluminium (foil)	0.04
Aluminium (polished)	0.05
Brass (polished)	0.10
Brick or concrete	0.95
Copper (polished)	0.03
Glass	0.92
Ice	0.97
Iron (polished)	0.06
Paper (black or white)	0.90
Paper (red)	0.76
Platinum (polished)	0.05
Porcelain	0.90
PVC	0.91
Rubber	0.90
Snow	0.80–0.90
Soil	0.93–0.96
Textiles	0.75–0.90
Vegetation	0.92–0.96
Water	0.96
Wood	0.80–0.90
Zinc (polished)	0.02

temperature fields. Some practical applications are also found in postharvest pest control research, such as: (i) temperature distribution over a cross-section of apples and oranges after RF and hot water treatments (Birla *et al.*, 2004); (ii) the surface temperature of walnuts after RF heating (Wang *et al.*, 2005b); and (iii) differential heating between honeybees and hornets (Ono *et al.*, 1995).

Figure 3.9 shows the temperature distribution over the cut section of an orange measured with the infrared thermal imaging technique after being heated with hot water at 53°C for 20 min (initial fruit temperature 20°C). The emissivity for the imaging was selected as 0.98. Good agreement is obtained for the orange core temperature measured by the infrared camera and thermocouple (Birla *et al.*, 2004). This type of thermal imaging allows direct assessment of special temperature distributions in commodities after thermal treatments.

Magnetic Resonance Imaging (MRI)

MRI temperature mapping has been developed for food processing research based on the sensitivity of water protons to temperature, including relaxation times, diffusion coefficients and chemical shifts (Nott and Hall, 1999). Relaxation times are related to the rotational mobility of water molecules,

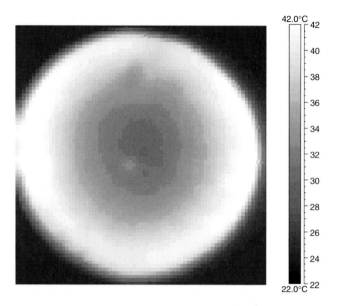

Fig. 3.9. Thermal imaging showing temperature distribution in oranges (9 cm) when subjected to hot water for 20 min (from Birla *et al.*, 2004).

diffusion coefficients are dependent on the random motion of water molecules and water proton chemical shifts vary with the fraction and nature of hydrogen bonds, all of which are temperature dependent.

MRI has been used to optimize industrial food processing that is related to thermal treatment, such as producing two-dimensional temperature mapping of a model food gel (Sun *et al.*, 1995) and mapping temperature distribution across potato, beef and carrot particles during ohmic heating (Ruan *et al.*, 1999).

Magnetic resonance imaging temperature mapping is a fast (e.g. 5 s is needed for 3-D mapping), non-invasive, three-dimensional technique. Temperature maps provide a spatial resolution of 0.94 mm and temporal resolution of 0.64 s with a temperature precision of about ± 1°C, but require highly specialized and expensive equipment. The cavity for sample holding in MRI measurement is relatively small and limited; currently, it is applied only in research laboratories.

3.3 Sensor Calibration, Precision and Response Time

Electrical transducers incorporating temperature sensors are not always stable and their performance may drift with time. Calibration of a given temperature probe is important to ensure reliable and accurate measurements and should be conducted on a regular basis. In general, two calibration methods are used. One method is to expose the sensor to an established fixed-point environment, such as the triple point of pure water and the freezing and boiling points of water. Another method is to compare readings with those of calibrated and traceable temperature sensors (e.g. pre-calibrated, mercury-in-glass thermometer by manufacturer) in the same thermal environment (e.g. water bath, oil bath or heating block). The output from the un-calibrated sensor can then be related to the temperature indicated by the calibrated sensor.

Calibration procedures are standardized for specific applications (ASTM, 1988). Special portable and laboratory temperature calibrators are commonly used in industrial applications and laboratory research for calibrating temperature sensors. Calibrators usually include automatic compensation of reference temperature in thermocouples, and can also be used for resistance sensors in two-, three- and four-wire configurations. The 744 Documenting Process calibrator and Fluke Corp 5500A calibrator (both Fluke Corporation, Everett, Washington, USA) are two good examples of commercial temperature calibrators with high precision.

Besides accuracy, precision and resistance to corrosion, an important consideration in selecting temperature sensors is response time, which is a measure of how quickly a sensor follows rapid temperature changes. The smaller the response time, the more closely the sensor follows the temperature change of the measured medium. For a step-change in the medium temperature, the temperature of the sensor $T(t)$ changes with time t following the relationship:

$$\frac{T_m - T(t)}{T_m - T_0} = \exp(-t / \tau) \tag{3.9}$$

where constant τ is the sensor response time (also referred to as the time constant), defined as the time for a sensor to reach 63.2% of a step-change in temperature $(T_m - T_0)$ (see Fig. 3.10a). The response time of the sensing element can be estimated from both the overall surface heat transfer coefficient, h (W/m^2 °C), and other physical properties:

$$\tau = \frac{\rho C_p V}{hA} \tag{3.10}$$

where A and V are the sensor's surface area (m^2) and volume (m^3), respectively, ρ is sensor density (kg/m^3), and C_p is specific heat (J/m^2/°C). It is clear from Eq. 3.10 that reducing the volume, V, of the sensing element (including the protective sheath) and increasing the heat transfer coefficient, h, can reduce sensor response time. That is why temperature probes are often made small.

For a particular sensor, however, the response time in water can be up to 100 times shorter than in air because of differences in the overall surface heat transfer coefficient. It is therefore critical that a temperature sensor (except for non-invasive ones) be in direct contact with the commodity to reduce response time and measurement errors, especially for rapid-heating processes. For many applications in which air temperature is to be measured, an aspirated temperature sensor is used to increase the surface heat transfer coefficient and reduce sensor response time.

Sensor response time can be used to assess measurement error in various applications. Figure 3.10b demonstrates the thermal lag of a temperature sensor when exposed to a medium with a temperature ramp of $\dfrac{dT}{dt}$. At steady state, the temperature lag is calculated by multiplying the thermal constant, τ, with the ramp rate, $\dfrac{dT}{dt}$. For example, when the medium temperature changes at 10°C/min and the time constant of the sensor is 30 s, the maximum thermal lag will be 5°C. Knowledge of the time constant of sensors is essential in designing effective dynamic temperature monitoring and control systems.

3.4 Application of Temperature Sensors

Sensor Selections

Temperature sensors are selected according to required accuracy, response time, initial investment, maintenance cost, ambient condition and stability of calibration. Typical applications of the temperature sensors discussed above are summarized in Table 3.2.

Thermistors are generally used for measuring temperature in a narrow range because of their high precision (± 0.1°C) in small ranges (Simpson *et al.*, 1991). Thermocouples are widely used because of acceptable precision, rapid response time and low cost. Resistance temperature devices, however, provide better accuracy and stability. Sonic thermometers are commonly used to measure air temperature in the turbulent environment of storage houses and bins (Liao *et al.*, 2004). Radiation thermometers are especially suitable for

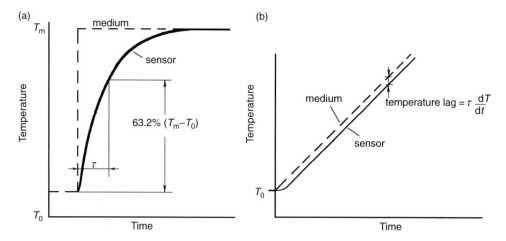

Fig. 3.10. Response time τ, defined under a step-change in medium temperature (a). Temperature lag of a sensor following a medium temperature ramp (b) (from Wang *et al.*, 2003).

Table 3.2. Properties and examples of typical applications of temperature sensors (from Wang *et al.*, 2003).

Sensor	Signal output	Measurement range (°C)	Accuracy	Example of typical applications
Thermistor	Resistance	–50~200	± 0.1%	Reference for data loggers
RTD	Resistance	–200~850	± 0.2°C	Precise and stable measurements
Thermocouple	Voltage	–250~400	± 0.5°C	Most food and agricultural applications
Sonic anemometer	Voltage	–10~80	± 0.4%	Airflow temperature in storage and bins
Far-infrared thermometer	Voltage	~4000	< 1%	Surface temperature
Fibre-optic thermometer	Voltage	–50~250	±0.5%	Microwave and radio frequency heating

measuring moving objects or objects inside vacuum or pressure vessels. Radiation sensors respond very quickly, but are more costly than thermocouples or RTDs, since their sighting paths and optical elements must be kept clean. Fibre-optic thermometers can be used in very strong electromagnetic fields with fast response time.

In selecting appropriate sensors for an application, environmental factors must be considered to reduce potential sources of error. It is also equally important to be familiar with the strategies that can be used to minimize environmental influences and maintain a desired level of temperature accuracy. Factors influencing measurement accuracy include sensor durability and range of operation, susceptibility to external noise influences, repeatability, handling/installation endurance, calibration method and unique requirements of certain sensors for electric, magnetic, radiation shielding or environmental protection. While direct exposure of the sensing element to the environment may increase sensor sensitive and reduce thermal time lags, in applications certain level of protection is always needed against corrosion, invasion of vapour, water, or mechanical abuse. Compromises need to be made with sound judgement and care.

The following illustrates an evaluation of insect mortality with regard to the effect of temperature distributions over a heating block (metal) surface that is heated electrically (Wang *et al.*, 2005a). Nine fast-response, ungrounded thermocouple temperature sensors with self-adhesive backing (SA1-T, Omega Engineering Inc., Stamford, Connecticut, USA) are uniformly distributed on a metal block surface (see Fig. 3.11).

Temperatures at nine different locations were monitored while the heating block temperature was raised from 22 to 50°C at heating rates of 10 and 15°C/min. The measured data from these nine sensors were recorded every 5 s by a data logger (DL2e, Delta-T Devices Ltd., Cambridge, UK). The hot (centre) and cold (four corners) spots were observed by this temperature measurement system. The overall maximum standard deviations for temperatures over the nine positions were 0.8 and 1.2°C for the heating rates of 10 and 15°C/min, respectively, when the mean block temperature reached the set point of 50°C (Wang *et al.*, 2005a).

Within 30 s of reaching 50°C, however, the temperature difference between the hottest location among the nine measured positions and the set point of

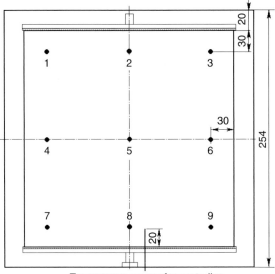

Temperature sensor for controller

Fig. 3.11. Surface mount thermocouple sensor distributions (1–9) on the bottom block together with the sensor from the controller (from Wang *et al.*, 2005a). All dimensions are in mm.

50°C were reduced to less than 0.6°C both for heating rates of 10 and 15°C/min (*see* Fig. 3.12). Thin-film RTDs with a measurement range –50 to +200°C and dimensions of $2.0 \times 2.3 \times 1.0$ mm (Omega Engineering Inc., Stamford, Connecticut, USA) can also be used for surface temperature measurements.

Selection of temperature probes also includes selection of compatible signal conditioning equipment. Common bench-top display units or PC-based equipment for a laboratory may be entirely unsuitable for use in processing. It is important to select probe styles that can be connected to suitable cables or enclosures for signal conditioners and transmitters that are protected from possible heat, moisture and chemical and mechanical damage. Transmitters should be selected according to the type of temperature sensor, precision, range and type of output required.

Although standards such as 4–20 milliamp DC analogue signals are still commonly used, modern temperature sensors are now being interfaced to measurement and process control systems using highly sophisticated technologies such as Ethernet, Field Bus and, most recently, wireless networks (Wang *et al.*, 2003).

Measurement Procedures and Locations for Fruit Temperature

To validate a measurement system, the loop resistance of a thermometer should first be checked, which depends on the length, type and diameter of the thermo-element and extension wires. If, on installation and at regular intervals,

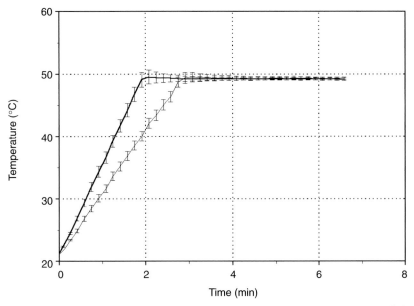

Fig. 3.12. Average and standard deviation values of temperature–time histories over nine positions on a bottom block surface from 22 to 50°C at the heating rates of 10 (—) and 15°C/min (—) (from Wang *et al.*, 2005a).

measurements are made of the loop resistance, a change in these values can be used to indicate wire thinning due to chemical attack, loose or corroded connections, increased contact resistance due to broken but touching wires or electrical shunting due to loss of insulation at some location along the wires (Childs, 2001).

The initial procedure is to check whether the temperature rises when fingers touch the sensor tip. The next step is to check if the sensor type matches the setting on the readout meter. A mis-metal (e.g. type T probe with type J setting) leads to error reading. The final step is to check the room and any other known medium temperatures against a standard thermometer.

Sudden changes in sensor output provide a diagnostic warning to the user. Fluctuations in the output temperatures exceeding the expected changes indicate that the thermometer wire is broken or that there is an unstable detection of the reference temperature.

Before placing into a water bath or hot air medium, fruit purchased from local stores should be kept at room condition until a thermal equilibrium is achieved throughout the fruit. Solid sensors such as thin thermocouples may be inserted directly into the required depth to measure the target temperature. Fibre-optic sensors or other non-strict sensors have to be inserted by a pre-inserted hole by using a slightly smaller metal needle. It is very important that the tip of the probe is in close contact with fruit tissue to eliminate possible thermal lag due to air pocket. For measuring the fruit surface temperature, the sensors need to reach the fruit surface through a side path. To avoid water ingress or heat conduction from hot water to the sensor tip, some insulation material such as silicon may be put around the probe on the fruit surface.

A typical temperature/time history of an apple core and surface when subjected to hot water heating is shown in Fig. 3.13. The apple core temperature takes a few minutes to demonstrate some small increases from its initial temperature at the beginning of heating, while the surface temperature reaches the set point very quickly. When the core temperature approaches the set point, the temperature increases again slowly due to a reduced temperature gradient in heat conduction.

The larger the fruit diameter, the longer the ramp time. Most fruit temperature profiles follow this trend in conventional heating. If there is any clear difference from the temperature trend shown in Fig. 3.13, one must check whether the measurement has been made correctly.

Measurement Errors and Improvement

Whenever a measurement is made, it is unlikely that the measured value will equal the true value or the exact value of the measured variable: this difference is the error, which is often caused by either: (i) the precision and resolution of the sensor itself; (ii) incorrect installation; (iii) malfunction; or (iv) unsuitable arrangement. The insertion of a transducer – or sometimes even thermal interactions between a remote sensor and an application – will result in disturbance to the temperature distribution. The magnitude of this disturbance will depend on the heat transfer process. Natural instabilities associated with the transducer and signal-processing device also contribute to deviation from

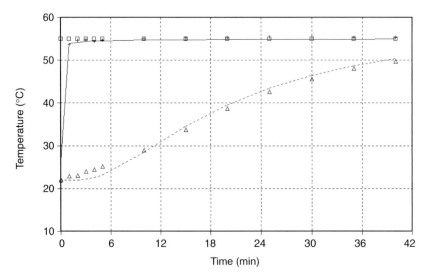

Fig. 3.13. Comparison of the surface and core temperatures between measurement and simulation for apples (7.2 cm diameter) in a water bath (55°C, 1 m/s). –, surface, simulated; - -, core, simulated; +, surface, measured; Δ, core, measured (from Wang *et al.*, 2001b).

the true value. Further deviations may be due to uncertainties arising from the calibration system.

General measurement for biological materials should be made a minimum of three times. If just one measurement is made, then a mistake could go unnoticed. If two measurements are made and the results are very different, one has to check the measurement process and know the reason for this difference. If three measurements are made and two agree, suspicion lies with the third. Persistent differences in the readings may be due to natural variations in the process. Statistically, more results are necessary to provide better confidence in the temperature measurement.

Thermal imaging is useful in measuring the temperature distribution over a subject surface. Measurement error may be caused by the precision of the sensor itself, the estimated emissivity of the material or the time lag between the treatment and measurement. Thermal imaging accuracy may be improved by adjusting the surface emissivity setting according to the agreement between the measured temperature with the thermal imaging system and that measured with a thermocouple. The time lag will result in the temperature drop around the boundary of the target object and the surface of interest due to heat loss to the surroundings.

Data Recorders

A measurement system comprises a transducer to convert a temperature-dependent phenomenon into a signal that is a function of: (i) temperature; (ii) a method to transmit and process the signal from the transducer; (iii) a display; and (iv) method of recording data. Of these functions, the data record is the final and most important process. This recording may be undertaken by the sensor system itself or by an external data acquisition system.

The most important group of indicating instruments are digital, based on LED and LCD technology. Digital meters are available as either portable or panel-mounted instruments. The displayed values can be converted into different temperature scales such as Fahrenheit, Celsius or Kelvin. Some multimeters also accept signals from temperature sensors at their input and are convenient tools for instant-action thermometry.

The problem with portable instruments is that they may only have one channel and cannot store temperature data. Once again a common mistake made by researchers who are not familiar with temperature measurements is a mismatch between the type of thermocouple used in the measurement and the type setting for a thermocouple digital read-out. For example, temperature measurement from a Type T thermocouple is read with a Type E meter.

Data loggers are widely used for measurement result recorders in laboratory and *in situ*. They are multi-channel instruments that accept signals from many different sensors (thermocouples, RTD and other resistance, voltage and current-based signals) to provide their versatile processing. The sampling rate can be set to as low as a few times per second, and the average result can be outputted and recorded from 1–15 min. The internal memory can store

32 kb or more data, which can be easily transferred to PC through an RS-232 serial port. Typical commercial data-logger systems are available from Keithley Instruments (Solon, Ohio, USA), Omega Inc. (USA), Fluke Corp. (Everett, Washington, USA), Campbell Scientific Inc. (Logan, Utah, USA) and Aqua Genesis (USA).

3.5. Temperature Control

Non-electric temperature controllers are often applied in thermostats or operate as the thermostat valves in water and steam heating systems; they are on-off controllers that provide simple and stable temperature control. Action is taken when the measured temperature is below or beyond the set point. The controller will remain in its previous state when the measured temperature is within the dead-band of the set point.

The most popular digital temperature controllers are usually based on typical algorithms and provide multistage conditioning and processing of the measured signal. According to algorithms, the systems divide into proportional (P), proportional-integral (PI) and proportional-integral-derivative (PID) controllers. The mathematical expression of the PID controllers is as follows:

$$r = k_1\theta + k_2\int \theta dt + k_3\frac{d\theta}{dt} \hspace{2cm} (3.11)$$

where r is the controller output, θ is the temperature difference between the measurement and the set-point and k_1, k_2 and k_3 are the constants for P, I and D actions, respectively. Many commercial products are available to apply the PID algorithm in multi-channel temperature controls. Proportional-integrative-derivative controllers (i/32 temperature & process controller, Omega Engineering Inc., Stamford, Connecticut, USA) have been used in heating block systems to provide accurate temperature output at different heating rates (Wang *et al.*, 2005a) via a solid-state relay. The block temperature is increased linearly to within ± 0.3°C of the desired set point.

3.6 Closing Remarks

Accurate temperature measurements are necessary in research and development of thermal treatments as well as in control of industrial treatment systems. Thermocouples and RTDs are most commonly used for thermal treatments with hot water or air, while fiber-optic sensors are needed when developing treatments based on radio frequency and microwave energy. Electronic sensors are less stable than temperature sensors that rely on thermal expansion, such as glass thermometers, but have much shorter response times and are more suited for rapid heating processes.

Electronic sensors need to be calibrated or checked on a regular basis against pre-calibrated thermometers to ensure that temperature readings are reliable and accurate. Infrared imaging methods are very effective in determining heating uniformity. But failure to calibrate against a direct temperature sensor and set correct emissivity values for the measured subject would result in erroneous readings and misleading information.

3.7 References

ASTM (1988) *Standard Guide for the Use in the Establishment of Thermal Process for Food Packaged in Flexible Containers. American Society for Testing Materials,* Philadelphia, Pennsylvania.

Baker, H.D., Ryder, E.A. and Baker, N.H. (1953) *Temperature Measurement in Engineering.* John Wiley & Sons, Inc., New York.

Belleville, C. and Duplain, G. (1993) White-light interferometric multimode fiber-optic strain sensor. *Optic Letters* 18, 78–80.

Birla, S.L., Wang, S., Tang, J. and Hallman, G. (2004) Improving heating uniformity of fresh fruits in radio frequency treatments for pest control. *Postharvest Biology and Technology* 33 (2), 205–217.

Boulard, T., Wang, S. and Haxaire, R. (2000) Mean and turbulent air flows and microclimatic patterns in a greenhouse tunnel. *Agricultural and Forest Meteorology* 100 (2–3), 169–181.

Campbell, G.S. and Unsworth, M.H. (1979) An inexpensive sonic anemometer for eddy correlation. *Journal of Applied Meteorology* 18, 1027–1077.

Childs, P.R.N. (2001) *Practical Temperature Measurement.* Butterworth-Heinemann, Oxford, UK.

Childs, P.R.N., Greenwood, J.R. and Long, C.A. (2000) Review of temperature measurement. *Review of Science Instrument* 71 (8), 2959–2978.

Coppin, P.A. and Taylor, K.J. (1983) A three-component sonic anemometer/thermometer system for general micrometeorological research. *Boundary-Layer Meteorology* 27, 27–42.

Doebelin, E.O. (1983) *Measurement Systems Application and Design.* McGraw-Hill Book Company, New York.

Follett, P.A. and Sanxter, S.S. (2001) Hot water immersion to ensure quarantine security for Cryptophlebia spp. (Lepidoptera: Tortricidae) in lychee and longan exported from Hawaii. *Journal of Economic Entomology* 94, 1292–1295.

Grattan, K.T.V. and Zhang, Z.Y. (1995) *Fiber Optic Fluorescence Thermometry.* Chapman and Hall, London.

Hansen, J., Wang, S. and Tang, J. (2004) A cumulated lethal time model to evaluate efficacy of heat treatments for codling moth Cydia pomonella (L.) (Lepidoptera: Tortricidae) in cherries. *Postharvest Biology and Technology* 33, 309–317.

Heather, N.W., Corcoran, R.J. and Kopittke, R.A. (1997) Hot air disinfestation of Australian 'Kensington' mangoes against two fruit flies (Diptera: Tephritidae). *Postharvest Biology and Technology* 10, 99–105.

Ikediala, J.N., Tang, J., Neven, L.G. and Drake, S.R. (1999) Quarantine treatment of cherries using 915 MHz microwaves: temperature mapping, codling moth mortality and fruit quality. *Postharvest Biology and Technology* 16, 127–137.

Kaimal, J.C. and Gaynor, J.E. (1991) Another look at sonic thermometry. *Boundary-Layer Meteorology* 56, 401–410.

Kyuma, K., Tai, S., Sawada, T. and Nunoshita, M. (1982) Fiber-optic instrument for temperature measurement. IEEE *Journal of Quantum Electronics* 18, 676–679.

Liao, T.L., Tasi, W.Y. and Huang, C.F. (2004) A new ultrasonic temperature measurement system for air conditioners in automobiles. *Measurement Science and Technology* 15 (2), 413–419.

Magison, E.C. (1990) *Temperature Measurement in Industry.* American Instrument Society. Pittsburgh, Pennsylvania.

Michalski, L., Eckersdorf, K., Kucharski, J. and McGhee, J. (2001) *Temperature Measure-*

ment 2nd edn. John Wiley & Sons Ltd, Chichester, UK.

Neven, L.G., Rehfield, L.M. and Shellie, K.C. (1996) Moist and vapour forced air treatment of apples and pears: effects on the mortality of fifth instar codling moth (Lepidoptera: Tortricidae). *Journal of Economic Entomology* 89, 700–704.

Nott, K.P. and Hall, L.D. (1999) Advances in temperature validation of foods. *Trends in Food Science and Technology* 10, 366–374.

Ono, M., Igarashi, T., Ohno, E. and Sasaki, M. (1995) Unusual thermal defence by a honeybee against mass attack by hornets. *Nature* 377, 334–336.

Ruan, R., Chen, P., Chang, K., Kim, H.J. and Taub, I.A. (1999) Rapid food particle temperature mapping during ohmic heating using Flash MRI. *Journal of Food Science* 64, 1024–1026.

Simpson, J.B., Pettibone, C.A. and Kranzler, G. (1991) Temperature. In: Henry, Z.A., Zoerb, G.C. and Birth, G.S. (eds) *Instrumentation and Measurement for Environmental Sciences.* American Society of Agricultural Engineers, pp. 601–617.

Sun, X.Z., Schmidt, S.J. and Litchfield, J.B. (1995) Temperature mapping in a potato using half Fourier transform MRI of diffusion. *Journal of Food Process Engineering* 17, 423–437.

Wang, S., Yernaux, M. and Deltour, J. (1999) A networked two-dimensional sonic anemometer system for measurement of air velocity in greenhouses. *Journal of Agricultural Engineering Research* 73 (2), 189–197.

Wang, S., Ikediala, J.N., Tang, J., Hansen, J.D., Mitcham, E., Mao, R. and Swanson, B. (2001a) Radio frequency treatments to control codling moth in in-shell walnuts. *Postharvest Biology and Technology* 22, 29–38.

Wang, S., Tang, J. and Cavalieri, R.P. (2001b) Modeling fruit internal heating rates for hot air and hot water treatments. *Postharvest Biology and Technology* 22, 257–270.

Wang, S., Tang, J., Johnson, J.A., Mitcham, E., Hansen, J.D., Cavalieri, R.P., Bower, J. and Biasi, B. (2002) Process protocols based on radio frequency energy to control field and storage pests in in-shell walnuts. *Postharvest Biology and Technology* 26, 265–273.

Wang, S., Tang, J. and Younce, F. (2003). Temperature measurement. In: Heldman, D.R. (ed.) *Encyclopedia of Agricultural, Food, and Biological Engineering.* Marcel Dekker, New York, pp. 987–993.

Wang, S., Johnson, J.A., Tang, J. and Yin, X. (2005a) Heating condition effects on thermal resistance of fifth-instar navel orangeworm (Lepidoptera: Pyralidae). *Journal of Stored Products Research* 41, 469–478.

Wang, S., Yue, J., Tang, J. and Chen, B. (2005b) Mathematical modelling of heating uniformity of in-shell walnuts in radio frequency units with intermittent stirrings. *Postharvest Biology and Technology* 35 (1), 97–107.

4   Physiological Responses of Agricultural Commodities to Heat Treatments

S. LURIE[1] AND E.J. MITCHAM[2]

[1]Department of Postharvest Science, Agricultural Research Organization, Volcani Center, Bet Dagan, Israel; e-mail: slurie43@volcani.agri.gov.il; [2]Department of Plant Sciences, University of California, Davis, California, USA; e-mail: ejmitcham@ucdavis.edu

4.1. Introduction

The chapters in this book reflect the wide interest in heat treatments for pest and disease control and maintenance of postharvest quality. Heat treatments include hot water dips, hot water brushing techniques, hot air treatments (vapour heat and forced air) and radio frequency (RF) heating. All have to be adapted to specific commodities to achieve direct benefits without causing damage. This chapter discusses briefly some of the effects that high temperatures have on harvested commodities, both beneficial and detrimental.

4.2 Effects on Physiology

High-temperature treatment of fruits or vegetables, whether applied for reasons of quarantine, decay control or to affect product physiology, has profound effects on tissue metabolism. These effects include changes in: (i) tissue respiration; (ii) hormone production – particularly through the effects of ethylene; (iii) enzyme activities; (iv) confirmation of macromolecules – including protein aggregation (discussed in Chapter 11, this volume); (v) membrane components that can lead to increased membrane leakage; and (vi) other changes that impact fruit and vegetable quality.

Respiration

The respiration rate of ripening fruit is initially increased by exposure to higher temperatures (Akamine, 1966; Maxie *et al.*, 1974; Inaba and Chachin, 1988;

Klein and Lurie, 1990; Lurie and Klein, 1990; Mitcham and McDonald, 1993). Following a heat treatment > 33°C for hours or days, the respiration rate of the commodity declines to near or below the level of the non-heated, control commodity. The magnitude of the respiratory response to temperature varies with the physiological age of the tissue. The temperature coefficient (Q_{10}) of apple fruit over the range of 10–30°C declines from 2.8 in young fruit to 1.6 in fruit 6 weeks later (Jones, 1942).

Heat treatments impact the subsequent climacteric respiratory rise in fruits and vegetables. For example, holding tomato fruit at temperatures of 33°C or higher for a number of days resulted in suppression of respiration in tomato fruit not completely reversed after returning the tomato fruit to ambient temperatures (Cheng et al., 1988; Inaba and Chachin, 1988). The increase between the initial preclimacteric level and the climacteric peak in avocado fruit was 250% at 20°C and only 30% at 40°C (Eaks, 1978).

During heat treatments at 43–48°C, an initial increase in respiration rate occurred in ripening apples (Lurie and Klein, 1990), papayas (Paull and Chen, 1990), mangoes (Mitcham and McDonald, 1997) and tomatoes (Lurie and Klein, 1992), and then declined to the same level or below in fruit that had received no heat treatment. The climacteric peak of heat-treated fruit may also be delayed (Mitcham and McDonald, 1993; Saftner et al., 2003). Klein and Lurie (1990) found a lower respiration in heated apples returned to ambient temperatures compared with non-heated fruit.

Cheng et al. (1988) found that the respiratory control ratio (the O_2 consumption compared to ATP production) is higher in tomato mitochondria isolated from fruit stressed for 3 days at 37°C than in fruit held at 20°C; mitochondrial ATPase activity was also very significantly reduced by the same stress. Chronic heat exposure led to a shift to the cyanide-insensitive pathway (Inaba and Chachin, 1988). This shift to alternative oxidative phosphorylation may reflect disruption of the thermoregulatory role of this pathway in order to maintain cellular stasis (Briedenbach et al., 1997). The high rate of respiration during heat treatment is a likely reason for the decrease in titratable acidity found in many studies of heat-treated commodities (Lurie, 1998), as the organic acids are utilized as respiratory substrates.

High rates of respiration also cause the internal atmospheres of fruits and vegetables to increase in CO_2 and decrease in O_2 (Shellie and Mangan, 2000), which may aid disinfestation. Insects also respire more quickly at high temperatures and are affected by high CO_2 and low O_2 in their surroundings (Neven, 2000). In many cases, applying controlled atmosphere to a heat treatment decreases the time needed for disinfestation treatments (see Chapter 10, this volume, for more details).

Ethylene and Ripening

The synthesis of ethylene, which synchronizes the ripening processes of climacteric fruits, is inhibited at temperatures near or above 35°C. In apples heated in air at 38°C, ACC oxidase activity decreases by 90% compared with that in untreated apples. The decline in apple ACC oxidase correlates well with ACC accumulation and the inhibition of ethylene produced by heated apples (Klein, 1989), an indication that a 38°C heat treatment inhibits ACC oxidase more than ACC synthase.

Roh *et al.* (1995) and Antunes and Sfakiotakis (2000) found that ACC synthase was inhibited by a 38°C heat treatment, but more slowly than ACC oxidase in both apples and kiwi fruit, respectively. Recovery following the inhibition of the ethylene pathway by heat is slow during apple storage and, upon removal from storage, ethylene production is often higher than from untreated fruit (Klein, 1989; Lurie and Klein, 1990). The heat-induced inhibition of ethylene synthesis may be due both to reduced synthesis of new enzymes and to direct inhibition of enzyme activity. The mRNA of ACC oxidase is strongly depressed at 38°C (Lurie *et al.*, 1996).

In climacteric fruits, which depend upon ethylene for coordinated ripening, the high-temperature inhibition of ethylene can inhibit many ripening processes, including fruit softening, colour changes and aroma development. These inhibitions, as in the case of heat-induced inhibition of ethylene synthesis, are reversible if the heat treatment is not too extended and does not cause damage. Volatile production, which was inhibited by 38°C hot air treatment of apples, or 42°C hot water immersion of tomatoes, recovered following the treatment when the fruit were held at ambient temperatures (Fallik *et al.*, 1997; McDonald *et al.*, 1998). Lycopene synthesis inhibited by heat treatment of tomatoes also recovered, and heat-treated tomatoes ripened normally (Lurie and Klein, 1992; McDonald *et al.*, 1998). In fact, if tomatoes are stored at low temperatures, such as 2°C, heat-treated fruit ripen normally after storage while control fruits decay as a result of chilling injury (Lurie and Klein, 1992).

Polygalacturonase, one of the enzymes involved in cell wall digestion that leads to fruit softening, is induced by ethylene. High temperature inhibits the activity of polygalacturonase and affects fruit softening in both mango (Ketsa *et al.*, 1998) and tomatoes (Mitcham and McDonald, 1992), and reduces mRNA abundance in tomato (Kagan-Zur *et al.*, 1995). High temperature also inhibits β-mannanase and α- and β-galactosidase activities in tomatoes (Sozzi *et al.*, 1997). Again, the inhibition of ripening-related processes such as colour development, volatile evolution and softening – as with ethylene synthesis – is at the levels of both enzyme activity and gene expression.

Despite inhibition of ethylene synthesis at high temperatures, heat-treated fruit do not respond to exogenous ethylene or propylene, suggesting inactivation of ethylene receptors (Yang *et al.*, 1990). Kiwi fruit are induced to ripen when treated with propylene at temperatures from 20 to 35°C. At 38°C and above, propylene does not induce ripening (Antunes and Sfakiotakis, 2000), while ripening is also inhibited in tomatoes treated at 38°C in the presence of ethylene (Yang *et al.*, 1990). High temperatures, i.e. above 35°C, appear to inhibit ripening by limiting the accumulation of ripening-related mRNAs due to temperature-inhibition of ethylene, but Kagan-Zur *et al.* (1995) found that exogenous ethylene treatment did not reverse 38°C inhibition of polygalacturonase mRNA accumulation.

Non-climacteric fruit also show a reduction in softening rate and colour development following heat stress. Strawberries, treated either with hot air for 3 h or hot water heated for 15 min at temperatures from 40 to 50°C and then held at 20°C for 3 days, had delayed colour development and reduced firmness loss compared to unheated berries (Garcia *et al.*, 1995; Civello *et al.*, 1997). Therefore, the inhibition of ripening may also be a consequence of a general inhibition of gene expression at heat stress-inducing temperatures (Vierling, 1991).

The inhibition of ripening in a particular fruit or vegetable after heat treatment results from a combination of factors: (i) preharvest environmental conditions; (ii) the physiological age of the commodity; (iii) the time and temperature of exposure; (iv) whether the commodity is removed from heat to storage or to ripening temperature; and (v) whether the heat treatment causes damage.

Ethylene and Senescence

Yellowing is one of the most important factors in the quality degradation of green vegetables. Broccoli senesces very rapidly and the florets turn yellow after harvest as a result of chlorophyll degradation. Reproductive structures within the florets produce high amounts of ethylene, which plays a role in broccoli senescence (Tian *et al.*, 1994). Chlorophyll breakdown is enhanced by heat treatment in many crops, including apples (Liu, 1978; Lurie and Klein, 1990), possibly due to enhanced chlorophyllase activity (Amir-Shapira *et al.*, 1986).

High-temperature treatment with 45°C water for 10 min (Tian *et al.*, 1996, 1997), 45°C for 14 min (Kazami *et al.*, 1991a, b), 50°C for 2 min (Forney, 1995) or 50°C hot air for 2 h (Terai *et al.*, 1999) delayed yellowing of harvested broccoli florets and reduced the rate of ethylene production. The difference in times and temperatures that were successful on this crop may be due to the fact that each study used different broccoli cultivars. The hot water treatments also slowed the loss of soluble proteins and ascorbic acid (Kazami *et al.*, 1991b; Tian *et al.*, 1997). The enzymes that promote chlorophyll degradation, chlorophyllase and peroxidase, were inhibited in the hot air-treated broccoli, while they increased in non-heated florets during storage (Funamoto *et al.*, 2002, 2003).

Similarly, hot moist air at 40 or 45°C from 30 to 60 min reduced the senescence of harvested kale and collard leaves (Wang, 1998). Yellowing of the leaves was delayed, losses of sugars and organic acid were slower and turgidity was maintained during storage after the heat treatments. However, the same treatments when applied to Brussels sprouts were not effective in delaying yellowing.

Other processes that involve ethylene in harvested vegetables, such as geotropic bending of asparagus (Paull and Chen, 1999) and extended growth of green onions (Hong *et al.*, 2000; Cantwell *et al.*, 2001), were also controlled by hot water treatments before storage. Hot water dips controlled sprout and root growth in garlic cloves (Cantwell *et al.*, 2003), in addition to sprouting and spoilage of potatoes (Ranganna *et al.*, 1998). The best treatments for inhibition of these physiological processes were 2–4-min dips in water ranging from 50 to 55°C. A 90 s dip in 50°C water reduced or prevented browning of the cut edges of both lettuce and celery petioles (Saltveit, 2000; Loaiza-Velarde *et al.*, 2003).

4.3 Types of Heat Damage

The goals of high-temperature treatments are to control microorganisms and/or insect pests whilst preventing commodity damage. Often, the difference between controlling a postharvest pest and causing damage to the commodity under treatment is a matter of only a few degrees (see Table 4.1). This small margin is one reason that a multitude of treatments exist, the result of research efforts to find a time–temperature regime that produces the desired effect without damaging the commodity – which can be both external and internal.

Most of the hypotheses that explain heat injury involve denaturation or disruption of protein synthesis. Many cellular processes decline during heat exposure as previously discussed, including chloroplast activity, chlorophyll levels, respiration rate, ethylene production and cell wall-degrading enzymes, while electrolyte leakage and heat shock protein synthesis increase (Paull, 1990).

External Damage

Damage to agricultural commodities can appear as peel browning (Kerbel *et al.*, 1987; Klein and Lurie, 1992; Lay-Yee and Rose, 1994; Woolf and Laing, 1996; Schirra and D'hallewin, 1997), pitting (Miller *et al.*, 1988; Jacobi and Gowanlock, 1995; Fig. 4.1) or yellowing of green vegetables such as zucchini (Jacobi *et al.*, 1996) or cucumber (Chan and Linse, 1989). One of the most common types of damage observed following heat treatment is surface scalding (see Fig. 4.2).

Table 4.1. Response of fruits to various exposure times and temperatures with respect to phytotoxic effects and aspects of ripening.

Fruit	Method	Time	Temperature (°C)	Phytotoxic symptoms	Softer/ firmer	Colour	Sweetness/ acidity
Apple	HA[a]	96 h	38	–	Firmer	Increased	Less acidic
	HA	96 h	43	–	Firmer	Increased	Less acidic
Oroblanco citrus	HA	4 h	46	+	Softer	Increased	No change
	HA	4 h	47	+	Softer	Increased	No change
Mango	VH[b]	15 min	46	–	No effect	No effect	No change
	VH	30 min	46	–	Softer	Increased	No change
	VH	30 min	47	+	Softer	Increased	No change

[a] Hot air; [b] vapour heat.

Fig. 4.1. Pitting in mango fruit (courtesy W. Miller, USDA ARS, Fort Pierce, Florida, USA).

Fig. 4.2. Scald in papaya fruit (courtesy R. Paull, University of Hawaii, Honolulu, Hawaii, USA).

Kerbel *et al.* (1987) found severe surface browning on 'Fuerte' avocados exposed to hot air at 43°C for 3.5–12 h. Anthony *et al.* (1989) found that the degree of external browning of stone fruit increased as the duration of the heat treatment at 52°C increased from 15 to 45 min, but was eliminated when the fruit were enclosed in plastic wrap, possibly due to reduced water loss and shriveling.

'Tommy Atkins' mangoes heated with 51.5°C air for 125 min showed some pitting on the skin (see Fig. 4.1) in areas where the fruit had previously been abraded or bruised (Miller *et al.*, 1991a). 'Manila' mangoes showed severe skin scalding when forced-air heated at temperatures of 45°C or higher, slight skin scalding from heating at 44°C but no damage at 43°C, indicating the presence of a threshold temperature for skin injury (Ortega-Zaleta and Yahia, 2000).

Hot water treatments can also result in damage to the epidermis of the commodity. Red ginger flowers heated in water at 49°C for 12–15 min developed necrotic tissue from injury to the inner bracts (Hara *et al.*, 1996), as well as an intensification of mechanical injury, emphasizing the importance of careful handling of materials destined for heat treatment. Surface browning of peaches exposed to hot water treatments increased with both time (1.5–5 min) and temperature (50–55°C) (Phillips and Austin, 1982). Fruit differed in susceptibility because of seasonal and maturity effects.

Many researchers have shown that heating with moist forced air is less damaging to fruit than hot water- or vapour-forced air. Shellie and Mangan (2000) closely monitored fruit surface temperatures during heating with hot water, water vapour pressure-deficit air (moist forced air), and water vapour-saturated air (vapour-forced air, see Fig. 4.3) and found the temperature of the fruit surface varied according to the heating medium used.

Fruit surface temperature was coolest when heated with water vapour pressure-deficit air (MFA) and reached only 81% of the temperature of the heating medium after 5 min of exposure. The surface of the fruit heated with vapour-saturated air or water reached 96% of the temperature of the heating medium within 5 min. During heating with vapour-saturated air, the surface temperature of the fruit exceeded the air temperature after about 20 min of heating and remained 1°C higher, while the surface temperature of fruit heated in vapour pressure-deficit air remained cooler than the air.

The authors suggest that the latent heat of condensation may be

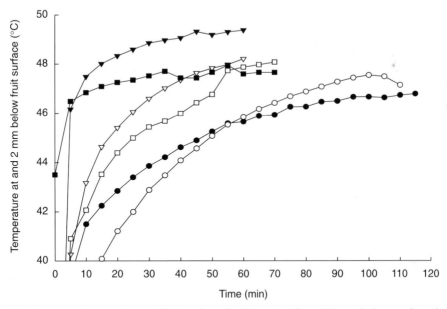

Fig. 4.3. Average temperature at fruit surface (solid symbols) and 2 mm below surface (open symbols) during heating via vapour pressure-deficit, forced air (circles), vapour-saturated air (triangles) and hot water (squares). Values represent the average of 12 fruit (grapefruit, orange, papaya and mango) over four treatment replications (from Shellie and Mangan, 2000).

responsible for the hotter surface temperature during vapour forced-air heating and that evaporative cooling may result in lower surface temperatures in moist forced-air heating where there is a vapour pressure deficit. These data indicate that the water vapour pressure during forced-air heating influences the surface heat transfer coefficient (Shellie and Mangan, 2000) and potential to develop skin scalding.

Increased Water Loss

Heat-treated commodities often lose water at an increased rate, such as musk melons exposed to 45°C hot air for 1.5 or 3 h (Lester et al., 1988), 'Tommy Atkins' mangoes treated with forced air at 51.5°C for 125 min (Miller et al., 1991a), strawberry fruit heated in air at 45°C for 3 h (Vicente et al., 2002) and peaches treated with hot water at 50–55°C for up to 5 min (Phillips and Austin, 1982). The mango fruit lost 1% more weight than their untreated counterpart during 3 weeks of storage at 12°C, and the strawberries had 2% higher weight loss at 20°C immediately after treatment.

Vapour heat treatment of table grapes had variable effects on weight loss following treatment (Lydakis and Aked, 2003). When fruit were heated at 52.5 or 55°C for 24 or 27 min, there was no effect on water loss, but when fruit were heated at 55 or 58°C for 21–30 min, there was a significant increase in water loss.

Heat treatments can cause cuticular wax to melt slightly and fill in surface micro-cracks and stomata (Schirra et al., 2000) in commodities, which may help to reduce transpiration and explain why some heat treatments have either no effect on water loss or even reduce it. Williams et al. (1994) found that Valencia oranges treated in 45°C water for 42 min lost significantly less moisture and remained firmer during storage than unheated control fruit. By contrast, Valencia oranges immersed in 53°C water for 12 min showed the greatest weight loss, suggesting that when heat treatment is more extreme, damage results that increases water loss, while a mild heat treatment can melt cuticular waxes and reduce water loss.

Lydakis and Aked (2003) observed that some grape berries heated in 58°C air showed fine splits. The authors postulated that this was due to a build-up of internal (turgor) pressure in the berry during the increase in temperature which exceeded the rupture strength of the skin, in part based on Lang and During's (1990) demonstration of a decrease in skin stiffness and strength with increasing temperature.

Internal Damage

Internal injury to fruit can also occur as a result of heat treatment, sometimes in the absence of any external damage. The most common symptoms include flesh darkening in avocado, citrus, lychee, nectarine and sapote mammey

(Jacobi *et al.*, 1993; Shellie *et al.*, 1993; Lay-Yee and Rose, 1994; Shellie and Mangan, 1994, 1996; Diaz-Perez *et al.*, 2001; Follett and Sanxter, 2003).

In mango and papaya, internal injury occurs as poor colour development, abnormal softening, a lack of starch breakdown and internal cavities (An and Paull, 1990; Jacobi and Wong, 1992; Mitcham and McDonald, 1993). Jacobi *et al.* (1995b) observed surface injuries to mango fruit from hot water treatment, but also internal cavities and starchy layers beneath the skin.

Heat treatments can modify the internal atmospheres of fruit during treatment. Mitcham and McDonald (1993) demonstrated that CO_2 increased to 13% and O_2 decreased to 6% in mangoes subjected to high-temperature forced air treatments and this was related to the development of internal cavities in the mango flesh (see Fig. 4.4).

Esquerra and Lizada (1990) showed a similar modification of the internal atmosphere of 'Carabao' mangoes following vapour heat treatment. Vapour heat-treated mangoes increased in internal cavitation from a range of 0–17% to 66–75% when heated to 46°C for 1.5 h (Esquerra and Lizada, 1990). Internal cavitation was also observed in 'Hayward' kiwi fruit heated at 40°C in a controlled atmosphere (0.4% O_2 + 20% CO_2) for 10 h when the fruit were not hydrocooled after treatment (Lay-Yee and Whiting, 1996). Internal cavities can also develop in mango fruit from exposure to high temperatures in the field (Gunjate *et al.*, 1982) or from exposure to modified atmospheres, especially at ambient temperatures in the tropics (Gautam and Lizada, 1984). Symptoms of cavitation include spongy tissues with air pockets, which do not appear until the fruit have ripened. Starch degradation in this tissue is inhibited.

Shellie and Mangan (1996) demonstrated that hot water treatments result in modification of a commodity's internal atmosphere, even more than during

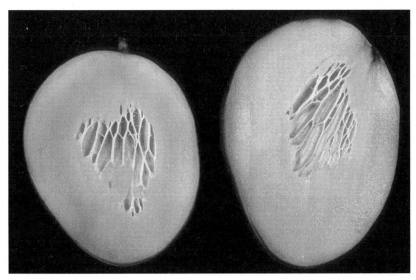

Fig. 4.4. Cavitation in mango fruit following hot air heating (courtesy E. Mitcham, University of California, Davis, California, USA).

a hot air treatment. During hot water heating of grapefruit at 44°C, internal CO_2 increased to 22% and O_2 decreased to 3% over a 3 h period. This significant internal atmosphere change may explain the greater quarantine efficiency of hot water over hot air treatments when the heating rate and product temperature are the same.

Decay

In many cases, a successful heat treatment can reduce the fungal inoculum, and also enhance resistance factors in the fruit or vegetable, making it harder for a pathogen to infect a commodity. These effects are discussed in Chapter 7, this volume. However, there are also many cases where the fruit or vegetable becomes more susceptible to decay as the result of a high-temperature treatment. This increase in susceptibility may occur even when there is no visible heat damage (Phillips and Austin, 1982).

Generally, the higher time–temperature conditions needed for disinfestations render a commodity more susceptible to decay than those conditions used to affect physiological processes or prevent decay. Miller *et al.* (1988) found an increase in *Penicillium* sp. decay on grapefruit that had been treated with hot water to control Caribbean fruit fly, and Hallman *et al.* (1990) found a similar increase in decay of grapefruit following a vapour heat treatment. Although the hot water quarantine treatment first used for fruit fly control in papaya fruit provided effective decay control, the hot air treatment that replaced it was ineffective in controlling postharvest diseases. However, hot forced-air treatments were found to be effective for anthracnose control in mango (Mitcham and McDonald, 1993; Fig. 4.5).

Tissue damage caused by heat can result in increased decay development (McGuire, 1991; Jacobi and Wong, 1992; Jacobi *et al.*, 1993; Lay-Yee and Rose, 1994). A 37°C hot air treatment of 'Tarocco' blood oranges enhanced secondary fungal infections such as *Phytophthora* rots following storage and shelf life (Schirra *et al.*, 2002), although a hot water dip of 50°C for 2 min decreased decay (Schirra *et al.*, 1997). An increase in decay susceptibility is part of the damage that can be caused by an inappropriate heat treatment. Higher decay in kiwi fruit heated in hot air could be caused by decay organisms invading areas injured by the treatment (Lay-Yee and Whiting, 1996).

Membrane Breakdown

Chen *et al.* (1982) demonstrated that electrolyte leakage is an effective measure of plant cell membrane thermostability, and other researchers (Martineau *et al.*, 1979; Ingram and Buchanan, 1981, 1984) reported a sigmoidal response in electrolyte leakage to treatment temperature for selected exposure times. A similar pattern of electrolyte leakage was observed in tomato disks exposed to 55°C for 34 min, 50°C for 105 min and 45°C for 166 min (Inaba and Crandall, 1988), but no visible signs of heat injury were observed.

Fig. 4.5. Control of anthracnose rot in mango fruit following forced hot air treatment at 46°C for 4 h (courtesy E. Mitcham, University of California, Davis, California, USA).

Loss of cellular membrane integrity with increasing temperatures is gradual, with repair and reversibility possible until a lethal temperature is reached. Mitcham and McDonald (1993) saw an increase in electrolyte leakage in the flesh of heat-treated mango fruit. Electrolyte leakage was higher in tissue from fruit treated at the higher temperature or for longer duration at a given temperature. Three days after treatment, electrolyte leakage had returned to control levels in all treatments except the most severe. Lurie and Klein (1990) observed 50% higher electrolyte leakage in fruit disks from apples held at 38°C for 3 days than in disks from unheated apples. The leakage returned to control levels after 2 days at 20°C. Electrolyte leakage may be an effect of heat treatment but not the cause of damage, since little change in electrolyte leakage has been reported during the first few hours of heat treatment (Paull and Chen, 1999).

Acclimation to high temperatures appears to involve changes in membrane components. Phospholipid content was higher and more saturated in heated apple fruit than in unheated fruit, and there was an increase in membrane microviscosity and sterols (Lurie *et al.*, 1995). Transgenic tobacco plants in which the chloroplast omega-3 fatty acid desaturase was silenced were better able to acclimate to higher temperatures than were wild-type plants (Murakami *et al.*, 2000). However, membrane lipid saturation is not correlated with heat tolerance in some systems (Gombos *et al.*, 1994).

Sensory Effects

Heat treatment can affect the sensory quality of fruit by altering flavour and texture. Volatile production was inhibited by 38°C hot air treatment of apples or

42°C hot water immersion of tomatoes, but recovered following the treatment, particularly if the fruits were stored (Fallik *et al.*, 1997; McDonald *et al.*, 1998). Volatiles and other compounds that contribute to off-flavours can accumulate in injured tissue (Schirra and D'halllewin, 1997; Schirra and Cohen, 1999). After 9 months of storage, 20% of sensory panellists indicated an off-flavour in raisins that had been heated with hot air at 55°C for 1 h before storage, but no difference was detected in prunes subjected to the same treatment (Johnson *et al.*, 1989).

Mango fruit subjected to hot air treatments showed an increase in ethanol and acetaldehyde concentrations, which could result in a fermented flavour (Mitcham and McDonald, 1993). This increases in ethanol and acetaldehyde were transient when the fruit was heated to 46°C for 3 or 4 h, but when the fruit was heated to 48°C for 5 h the concentrations were much higher and continued to increase for several hours after treatment. The increase in these fermentative metabolites was linked to a modification of the fruit's internal atmosphere during the heat treatment. Greater demand for O_2 can be expected at elevated temperatures, which may increase the O_2 concentration at which fermentative metabolism is initiated (Zhou *et al.*, 2000).

In stone fruit, such as peaches and nectarines, heat treatment may enhance the development of flesh mealiness following storage (Obenland and Carroll, 2000) and, in sweet cherries, hot water treatments often resulted in a 'cooked' texture and flavour according to informal sensory panels (E. Mitcham, unpublished data).

4.4 Responses of Dried Commodities to Heat Treatment

Water Content

The moisture content of dried commodities is carefully regulated to prevent microorganism growth and the development of rancidity, while avoiding over-drying (Rockland, 1962; Thompson *et al.*, 1998). Walnut kernels are most stable at 3.5% equilibrium relative humidity; darkening and rancidity occurred at an accelerated rate after storage at either 0.5% higher or lower moisture levels (Rockland, 1962). When heat is applied to dry commodities, it is often in the form of non-humidified air to prevent undesirable increases in product moisture content. However, this can result in excessive drying of the commodity. Dry hot air heating of walnut and almond kernels reduced their moisture content in relation to treatment temperature and duration (Buransompob *et al.*, 2003). Walnut kernel moisture content decreased by 25–30% after hot air treatments at 55°C for 2 min or 60°C for 10 min, compared to 5.5–11.6% weight loss in almond kernels; initial moisture contents were 4.4 and 5.7%, respectively.

Wahab *et al.* (1984) reported a decrease in pecan kernel moisture content as a result of pre-storage heating. Wang *et al.* (2002) found that using radio frequency (RF) energy to heat in-shell walnuts to 55°C for 5 or 10 min had no effect on shell moisture, but significantly reduced kernel moisture content relative to the unheated walnuts. The sensory panellists did not detect any differences between control and heat-treated walnut kernel flavour or force required to crack the kernels. The authors suggest that the drying effect from RF heating might be beneficial for the tree nut industry.

Rancidity

Walnuts and almonds contain unsaturated fatty acids that are susceptible to oxidative rancidity when subjected to high temperatures, prompting processors to avoid temperatures greater than 43°C for drying (Rockland, 1962; Olson et al., 1978). Almonds are less susceptible to rancidity than walnuts because of smaller concentrations of polyunsaturated fatty acids and larger amounts of tocopherol antioxidants (Young and Cunningham, 1991; Macrae et al., 1993).

According to sensory panels, almonds, macadamia nuts and pistachios were acceptable after 6–9 months' storage at 38°C, while cashews, hazelnuts, peanuts, pecans and walnuts were acceptable only after 2–5 months' storage at 38°C before becoming rancid (Young and Cunningham, 1991). In one study, walnuts were heated with hot air at 40°C for 48 h or 45°C for 4 h. After 8–9 months of storage, the walnuts became rancid, as related to the severity of the heat treatment (Johnson et al., 1989). However, short-time heat treatments of 55°C for 2 min did not promote rancidity of almonds or walnuts (Buransompob et al., 2003).

Nut quality is usually determined by the presence of off-flavours, most commonly due to oxidative rancidity (Fourie and Basson, 1989; Zacheo et al., 2000). Peroxide value (PV) is measured as an indication of oxidative rancidity and is reported as milliequivalents (meq) of peroxide per kg of oil (Hui, 1992). Pecan kernels heated at 71 or 82°C for 5 or 10 min exhibited increasingly lower PVs than unheated pecan kernels during storage at room temperature (Wahab et al., 1984).

Reduced peroxide levels were also seen in walnut and almond kernels following short-time hot air treatments of 55°C for 2 min (Buransompob et al., 2003; 2007) and for walnut kernels heated to 55°C using RF heating and held for 5 or 10 min during accelerated storage of up to 20 days at 35°C (representing 2 years at 4°C based on a Q_{10} value of 3.4 at 35°C; Taoukis et al., 1997; Wang et al., 2002; Fig. 4.6).

Zacheo et al. (2000) reported that lipoxygenase activity increased during ageing of almond kernels, but heating almond kernels at 80°C for 10 min eliminated it. Buransompob (2001) also demonstrated that heating almonds and walnuts in hot air at 55 or 60°C for 2 to 10 min inactivated the lipoxygenase activity retarded the development of the oxidative rancidity and extended shelf-life of almonds and walnuts. When soybean lipoxygenase was added to a crude extract of walnuts or almonds, a higher activity was exhibited in walnut extract. This difference was attributed to the higher amount of alpha-tocopherol (24.0 mg/100 g) in almonds as compared with walnuts (2.6 mg/100 g) (Buransompob, 2001).

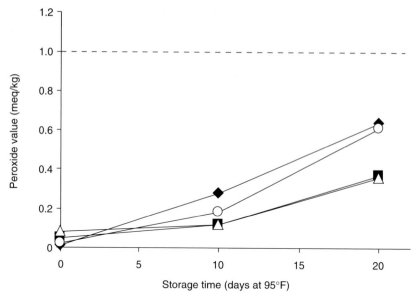

Fig. 4.6. Changes in peroxide values in walnut kernels during accelerated storage at 35°C for 20 d, simulating 2 years at 4°C. ♦, control; ■, RF plus 5 min storage; △, RF plus 10 min storage; □, RF plus 10 min storage plus hot air (from S. Wang and J. Tang, Washington State University, Pullman, Washington, USA).

Hydrolytic rancidity results from the hydrolysis of triacylglycerols in the presence of lipases and moisture, and is assessed by measuring fatty acids in oils by titration (Buransompob *et al.*, 2003). Heat treatments reportedly reduce hydrolytic rancidity: for example, when unshelled pecans were heated at 80°C for 15 min, a decrease in rancidity was observed along with inhibition of enzyme activity (McGlammery and Hood, 1951). Reducing moisture content and the inactivation of lipase and esterase activity by heat treatment may also inhibit the onset of hydrolytic rancidity (Allen and Hamilton, 1999). Ekstrand *et al.* (1993) reported an increase in fatty acid values in stored pecans, but a less pronounced increase in heated pecans. However, Wang *et al.* (2002) reported no effect on fatty acid values in walnut kernels heated to 55°C using RF and held for 5 or 10 min compared with unheated kernels during accelerated storage of up to 20 days at 35°C, suggesting that a higher temperature is required for inhibition of hydrolytic rancidity than for oxidative rancidity.

4.5 Factors Affecting Response to Heat Treatment

The maximum heat tolerance of plants, normally in the range of 42–60°C, correlates roughly with the original evolutionary habitat (Kappen, 1981). Lethal temperatures are reported at 45°C for tomatoes, 63°C for grapes and 49–52°C for apples (Huber, 1935). This variation in lethal temperatures can be traced, in part, to exposure duration (Paull and Chen, 1999). A short exposure to near-lethal temperatures is a 'crisis' situation related to survival and the capacity to recover, while longer exposures to lower temperatures are more difficult to quantify.

Exposure of pears (Maxie *et al.*, 1974), papaya (An and Paull, 1990) and tomatoes (Biggs *et al.*, 1988; Picton and Grierson, 1988) to 30°C for 48 h or more results in a delay of various aspects of ripening, including pigmentation, softening and ethylene production. Couey (1989) grouped fruit into heat-tolerant and heat-sensitive types based on their exposure to a hot water treatment (see Table 4.2). Among the heat-tolerant types were banana, papaya and mango. Differences may reflect actual thermal adaptation; however, many other factors influence the response of fruit to heat treatments.

Growing Conditions

Postharvest heat treatments can modify postharvest responses of fruits and vegetables, and serve as both antifungal and quarantine treatments (Lurie, 1998), but preharvest growing conditions – including water stress, high light intensity and field temperatures – may influence the response of the commodity to a heat treatment. Preharvest cultural factors such as rootstock, production practices (e.g. the use of growth regulators before harvest that alter maturation of the fruit or vegetable) and climatic conditions during physiological development of the fruit also need to be considered when deciding on a heat treatment.

If a commodity has a long harvest season, such as citrus fruits, the season of harvest may play a role in the response of fruit to a heat treatment. Early-season fruit are more susceptible to injury (Miller and McDonald, 1992), possibly due to degreening ethylene treatment. Two grapefruit cultivars, 'Marsh' – a white-fleshed variety and 'Ruby Red' – a thinner peel, red-fleshed variety, were harvested at different times in the season and tested for their ability to withstand a vapour heat disinfestation protocol of 43.5°C for 30 min (Hallman *et al.*, 1990; Miller and McDonald, 1991; Miller *et al.*, 1991b). The

Table 4.2. Fruits that are heat-tolerant or heat-sensitive according to their response to hot water treatment (adapted from Couey, 1989, with additions from Lurie, unpublished).

Heat-tolerant	Heat-sensitive
Apple	Bell pepper
Banana	Cantaloupe
Lychee	Citrus
Mango	Peach
Papaya	Raspberry
Pear	

disinfestation treatment did not injure 'Marsh' or 'Ruby Red' fruit from mid- or late-season harvests (Miller and McDonald, 1991), but 'Ruby Red' fruit was more sensitive to heat treatment than 'Marsh' grapefruit when harvested early (Miller and McDonald, 1992).

In another study on 'Marsh' grapefruit, with 46°C forced hot air for a quarantine treatment (McGuire and Reeder, 1992), the most resistant fruit to heat damage were those from mid-season harvests. Both early- and late-season fruit were more easily damaged by temperatures of 48°C or higher than were the mid-season fruit. The peel of the early-season fruit is immature (Benschoter, 1979), while fruit harvested late in the season has begun to senesce (McGuire and Reeder, 1992). Both immature and senescence peel were more delicate than mid-season peel.

One treatment of grapefruit with the growth regulator gibberellin to retard peel yellowing maintained increased peel resistance to ovipositioning by fruit flies, by maintaining the peel in an immature condition that did not attract the flies (Greany et al., 1995). Gibberellin-treated fruits, when subjected to vapour heat treatment, had higher quality after storage than fruit with no gibberellin treatment (Miller and McDonald, 1997), including reduced levels of postharvest decay. In contrast, degreening treatments with ethylene on early-season citrus fruits increased the susceptibility of the fruit to heat damage during quarantine treatments (Miller and McDonald, 1992).

Smith and Lay-Yee (2000) tested apples for their tolerance to hot water quarantine treatments in the temperature range of 44–46°C for 35–45 min. The incidence of heat damage to 'Royal Gala' apples harvested in five orchards in two regions of New Zealand varied between growing regions, harvest dates and orchards. Early-season fruit had lower levels of both external (skin scalding) and internal (flesh breakdown) damage than did mid- and late-season fruit. Fruit from all harvests tolerated 44°C for 35 min followed by cold storage, but there was a relatively small margin between this treatment and a higher temperature of 47°C that caused damage. The difference in behaviour of apples from different growing regions and harvests may be partially determined by the heat units experienced in the orchard close to harvest. Flesh temperatures in fruit exposed to sunlight can be as much as 15°C above ambient air temperatures (Ferguson et al., 1998; Woolf and Ferguson, 2000). Preharvest high temperatures may have a protective effect in increasing tolerance of fruit to both low- and high-temperature postharvest treatments. In apples, increased fruit temperature led to the accumulation of heat shock proteins, which led to thermotolerance to higher temperatures (Vierling, 1991) such as those given in postharvest heat treatments.

In tropical and subtropical fruit, high flesh temperatures may arise as much from the ambient air temperature as from direct exposure to the sun. Papaya suffered from impaired ripening (i.e. areas of the flesh did not soften) after a hot water disinfestation treatment (Paull and Chen, 1990). This problem was associated with harvest period, winter-harvested fruit being the most sensitive to damage. Paull (1995) found a correlation between the minimum temperature during the 3 days before harvest and development of the disorder after quarantine treatment: if the minimum temperature was > 22.4°C, fruit injury did not occur.

Other subtropical fruit such as avocados and mangoes react differently to disinfestation heat treatments depending on their stage of maturity and preharvest temperature history. Immature 'Carabao' mangoes suffered physiological damage after vapour heat treatments (Esquerra and Lizada, 1990; Esquerra *et al.*, 1990), while 'Kensington Pride' mangoes showed no adverse reaction at any stage of maturity to a vapour heat treatment of 46.5°C for 10 min (Jacobi *et al.*, 1995a). However, 'Kensington Pride' mango did show heat injury when the disinfestation procedure was conducted in hot water at 46.5°C (Jacobi and Wong, 1992).

Avocados growing in direct sunlight in both New Zealand and Israel showed a greater tolerance than shade-grown fruit to both high and low postharvest temperatures. In both cases fruit had flesh temperatures routinely > 40°C (Woolf *et al.*, 1999, 2000). Better tolerance to a 50°C hot water treatment was found in sun-exposed fruit from all avocado cultivars tested, except 'Ettinger', the earliest cultivar (Woolf *et al.*, 2000; Fig. 4.7).

The above studies demonstrate that growing conditions can influence the sensitivity or resistance of different fruit to a high-temperature treatment, as can the maturity of the fruit and genetic influence of the species and cultivar. However, it appears that by varying the type of heat treatment applied – choosing between vapour heat, forced hot air heating, hot water dips or RF heating – and determining the time–temperature treatment needed for the particular application (disinfestation or disinfection), a high-temperature treatment can be developed with minimum damage to a commodity.

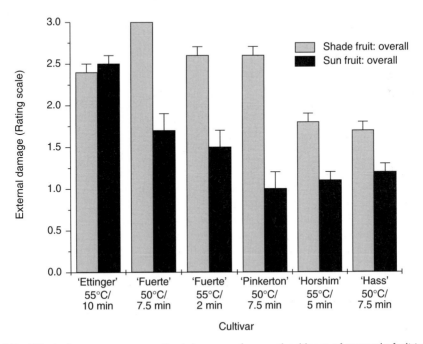

Fig. 4.7. Effect of sun exposure on the tolerance of several cultivars of avocado fruit to hot water treatment (from A. Woolf, HortResearch, Auckland, New Zealand).

Preconditioning

Just as fruits and vegetables exposed to high temperatures in the field acquire thermotolerance, a harvested fruit or vegetable can be given a pretreatment at a moderate elevated temperature to allow it to become resistant to heat damage from a higher disinfestation or disinfection treatment. This pretreatment to help in the development of resistance to more extreme conditions has been examined for development of resistance to high temperature, low temperature and irradiation quarantine treatments. Temperatures of 35–40°C have generally been used to develop tolerance, but this also induces the synthesis of heat shock proteins (Vierling, 1991).

In the previous section we mentioned disinfestation treatments for mangoes that had been either hot water- or vapour heat-based. 'Kensington Pride' mangoes tolerated a vapour heat treatment better than did hot water. However, Joyce and Shorter (1994) found that a pretreatment in air at 37°C could induce tolerance to a hot water immersion treatment of 47°C for 25 min that prevented pulp injury, but did not completely prevent peel scalding.

In another study with 'Kensington Pride' mango, the pretreatment was conducted in air at 39°C between 4 and 12 h and then a 46°C hot water treatment for 30 min was given (Jacobi et al., 1995b). The pretreatment alleviated both external and internal injuries. In both studies, the most effective pretreatment times were 8–12 h. The tolerance to high temperatures induced by the pretreatment was decreased if the mangoes were held at 22°C for periods of 16 h or longer before being given the hot water disinfestation treatment (Jacobi et al., 2001), twice as long as that required to induce the heat tolerance.

'Keitt' mangoes also became tolerant to a 47°C hot water disinfestation treatment after heated to 40°C in hot water before the higher-temperature treatment (McGuire and Sharp, 1997). Similar results were found in a study where 'Keitt' and 'Tommy Atkins' mangoes were held 60 min in 36.5°C water before a 46.5°C hot water disinfestation treatment (Nyanjage et al., 1998).

Other subtropical fruits also develop thermotolerance when pretreated at temperatures < 40°C before a higher disinfestation treatment. Papayas held for 4 h at 38°C developed thermotolerance to a 49°C treatment for 70 min (Paull and Chen, 1990). Holding 'Hass' avocadoes for 60 min in 38°C water gave them thermotolerance to a 10 min, 50°C hot water treatment (Woolf and Lay-Yee, 1997). The thermotolerance in avocados induced by a 38°C pretreatment was maintained if the higher heat treatment was delayed for up 5 days, although if the hot water pretreatment was shorter, the thermotolerance disappeared sooner (Woolf et al., 2004).

Numerous studies show that a high-temperature preconditioning treatment can alleviate chilling injury (Lurie, 1998) from cold storage or cold disinfestation treatments. In citrus, for example, a hot water dip at 53°C for 2 min, a hot water brush at 60°C for 30 s and a hot air treatment at 36°C for 3 days all alleviated chilling injury in 'Star Ruby' grapefruit (Porat et al., 2000). Lurie and Klein (2004) list similar studies on many other fruits and vegetables.

It therefore appears that high-temperature preconditioning can help prevent damage resulting from more extreme temperature treatments (heat or cold) used for quarantine protection. Effective treatment depends upon the commodity and method of preconditioning chosen.

4.6 Conclusions

Application of heat treatments to plant materials as a means of controlling pests provides a non-chemical method of control. However, the tolerance to such treatments must be carefully evaluated. Heat treatment can have beneficial effects beyond pest control such as reducing susceptibility to chilling injury and reducing the rate of ripening. Heat damage may be immediate or develop after a period of storage. Tolerance to heat exposure is influenced by species, cultivar, harvest maturity, growing conditions and handling between harvest and treatment. In addition, the method used to apply heat (air, water, RF energy) can greatly influence product tolerance. In some cases, special treatments can be applied to increase product tolerance to a heat treatment, but one must also consider if this will influence treatment efficacy against insect pests and pathogens.

4.7 References

Akamine, E.K. (1966) Respiration of fruits of papaya (*Carica papaya* L. var. Solo) with reference to the effect of quarantine disinfestation treatments. *Proceedings of the American Society of Horticultural Science* 89, 231–236.

Allen, J.R. and Hamilton, R.J. (1999) *Rancidity in Foods*. Aspen Publishers, Gaithersburg, Maryland, pp. 1–33.

Amir-Shapira, D., Goldschmidt, E. and Altman, A. (1986) Autolysis of chlorophyll in aqueous and detergent suspensions of chlorophyll fragments. *Plant Science* 43, 201–206.

An, J.F. and Paull, R.E. (1990) Storage temperature and ethylene influence on ripening of papaya fruit. *Journal of the American Society of Horticultural Science* 115, 949–955.

Anthony, B.R., Phillips, D.J., Badr, S. and Aharoni, Y. (1989) Decay control and quality maintenance after moist air heat treatment of individually plastic-wrapped nectarines. *Journal of the American Society of Horticultural Science* 114, 946–949.

Antunes, M.D.C. and Sfakiotakis, E.M. (2000) Effect of high temperature stress on ethylene biosynthesis, respiration and ripening of 'Hayward' kiwi fruit. *Postharvest Biology and Technology* 20, 251–258.

Benschoter, C.A. (1979) Seasonal variation in tolerance of Florida 'Marsh' grapefruit to a combination of methyl bromide and cold storage. *Proceedings of the Florida State Horticulture Society* 92, 166–167.

Biggs, M.S., Woodson, W.R. and Handa, A.K. (1988) Biochemical basis of high temperature inhibition of ethylene biosynthesis in ripening tomato fruits. *Physiologia Plantarum* 72, 572–578.

Breidenbach, R.W., Saxton, R.J., Hansen, L.D. and Criddle, R.S. (1997) Heat generation and dissipation in plants: can the alternative oxidative phosphorylation pathway serve a thermoregulatory role in plant tissue other than specialized organs? *Plant Physiology* 114, 1137–1140.

Buranasompob, A. (2001) Rancidity and lipoxygenase activity of walnuts and almonds. MSc Thesis, Department of Food Science and Human Nutrition, Washington State University, Pullman, Washington.

Buranasompob, A., Tang, J., Powers, J.R., Reyes, J., Clark, S. and Swanson, B.G. (2007) Lipoxygenase activity in walnuts and almonds. *LWT – Food Science and Technology* 40, 893-899.

Buranasompob, A., Tang, J., Runsheng, M. and Swanson, B.G. (2003) Rancidity of walnuts and almonds affected by short time heat treatments for insect control. *Journal of Food Processing and Preservation* 27, 445-464.

Cantwell, M.I., Hong, G. and Suslow, T.V. (2001) Heat treatments control extension growth and enhance microbial disinfection of minimally processed green onions. *HortScience* 36, 732-737.

Cantwell, M.I., Kang, J. and Hong, G. (2003) Heat treatments control sprouting and rooting of garlic cloves. *Postharvest Biology and Technology* 30, 57-65.

Chan, H.T. and Linse, E. (1989) Conditioning cucumbers for quarantine treatments. *HortScience* 24, 985-989.

Chen, H.H., Shen, Z.Y. and Li, P.H. (1982) Adaptability of crop plants to high temperature stress. *Crop Science* 22, 719-725.

Cheng, T.S., Floros, J.D., Shewfelt, R.L. and Chang, C.J. (1988) The effect of high temperature stress on ripening of tomatoes (*Lycopersicun esculentum*). *Journal of Plant Physiology* 132, 459-464.

Civello, P.M., Martinez, G.A., Chaves, A.R. and Anan, M.C. (1997) Heat treatments delay ripening and postharvest decay of strawberry fruit. *Journal of Agricultural Food Chemistry* 45, 4589-4596.

Couey, H.M. (1989) Heat treatment for control of postharvest diseases and insect pests of fruits. *HortScience* 24, 198-201.

Diaz-Perez, J.C., Mejia, A., Bautista, S., Zaveleta, R., Villanueva, R. and Gomez, R.L. (2001) Response of sapote mamey (*Pouteria sapota* (Jacq.) H.E. Moore & Stearn) fruit to hot water treatments. *Postharvest Biology and Technology* 22, 159-167.

Eaks, I.L. (1978) Ripening, respiration, and ethylene production of 'Hass' avocado fruits at 20°C and 40°C. *Journal of the American Society of Horticultural Science* 103, 576-578.

Ekstrand, B., Gangby, J., Akesson, G., Stollman, U., Lingert, H. and Dahl, S. (1993) Lipase activity and development of rancidity in oats and oat products related to heat treatment during processing. *Journal of Cereal Science* 17, 247-254.

Esquerra, E.B. and Lizada, M.C.C. (1990) The postharvest behaviour and quality of 'Carabao' mangoes subjected to vapour heat treatment. *ASEAN Food Journal* 5, 6-11.

Esquerra, E.B., Brena, S.R., Reyes, M.U. and Lizada, M.C.C. (1990) Physiological breakdown in vapour heat treated 'Carabao' mango. *Acta Horticulturae* 269, 425-434.

Fallik, E., Douglas, A., Hamilton-Kemp, T.R., Loughrin, J.H. and Collins, R.W. (1997) Heat treatment temporarily inhibits aroma volatile compound emission from Golden Delicious apples. *Journal of Agricultural Food Chemistry* 45, 4038-4044.

Ferguson, I.B., Snelgar, W., Lay-Yee, M., Watkins, C.B. and Bowen, J.H. (1998) Expression of heat shock protein genes in apple fruit in the field. *Australian Journal of Plant Physiology* 25, 155-163.

Follett, P.A. and Sanxter, S.S. (2003) Lychee quality after hot water immersion and X-ray irradiation quarantine treatments. *HortScience* 38, 1159-1162.

Fourie, P.C. and Basson, D.S. (1989) Predicting occurrence of rancidity in stored nuts by means of chemical analyses. *Leben-Wissen. Technology* 22 (5), 251-253.

Forney, C.F. (1995) Hot water dips extend the shelf life of fresh broccoli. *HortScience* 30, 1054-1057.

Funamoto, Y., Yamauchi, N., Shigenaga, T. and Shigyo, M. (2002) Effects of heat treatment on chlorophyll degrading enzymes in stored broccoli (*Brassica oleracea* L.) *Postharvest Biology and Technology* 24, 163-170.

Funamoto, Y., Yamauchi, N. and Shigyo, M. (2003) Involvement of peroxidase in chlorophyll degradation in stored broccoli (*Brassica oleracea* L.) and inhibition of the activity by heat treatments. *Postharvest Biology and Technology* 28, 39–46.

Garcia, J.M., Aguilera, C. and Albi, M.A. (1995) Postharvest heat treatment on Spanish strawberry (*Fragaria* × *ananassa* cv. Tulda). *Journal of Agricultural Food Chemistry* 43, 1489–1495.

Gautam, D.M. and Lizada, M.C.C. (1984) Internal breakdown in 'Carabao' mango subjected to modified atmospheres. I. Storage duration and severity of symptoms. *Postharvest Research Notes* 1, 28–30.

Gombos, Z., Wada, H., Hideg, E. and Murata, N. (1994) The unsaturation of membrane lipids stabilizes photosynthesis against heat stress. *Plant Physiology* 104, 563–567.

Greany, P.D., McDonald, R.E., Schroeder, W.J., Shaw, P.E., Aluja, M. and Malavasi, A. (1995) Use of gibberellic acid to reduce citrus fruit susceptibility to fruit flies. In: Hedin, P.A. (ed.) *Bioregulators for Crop Protection and Pest Control*. American Chemical Society, Washington, DC, pp. 40–48.

Gunjate, R.T., Walimbe, D.P., Lad, B.L. and Limaye, V.P. (1982) Development of internal breakdown in 'Alphonso' mango by postharvest exposure of fruits to sunlight. *Science and Culture* 48, 188–190.

Hallman, G.J., Gaffney, J.J. and Sharp, J.L. (1990) Vapor heat treatment for grapefruit infested with Caribbean fruit fly (Diptera: Tephritida). *Journal of Economic Entomology* 83, 1475–1478.

Hara, A.H., Hata, T.Y., Tenbrink, V.L., Hu, B.K.-S. and Kandko, R.T. (1996) Postharvest heat treatment of red ginger flowers as a possible alternative to chemical insecticidal dip. *Postharvest Biology and Technology* 7, 137–144.

Hong, G., Peiser, G. and Cantwell, M.I. (2000) Use of controlled atmospheres and heat treatment to maintain quality of intact and minimally processed green onions. *Postharvest Biology and Technology* 20, 53–61.

Huber, H. (1935) Der warmehaushalt der pflanzen. *Naturwissenshaften* 17, 1–148.

Hui, Y.H. (1992) *Encyclopedia of Food Science Technology*, Volume 2. John Wiley & Sons, New York, pp. 825–831.

Inaba, M. and Chachin, K. (1988) Influence of and recovery from high temperature stress on harvested mature green tomatoes. *HortScience* 23, 190–192.

Inaba, M. and Crandall, P.G. (1988) Electrolyte leakage as an indicator of high temperature injury to harvested mature green tomatoes. *Journal of the American Society of Horticultural Science* 113, 96–99.

Ingram, D.L. and Buchanan, D. (1981) Measurement of direct heat injury of roots of three woody plants. *HortScience* 16, 769–771.

Ingram, D.L. and Buchanan, D. (1984) Lethal high temperatures for roots of three citrus rootstocks. *Journal of the American Society for Horticultural Science* 109, 189–193.

Jacobi, K.K. and Gowanlock, D. (1995) Ultrastructural studies of 'Kensington' mango (*Mangifera indica* Linn.) heat injuries. *HortScience* 30, 102–105.

Jacobi, K.K. and Wong, L.S. (1992) Quality of 'Kensington' mango (*Mangifera indica* Linn.) following hot water and vapour heat treatments. *Postharvest Biology and Technology* 1, 349–359.

Jacobi, K.K., Wong, L.S. and Giles, J.E. (1993) Lychee (*Lichi chinensis* Sonn.) fruit quality following vapour heat treatment and cool storage. *Postharvest Biology and Technology* 3, 111–117.

Jacobi, K.K., Wong, L.S. and Giles, J.E. (1995a) Effect of fruit maturity on quality and physiology of high humidity hot air treated 'Kensington' mango (*Mangifera indica* Linn.) *Postharvest Biology and Technology* 5, 149–159.

Jacobi, K.K., Giles, J., MacRae, E.A. and Wegrzyn, T. (1995b) Conditioning 'Kensington' mango with hot air alleviates hot water disinfestations injuries. *HortScience* 30, 562–565.

Jacobi, K.K., Wong, L.S. and Giles, J.E. (1996) Postharvest quality of zucchini (*Cucurbita pepo* L.) following high humidity hot water disinfestation treatments and cool storage. *Postharvest Biology and Technology* 7, 309–315.

Jacobi, K.K., MacRae, E.A. and Hetherington, S.E. (2001) Loss of heat tolerance in 'Kensington' mango fruit following heat treatment. *Postharvest Biology and Technology* 21, 321–330.

Johnson, J.A., Boling, H.R., Fuller, G. and Thompson, J.F. (1989) Efficacy of temperature treatments for insect disinfestations of dried fruits and nuts. Prune Research Reports.

Jones, W.W. (1942) Respiration and chemical changes of the papaya fruit in relation to temperature. *Plant Physiology* 17, 481–486.

Joyce, D.C. and Shorter, A.J. (1994) High-temperature conditioning reduces hot water treatment injury of 'Kensington Pride' mango fruit. *HortScience* 29, 1047–1051.

Kagan-Zur, V., Tieman, D., Marlow, S.J. and Handa, A. (1995) Differential regulation of polygalacturonase and pectin methylesterase gene expression during and after heat stress in ripening tomato (*Lycopersicon esculentum* Mill.) fruits. *Plant Molecular Biology* 29, 1101–1109.

Kappen, L. (1981) Ecological significance of resistance to high temperatures. In: Lange, O.L., Nobel, P.S., Osmond, C.B. and Ziegler, H. (eds) *Encyclopedia of Plant Physiology*, Vol. 12 A. Springer, New York, pp. 440–474.

Kazami, D., Sato, T., Nakagawa, H. and Ogura, N. (1991a) Effect of preharvest hot water dipping of broccoli heads on shelf life and quality during storage. *Journal of the Agriculture and Chemical Society of Japan* 65, 19–26.

Kazami, D., Sato, T., Nakagawa, H. and Ogura, N. (1991b) Effect of preharvest hot water dipping of broccoli heads on soluble protein, free amino acid contents and protease activities during storage. *Journal of the Agricultural and Chemical Society of Japan* 65, 27–33.

Kerbel, E.L., Mitchell, G. and Mayer, G. (1987) Effect of postharvest heat treatment for insect control on the quality and market life of avocados. *HortScience* 22, 92–96.

Ketsa, S., Chidragool, S., Klein, J.D. and Lurie, S. (1998) Effect of heat treatment on changes in softening, pectic substances and activities of polygalacturonase, pectinesterase and β-galactosidase of ripening mango. *Journal of Plant Physiology* 153, 457–461.

Klein, J.D. (1989) Ethylene biosynthesis in heat treated apples. In: Clijsters, M. (ed.) *Biochemical and Physiological Aspects of Ethylene Production in Lower and Higher Plants*. Kluwer Academic Press, Dordrecht, Netherlands, pp. 184–190.

Klein, J.D. and Lurie, S. (1990) Prestorage heat treatment as a means of improving poststorage quality of apples. *Journal of the American Society of Horticultural Science* 115, 265–269.

Klein, J.D. and Lurie, S. (1992) Prestorage heating of apple fruit for enhanced postharvest quality: interaction of time and temperature. *HortScience* 27, 326–328.

Lang, A. and During, H. (1990) Grape berry splitting and some mechanical properties of the skin. *Vitis* 29, 61–70.

Lay-Yee, M. and Rose, K.J. (1994) Quality of 'Fantasia' nectarines following forced air heat treatments for insect disinfestations. *HortScience* 29, 663–555.

Lay-Yee, M. and Whiting, D.C. (1996) Response of 'Hayward' kiwi fruit to high-temperature controlled atmosphere treatments for control of two-spotted spider mite (*Tetranychus urticae*). *Postharvest Biology and Technology* 7, 73–81.

Lester, G., Dunlap, J. and Lingle, S. (1988) Effect of postharvest heating on electrolyte leakage and fresh weight loss from stored muskmelon fruit. *HortScience* 23, 407.

Liu, F.W. (1978) Modification of apple quality by high temperature. *Journal of the American Society of Horticultural Science* 103, 584–587.

Loaiza-Velarde, J.G., Mangrich, M.E., Campos-Vargas, R. and Saltveit, M.E. (2003) Heat shock reduces browning of fresh-cut celery petioles. *Postharvest Biology and Technology* 27, 305–311.

Lurie, S. 1998. Postharvest heat treatments. *Postharvest Biology and Technology* 14, 257–269.

Lurie, S. and Klein, J.D. (1990) Heat treatment of ripening apples: differential effects on physiology and biochemistry. *Physiologia Plantarum* 78, 181–186.

Lurie, S. and Klein, J.D. (1992) Ripening characteristics of tomatoes stored at 12°C and 2°C following a prestorage heat treatment. *Scientia Horticulturae* 51, 55–64.

Lurie, S. and Klein, J.D. (2004) Prestorage temperature manipulations. In: *The Commercial Storage of Fruits, Vegetables, and Florist and Nursery Stocks*. USDA Handbook 66, Government Printing Office, Washington, DC (http://www.ba.ars.usda.gov/hb66/).

Lurie, S., Othman, S. and Borochov, A. (1995) Effects of heat treatment of plasma membrane of apple fruit. *Postharvest Biology and Technology* 5, 29–38.

Lurie, S., Handros, A., Fallik, E. and Shapira, R. (1996) Reversible inhibition of tomato fruit gene expression at high temperature. *Plant Physiology* 110, 1207–1214.

Lydakis, D. and Aked, J. (2003) Vapour heat treatment of Sultanina table grapes. II. Effects of postharvest quality. *Postharvest Biology and Technology* 27, 117–126.

Macrae, R., Robinson, R.K. and Sadler, M.J. (1993) *Encyclopedia of Food Science, Food Technology and Nutrition*. Academic Press, New York, p. 4832.

Martineau, J.R., Specht, J.E., Williams, J.H. and Sullivan, C.Y. (1979) Temperature tolerance in soybeans. I. Evaluation of techniques for assessing cellular membrane thermostability. *Crop Science* 19, 76–78.

Maxie, E.C., Mitchell, F.G., Sommer, N.F., Snyder, R.G. and Rae, H.L. (1974) Effects of elevated temperatures on ripening of 'Bartlett' pears *Pyrus communis* L. *Journal of the American Society of Horticultural Science* 99, 344–349.

McDonald, R.E., McCollum, T.G. and Baldwin, E.A. (1998) Heat treatment of mature-green tomatoes: differential effects of ethylene and partial ripening. *Journal of the American Society of Horticultural Science* 123, 457–462.

McGlammery, J.B. and Hood, M.P. (1951) Effect of two heat treatments on rancidity development in unshelled pecans. *Food Technology* 26, 80.

McGuire, R.G. (1991) Concomitant decay reductions when mangoes are treated with heat to control infestations of Caribbean fruit flies. *Plant Disease* 75, 946–949.

McGuire, R.G. and Reeder, W.F. (1992) Predicting market quality of grapefruit after hot air quarantine treatment. *Journal of the American Society for Horticultural Science* 117, 90–95.

McGuire, R.G. and Sharp, J.L. (1997) Quality of colossal Puerto Rican 'Keitt' mangoes after quarantine treatment in water at 48°C. *Tropical Science* 37, 154–159.

Miller, W.R. and McDonald, R.E. (1991) Quality of stored 'Marsh' and 'Ruby Red' grapefruit after high-temperature, forced-air treatment. *HortScience* 26, 1188–1191.

Miller, W.R. and McDonald, R.E. (1992) Postharvest quality of early season grapefruit after forced-air vapour heat treatment. *HortScience* 27, 422–424.

Miller, W.R. and McDonald, R.E. (1997) Comparative responses of preharvest GA-treated grapefruit to vapour heat and hot water treatment. *HortScience* 32, 275–277.

Miller, W.R., McDonald, R.E., Hatton, T.T. and Ismail, M. (1988) Phytotoxicity to grapefruit exposed to hot water immersion treatment. *Proceedings of the Florida State Horticultural Society* 101, 192–201.

Miller, W.R., McDonald, R.E. and Sharp, J.L. (1991a) Quality changes during storage and ripening of 'Tommy Atkins' mangoes treated with heated forced air. *HortScience* 26, 395–397.

Miller, W.R., McDonald, R.E., Hallman, G. and Sharp, J.L. (1991b) Condition of Florida grapefruit after exposure to vapour heat quarantine treatment. *HortScience* 26, 42–44.

Mitcham, E.J. and McDonald, R.E. (1992) Effect of high temperature on cell wall metabolism associated with tomato (*Lycopersicon esculentum* Mill.) fruit ripening. *Postharvest Biology and Technology* 1, 257–264.

Mitcham, E.J. and McDonald, R.E. (1993) Respiration rate, internal atmosphere, and ethanol and acetaldehyde accumulation in heat treated mango fruit. *Postharvest Biology and Technology* 3, 77–84.

Mitcham, E.J. and McDonald, R.E. (1997) Effects of postharvest heat treatment on inner and outer tissue of mango fruit. *Tropical Science* 37, 193–205.

Murakami, Y., Tsuyama, M., Kobayashi, Y., Kodama, H. and Iba, K. (2000) Tienoic fatty acids and plant tolerance of high temperature. *Science* 287, 476–479.

Neven, L.G. (2000) Physiological responses of insects to heat. *Postharvest Biology and Technology* 21, 103–112.

Nyanjage, M.O., Wainwright, H. and Bishop, C.F.H. (1998) The effects of hot water treatments in combination with cooling and/or storage on the physiology and disease of mango fruits (*Mangifera indica* Linn.). *Journal of Horticultural Science and Biotechnology* 73, 589–597.

Obenland, D.M. and Carroll, T.R. (2000) Mealiness and pectolytic activity in peaches and nectarines in response to heat treatment and cold storage. *Journal of the American Society of Horticultural Science* 125, 723–728.

Olson, W.H., Sibbett, G.S. and Martin, G.C. (1978) *Walnut Harvesting and Handling in California.* Leaflet #21036, Division of Agricultural Science, University of California, Davis, California.

Ortega-Zaleta, D. and Yahia, E.M. (2000) Tolerance and quality of mango fruit exposed to controlled atmospheres at high temperatures. *Postharvest Biology and Technology* 20,195–201.

Paull, R.E. (1990) Postharvest heat treatments and fruit ripening. *Postharvest News and Information* 1, 355–363.

Paull, R.E. (1995) Preharvest factors and the heat sensitivity of field-grown ripening papaya fruit. *Postharvest Biology and Technology* 6, 167–175.

Paull, R.E. and Chen, N.J. (1990) Heat shock response in field-grown, ripening papaya fruit. *Journal of the American Society of Horticultural Science* 115, 623–631.

Paull, R.E. and Chen, N.J. (1999) Heat treatment prevents postharvest geotropic curvature of asparagus spears (*Asparagus officinalis* L.) *Postharvest Biology and Technology* 16, 37–41.

Phillips, D.J. and Austin, R.K. (1982) Changes in peaches after hot water treatment. *Plant Disease* 66, 487–488.

Picton, S. and Grierson, D. (1988) Inhibition of expression of tomato ripening genes at high temperature. *Plant Cell and Environment* 11, 265–272.

Porat, R., Pavoncello, D., Peretz, J., Ben Yehoshua, S. and Lurie, S. (2000) Effects of various heat treatments on the induction of cold tolerance and on the postharvest qualities of 'Star Ruby' grapefruit. *Postharvest Biology and Technology* 18, 159–165.

Ranganna, B., Raghavan, G.S.V. and Kushalappa, A.C. (1998) Hot water dipping to enhance storability of potatoes. *Postharvest Biology and Technology* 13, 215–223.

Rockland, L.B. (1962) Studies on the processing of shelled walnuts. *California Macadamia Society Yearbook* 8, 30–34.

Roh, K.A., Lee, B.M., Lee, D.C. and Park, M.E. (1995) Effects of temperature on ethylene biosynthesis of apples. *Journal of Agriculture, Science and Farm Management* 37, 696–700.

Saftner, R.A., Abbott, J.A., Conway, W.S. and Barden, C.L. (2003) Effects of 1-Methylcyclopropene and heat treatments on ripening and postharvest decay in 'Golden Delicious' apples. *Journal of the American Society of Horticultural Science* 128, 120–127.

Saltveit, M.E. (2000) Wound induced changes in phenolic metabolism and tissue browning are altered by heat shock. *Postharvest Biology and Technology* 21, 61–69.

Schirra, M. and Cohen, E. (1999) Long-term storage of 'Olind' oranges under chilling and intermittent warming temperatures. *Postharvest Biology and Technology* 16, 63–70.

Schirra, M. and D'hallewin, G. (1997) Storage performance of Fortune mandarins following hot water dips. *Postharvest Biology and Technology* 10, 229–236.

Schirra, M., Agabbio, M., D'hallewin, G., Pala, M. and Ruggio, R. (1997) Response of Tarocco oranges to picking date, postharvest hot water dips, and chilling storage temperature. *Journal of Agricultural Food Chemistry* 45, 3216–3220.

Schirra, M., D'hallewin, G., Ben-Yehoshua, S. and Fallik, E. (2000) Host–pathogen interactions modulated by heat treatment. *Postharvest Biology and Technology* 21, 71–85.

Schirra, M., Cabras, P., Angioni, A., D'hallewin, G. and Pala, M. (2002) Residue uptake and storage responses of Tarocco blood oranges after preharvest thiabendazole spray and postharvest heat treatment. *Journal of Agricultural Food Chemistry* 50, 2293–2296.

Shellie, K.C. and Mangan, R.L. (1994) Postharvest quality of 'Valencia' orange after exposure to hot, moist forced air for fruit fly disinfestations *HortScience* 29, 1524–1528.

Shellie, K.C. and Mangan, R.L. (1996) Tolerance of red fleshed grapefruit to a constant or stepped temperature, forced air quarantine heat treatment. *Postharvest Biology and Technology* 7, 151–160.

Shellie, K.C. and Mangan, R.L. (2000) Postharvest disinfestation heat treatment: response of fruit and fruit fly larvae to different heating media. *Postharvest Biology and Technology* 21, 51–60.

Shellie, K.C., Firko, M.J. and Mangan, R.L. (1993) Phytotoxic response of 'Dancy' tangerines to high temperature, moist, forced air treatment for fruit fly disinfestations. *Journal of the American Society for Horticultural Science* 118, 481–489.

Smith, K.J. and Lay-Yee, M. (2000) Response of 'Royal Gala' apples to hot water treatment for insect control. *Postharvest Biology and Technology* 19, 111–122.

Sozzi, G.O., Cascone, O. and Fraschina, A.A. (1997) Effect of high temperature stress on endo-β-mannanase and α- and β-galactosidase activities during tomato fruit ripening. *Postharvest Biology and Technology* 9, 49–63.

Taoukis, P.S., Labuza, T.P. and Sagus, I.S. (1997) Kinetics of food deterioration and shelf-life prediction. In: Valentas, K.J., Rotstein, E. and Singh, R.P. (eds) *Handbook of Food Engineering Practice*. CRC Press, Boca Raton, Florida, p. 374.

Terai, H., Kanou, M., Mizuno, M. and Tsuchida, M. (1999) Inhibition of yellowing and ethylene production in broccoli florets following high temperature treatments with hot air. *Food Preservation Science* 25, 221–227.

Thompson, J.F., Rumsey, T.R. and Grant, J.A. (1998) Dehydration. In: Ramos, D.E. (ed.) *Walnut Production Manual*. Agricultural Natural Resource Publication #3373, University of California, Oakland, California.

Tian, M.S., Downs, C.G., Lill, R.E. and King, G.A. (1994) A role for ethylene in the yellowing of broccoli after harvest. *Journal of the American Society of Horticultural Science* 119, 276–281.

Tian, M.S., Woolf, A.B., Bowen, J.H. and Ferguson, I.B. (1996) Changes in color and chlorophyll fluorescence of broccoli florets following hot water treatment. *Journal of the American Society of Horticultural Science* 121, 310–313.

Tian, M.S., Islam, T., Stevenson, D.G. and Irving, D.E. (1997) Color, ethylene production, respiration and compositional changes in broccoli dipped in hot water. *Journal of the American Society of Horticultural Science* 122, 112–116.

Vicente, A.R., Martínez, G.A., Civello, P.M. and Chaves, A.R. (2002) Quality of heat-treated strawberry fruit during refrigerated storage. *Postharvest Biology and Technology* 25, 59–71.

Vierling, E. (1991) The roles of heat shock proteins in plants. *Annual Review of Plant Physiology and Plant Molecular Biology* 42, 579–531.

Wahab, F.K. Hamady, A.M., Nabawy, S.M., Abou, R.M. and Hagagg, L.F. (1984) Effect of storage and prestorage treatments on pecan fruit delay and oil properties. *Gartenbauwissenschaft* 49, 61–64.

Wang, C.Y. (1998) Heat treatment affects postharvest quality of kale and collard, but not of Brussels sprouts. *HortScience* 33, 881–883.

Wang, S., Tang, J., Johnson, J.A., Mitcham, E., Hansen, J.D., Cavalieri, R.P., Bower, J. and Biasi, B. (2002) Process protocols based on radio frequency energy to control field and storage pests in in-shell walnuts. *Postharvest Biology and Technology* 26, 265–273.

Williams, M.H., Brown, M.A., Vesk, M. and Brady, C. (1994) Effect of postharvest heat treatments on fruit quality, surface structure, and fungal disease in Valencia oranges. *Australian Journal of Experimental Agriculture* 34, 1183–1190.

Woolf, A.B. and Ferguson, I.B. (2000) Postharvest responses to high fruit temperatures in the field. *Postharvest Biology and Technology* 21, 7–20.

Woolf, A.B. and Laing, W.A. (1996) Avocado skin fluorescence following hot water treatments and pretreatments. *Journal of the American Society of Horticultural Science* 121, 147–155.

Woolf, A.B. and Lay-Yee, M. (1997) Pretreatments at 38°C of 'Hass' avocado confer thermotolerance to 50°C hot water treatments. *HortScience* 34, 705–708.

Woolf, A.B., Bowen, J.H. and Ferguson, I.B. (1999) Preharvest exposure to the sun influences postharvest responses of 'Hass' avocado fruit. *Postharvest Biology and Technology* 15, 143–153.

Woolf, A.B., Weksler, A., Prusky, D., Kobiler, E. and Lurie, S. (2000) Direct sunlight influences postharvest temperature responses and ripening of five avocado cultivars. *Journal of the American Society of Horticultural Sciences* 125, 370–376.

Woolf, A.B., Bowen, J.H., Ball, S., Durant, S., Laidlaw, W.G. and Ferguson, I.B. (2004) A delay between a 38°C pretreatment and damaging high and low temperature treatments influences pretreatment efficacy in 'Hass' avocados. *Postharvest Biology and Technology* 34, 143–153.

Yang, R.F., Cheng, T.S. and Shewfelt, R.L. (1990) The effect of high temperature and ethylene treatment on the ripening of tomatoes. *Journal of Plant Physiology* 136, 368–375.

Young, C.K. and Cunningham, S. (1991) Exploring the partnership of almonds with cereal foods. *Cereal Foods World* 36 (5), 412–418.

Zacheo, G., Cappello, M.S., Gallo, A., Sanito, A. and Cappello, A.R. (2000) Changes associated with postharvest ageing in almond seeds. *Lebens.-Wissen. Technology* 33 (6), 415–423.

Zhou, S., Criddle, R.S. and Mitcham, E.J. (2000) Metabolic response of *Platynota stultana* pupae to controlled atmospheres and its relation to insect mortality response. *Journal of Insect Physiology* 46, 1375–1385.

5 Experimental and Simulation Methods of Insect Thermal Death Kinetics

S. Wang,[1]* J. Tang[1]** and J.D. Hansen[2]

[1]Department of Biological Systems Engineering, Washington State University, Pullman, Washington, USA; e-mail: *shaojin_wang@wsu.edu **jtang@wsu.edu; [2]USDA-ARS Yakima Agricultural Research Laboratory, Wapato, Washington, USA; e-mail: jimbob@yarl.ars.usda.gov

5.1 Introduction

Knowledge of the thermal death kinetics of insect pests, which vary widely in tolerance to heat treatments, will facilitate successful development of thermal treatments to control pests in nuts and fruits as an alternative to chemical fumigation (Yokoyama et al., 1991; Neven et al., 1996; Mangan et al., 1998; Tang et al., 2000; Wang et al., 2002a, b; Hansen et al., 2004). With this information, appropriate means of delivering the desired lethal energy to different insects can be achieved.

The heat limitation to which insects – such as the codling moth (Cydia pomonella (L.) [Lepidoptera: Tortricidae]) – can tolerate has been studied with a variety of thermal methods (Yokoyama et al., 1991; Neven, 1994; Neven and Rehfield, 1995; Neven and Mitcham, 1996; Ikediala et al., 1999; Wang et al., 2002a). As shown in Table 5.1, the observed time to reach 100% mortality of codling moth is not consistent, probably caused by the differences in the test conditions and heating rates. Reported information on thermal tolerance for codling moth and other insects is often confounded by variations in heat transfer in the materials and heat application methods employed.

Heating rates are believed to significantly affect insect metabolism and physiological adjustment to heat treatments (Evans, 1986; Neven, 1998a, b). Neven (1998a) reported that codling moth larvae in fruits might experience thermal conditioning and acclimation at heating rates of 0.13–0.20°C/min in a controlled atmosphere (CA) chamber based on experimental data. Consequently, a longer holding time is required at a final temperature in order to achieve the same mortality at a slower heating rate.

For conventional heating, the rates in the interior of commodities range

Table 5.1. Thermal tolerance of fifth-instar codling moths (from Tang *et al.*, 2000).

Temp. (°C)	Medium	Heating method	Time for 100% mortality (min)	Heating rate (°C/min)	Source
45	15 ml vials	Water bath	> 55	NR	Yokoyama *et al.* (1991)
	0.48 l jars	Walk-in chamber	1440	NR	Soderstrom *et al.* (1996)
	Infested cherry	Test chamber	60	4.4	Neven and Rehfield (1995)
	Infested cherry	Test chamber	124	1.66	Neven and Mitcham (1996)
46	15 ml vials	Water bath	> 55	NR	Yokoyama *et al.* (1991)
	Apples	Test chamber	480	4.0	Neven and Rehfield (1995)
	7.5 vials	Water bath	97	NR	Neven *et al.* (1996)
47	15 ml vials	Water bath	25	NR	Yokoyama *et al.* (1991)
48	15 ml vials	Water bath	10	NR	Yokoyama *et al.* (1991)
	7.5 ml vials	Water bath	177	NR	Neven *et al.* (1996)
	7.5 ml vials	Water bath	73	NR	Neven *et al.* (1996)
49	15 ml vials	Water bath	7.5	NR	Yokoyama *et al.* (1991)
50	15 ml vials	Water bath	4.0	NR	Yokoyama *et al.* (1991)
51	15 ml vials	Water bath	3.0	NR	Yokoyama *et al.* (1991)
52	Metal surface	Heating block	2.0	20	Ikediala *et al.* (2000)
52	Metal surface	Heating block	2.0	18	Wang *et al.* (2002a)

NR, not reported.

between 0.05 and 2.00°C/min, depending on heating methods, type and size of commodity and the end temperature (see Chapter 2, this volume). In addition, under a constant treatment condition the heating rate at the interior of a commodity decreases with time due to decreasing temperature gradients between the heating medium and fruit. As a result, conventional heat treatments require the long times to achieve efficacy against insects.

Only limited numbers of temperature–time combinations and relatively small insect sample sizes can be used for experimentation due to labour costs and time limitations. But carefully planned experiments can lead to development of empirical and fundamental thermal death kinetic models that in turn can be used to estimate time required to kill insects at a wide range of population levels and conditions. In particular, the thermal death kinetic information can be used in combination with engineering theory for heat transfer to design new treatment protocols for delivering the desired lethal energy to the insects (Tang *et al.*, 2000; Wang and Tang, 2004).

During experimentation, evaluation of insect mortality generally includes the acute mortality obtained relatively soon after treatment, which is helpful for quarantine purposes, as well as long-term chronic mortality useful for phytosanitation, pest management programmes and quarantines such as irradiation (Hansen and Sharp, 1998).

This chapter reviews the different experimental methods recorded in the literature to obtain thermal death kinetic data of insect pests and various models developed based on the experimental data, including a detailed discussion on their respective advantages and disadvantages. Finally, an application of the kinetic models is proposed to predict the thermal mortality of insects in commodities when subjected to practical thermal treatments.

5.2 Experimental Methods for Obtain Thermal Kinetic Response Information

Ideal test conditions for obtaining information on kinetic responses of insects to heat treatments should simulate the real environment for the infesting insects in the commodities. These include suitable humidity, oxygen level and movement space. One method of capturing the effect of temperature and time combinations on insect thermal mortality is to provide a step-functional temperature profile via instantaneous heating and cooling (see Fig. 5.1), thus avoiding the confounded effect of thermal lags experienced by the insects. In doing so, heat should be uniformly delivered to the whole insect population of a sample within a defined space. In reality, however, the temperatures experienced by insects in experimentation may be different from an ideal step-function. These deviations depend on the heating and cooling methods, heating medium and size of the heating object.

Experimental methods for characterizing the temperature–time effect on insect mortality include: (i) directly exposing insects in a water bath for specific times; (ii) heating insects in tubes which in turn are submerged in a water bath;

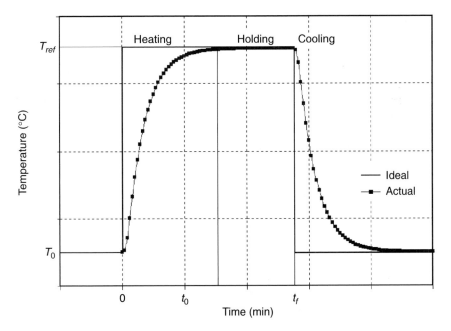

Fig. 5.1. Typical ideal and actual temperature–time histories of the targeted part for thermal treatment during heating, holding and cooling periods. T_0 and T_{ref}, initial and target temperatures; t_0 and t_f, starting and ending times).

(iii) heating insects in fruits, fruit pulp or other substrates (Hansen *et al.*, 1990; Yokoyama *et al.*, 1991; Thomas and Mangan, 1997; Hansen and Sharp, 2000; Waddell *et al.*, 2000); (iv) heating insects in a uniform heating environment in a heating block system (Ikediala *et al.*, 2000; Wang *et al.*, 2002a, b). These experimental methods will be discussed in detail in the following sections.

Infested Commodities Subjected to Heating

One way to determine thermal mortality characteristics of insects is to heat infested commodities in a water bath or hot air chamber set at a selected temperature for predetermined times (Hayes *et al.*, 1987; Sharp *et al.*, 1989; Neven *et al.*, 1996; Follett and Sanxter, 2001; Hansen *et al.*, 2004), which is similar to the situation found in a natural infestation. This method works well as a confirmation for conventional heat treatments, but is not suited for kinetic studies because the insects can be randomly distributed in the infested commodities, whereas heat transfer starts from the surface. As a result, the insects are not exposed to the same thermal condition during the tests.

In Fig. 3.9, a cross-section thermal image of an orange is shown after submersion in hot water at 53°C for 20 min. A large temperature gradient is observed between the surface (53°C) and the core (33°C), which would result in a corresponding significantly higher insect mortality close to the surface than at the core. Mortality variations increase with the size of the commodity, from small commodities such as cherries and lychees (Follett and Sanxter, 2001; Hansen *et al.*, 2004) to large commodities such as apples, oranges and grapefruits (Hallman, 1996; Neven *et al.*, 1996; Mangan *et al.*, 1998).

The long heating time needed to bring a commodity centre to a desired time–temperature level adds another complication to the interpretation and analyses of kinetic data. As noted before and demonstrated in Table 5.1, information from such experiments is only applicable to the tested products and specific test conditions and so cannot be used directly to develop thermal death kinetic models.

Direct Immersion of Insects in Water Baths

Direct hot water bath immersion is a common technique for the study of thermal kinetic behaviour of insect pests (Hayes *et al.*, 1984; Jang, 1986; Yokoyama and Miller 1987; Jones and Waddell, 1997; Thomas and Mangan, 1997). For example, Jang (1986) and Thomas and Mangan (1997) placed fruit fly larvae in mesh bags (150 or 30 ml) and then immersed the bags in a water bath at selected temperatures. Jang (1986) determined the thermal tolerance of eggs and first-instars of Mediterranean fruit fly (*Ceratitis capitata* (Wiedemann) [Diptera: Tephritidae]), oriental fruit fly (*Bactrocera dorsalis* (Hendel) [Diptera: Tephritidae]) and melon fly (*Bactrocera cucurbitae* (Coquillett) [Diptera: Tephritidae]) by using a permeable container immersed in circulating hot water.

Hallman (1990) used a similar method to expose third-instars of Caribbean fruit fly (*Anastrepha suspense* (Loew) [Diptera: Tephritidae]) to heat. Although this method provides an almost ideal step-functional temperature profile where all insects are exposed to the same test conditions, it also results in instantaneous wet-heating of the insects, which may not be similar to the conditions that insects experience during hot air treatments, or even water dip treatments of actual commodities where air pockets may exist around the insects.

Direct immersion of insects in water at elevated temperatures may lead to drowning, especially in heated water for an extended period of time (Hansen and Sharp, 2000). Because of the long treatment duration, this method may cause insect suffocation when conducted under anaerobic conditions. In addition, ion concentrations, pH and dissolved O_2 levels in the immersion water are all thought to influence the results of thermal mortality tests, making this a difficult method to yield repeatable and reliable experimental data that truly represent the inherent thermal tolerance of the studied insects (Paull, 1994; Shellie *et al.*, 2001).

Direct hot water immersion of Mediterranean fruit fly life stages (Jang, 1986; Waddell *et al.*, 1997; Jang *et al.*, 1999) shows that the egg is the most heat-tolerant stage, while the third-instar is more susceptible to heat than the first-instar. However, when fruit infested with fruit flies is subjected to forced hot air treatments (Heather *et al.*, 1997), the opposite result is obtained, where the egg is the most susceptible to heat. Hansen and Sharp (1998, 2000) suggested that water immersion restricts aerobic oxygen from the heated fruit fly larvae, and thus synergizes their mortality.

Glass Vials and Metal Tubes in Water Baths

To avoid the problem associated with directly immersing insects in heated water, many researchers use an indirect immersion method. For example, Yokoyama *et al.* (1991) and Neven (1994) placed codling moth larvae in glass vials (1.5 and 2 cm in diameter, respectively), and Thomas and Shellie (2000) placed Mexican fruit fly (*Anastrepha ludens* (Loew) [Diptera: Tephritidae]) larvae in metal tubes (internal diameter 2.5 cm) containing papaya pulp before heating in water baths. Hansen and Sharp (1998, 2000) heated third-instars of the Caribbean fruit fly using a rearing diet and fruit pulp in metal tubes to transfer the heat and observed that the mortality rate was significantly lower compared with the use of permeable cloth sleeves.

In a recent study, Lurie *et al.* (2004) heated medfly larvae in glass test tubes (2 cm outer diameter) placed on a plastic rack and immersed in a circulating water bath. The water level in the tubes was 1–2 cm below the bath's water level. A temperature sensor in one of the tubes monitored the internal temperature. The treatment started when the internal temperature reached the set point. At the end of the exposure period, tubes were transferred immediately into circulating water at a temperature of 22°C. One test tube containing insects in water at room temperature (22°C) was used as a control.

The treated pests were transferred to a larval diet to complete their development.

Mortality rates (%) were calculated based on egg hatching, pupation or adult emergence depending on the targeted insect life stages. They observed that medfly eggs were more heat resistant than larvae at 43°C, and the times needed to generate complete kill were different from those found in fruits treated in the Controlled Atmosphere/Temperature Treatment System (CATTS) (Lurie *et al.*, 2004). This method maintained the temperature gradient between the surface and centre of the tube due to slow heat conduction within the tube. The magnitude of the temperature gradient increased with the dimensions and depended on the materials of the tube and medium for the insects. Locations of insects and the surrounding environment in the tube (e.g. air, diet, pulp or water) also influenced the test results.

In a similar test, Hansen and Heidt (2006) thermally treated fifth-instar codling moths in glass flasks in a 50°C water bath with an interior atmosphere of flowing air, static air, nitrogen, a vacuum or water. After 5 min, 96% of the larvae survived in the flowing air, 24% in static air, 16% in the vacuum, 8% in nitrogen but none in water. These data demonstrate that anaerobic conditions greatly enhance thermal pest treatments, with water submersion having additional effects.

Heating Block Systems

A unique experimental heating block system (HBS) was developed at Washington State University to study intrinsic thermal death kinetics of insect pests (Ikediala *et al.*, 2000; Wang *et al.*, 2002b; Fig. 5.2). The HBS directly and uniformly heats each of the tested insects in any desired environment (e.g. humid air, controlled atmospheres, etc.). This system consists of top and bottom aluminium blocks (25.4 × 25.4 cm), heating pads, controlled atmosphere circulating channels and a data acquisition/Proportional-Integral-Derivative (PID) control unit (i/32 temperature and process controller, Omega Engineering Inc., Stamford, Connecticut, USA). The heating blocks are made of aluminium alloys with low thermal capacitance (903 J/kg/°C) and high thermal conductivity (234 W/m/°C).

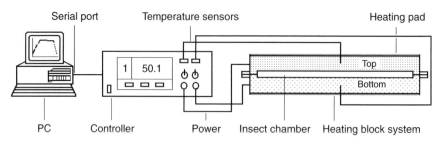

Fig. 5.2. Diagram of a heating block system (not to scale) (from Wang and Tang, 2004).

When the two blocks are assembled, one on top of the other, a 3-mm gap machined in the blocks forms a closed chamber for the insects. Eight custom-made electric heating pads (250 W, 120 V, Heat-Con., Inc., Seattle, Washington, USA) are glued to the top and bottom block surfaces, providing a maximum heating flux density of 15,500 W/m^2. The HBS is capable of heating insects at any rate between 0.2 and 18°C/min (Wang et al., 2002b). Calibrated type-T thermocouples inserted through sensor holes near the centre of each block are used to monitor the temperatures of the top and bottom blocks, in which PID controllers maintain via solid relays to linearly increase within ± 0.3°C of the desired set point (see Fig. 5.3).

The temperature profiles at the heating rates of 15–18°C/min in the HBS are close to the step-function shown in Fig. 5.1, with the temperature in the chamber very uniform. For example, in one study, we determined that the overall maximum standard deviations for temperatures over the nine positions uniformly distributed on the HBS were 0.8 and 1.2°C for heating rates at 10 and 15°C/min, respectively, when the mean block temperature had reached 50°C (Wang et al., 2005). Within 30s of reaching 50°C, however, the temperature difference between the hottest location among the nine measured positions and the set point of 50°C was reduced to < 0.6°C for both 10 and 15°C/min heating rates.

Researchers have used HBS applications to determine the thermal death kinetics of many insect pests. Thermal mortality kinetic data were obtained for fifth-instar codling moth, Indianmeal moth (*Plodia interpunctella* (Hübner) [Lepidoptera: Pyralidae]), navel orangeworm (*Amyelois transitella* (L.) [Lepidoptera: Pyralidae]) (Ikediala et al., 2000; Wang et al., 2002a, b; Johnson et al., 2003) as well as third-instar Mediterranean (Gazit et al., 2004) and

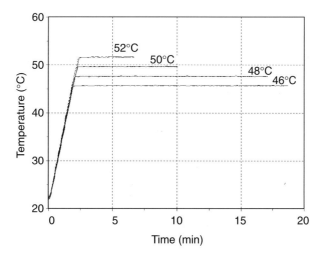

Fig. 5.3. Top and bottom plate temperatures of a heating block system when heated to four different final temperatures at a rate of 15°C/min. (—), bottom plate; (- -), top plate.

Mexican fruit flies (Hallman *et al.*, 2005). The HBS results were confirmed with radio frequency (RF) heat treatments for walnuts infested with fifth-instar codling moth (Wang *et al.*, 2001a) and navel orangeworm (Wang *et al.*, 2002c, Mitcham *et al.*, 2004) and hot water treatments for cherries infested with third-instar codling moth (Feng *et al.*, 2004; Hansen *et al.*, 2004).

5.3 Insect Mortality Models

Models used in the literature to describe the thermal mortality kinetic data for insects range from fundamental kinetic models that include the influence of temperature to semi- or purely empirical models that are specific to fixed temperatures. This section introduces some of the most often-used models.

Fundamental Kinetic Model

Reduction in an insect population N as a function of time t (min) at an elevated temperature can be described by the following fundamental kinetic model:

$$\frac{dN}{dt} = -KN^n \tag{5.1}$$

where K is the rate constant (min^{-1}) and n is the kinetic order of the reaction. This equation can be written for changes in survival ratio:

$$\frac{dS}{dt} = -kS^n \tag{5.2}$$

where S is the ratio of the surviving numbers of insects (N) at time t to the initial number of the population (N_0) expressed as a percentage. That is:

$$S = \frac{N}{N_0} \times 100 \tag{5.3}$$

Integrating Eq. 5.2 yields:

$$\begin{align} \ln S &= -kt + c \quad (n = 1) \\ S^{1-n} &= -kt + c \quad (n \neq 1) \end{align} \tag{5.4}$$

where c is a constant.

A linear regression analysis can be performed on experimentally obtained survival ratios of an insect population as a function of time at a fixed temperature using zero-order ($n = 0$), half-order ($n = 0.5$), first-order ($n = 1$), 1.5th-order ($n = 1.5$) or second-order ($n = 2$) reactions. The highest correlation determines the order of the kinetic reaction.

As an example, experimental data (time and survival ratio) reported by Gazit *et al.* (2004) for third-instar Mediterranean fruit fly at 46°C are listed in Table 5.2. The survival ratios were converted according to Eq. 5.4 for different reaction orders, and also listed in Table 5.2. Linear regression can then be

Table 5.2. Survival rate (N/N_0) of third-instar Mediterranean fruit fly at different reaction orders (n) versus time at 46°C (from Gazit et al., 2004).

Time (min)	N/N_0	$n = 0$ $(N/N_0)^{1.0-0.0}$	$n = 0.5$ $(N/N_0)^{1.0-0.5}$	$n = 1$ Ln (N/N_0)	$n = 1.5$ $(N/N_0)^{1.0-1.5}$	$n = 2$ $(N/N_0)^{1.0-2.0}$
0	1.000	1.000	1.000	0.000	1.000	1.000
5	0.959	0.959	0.979	−0.042	1.021	1.043
10	0.697	0.697	0.835	−0.361	1.198	1.434
20	0.585	0.585	0.765	−0.535	1.307	1.708
30	0.269	0.269	0.519	−1.312	1.927	3.713
50	0.043	0.043	0.207	−3.146	4.821	23.238
60	0.001	0.001	0.032	−6.908	31.623	1000.000

performed between the converted data sets and exposure times at 46°C. Similar analyses can also be performed with data at other temperatures.

The best-fitted lines are determined by comparing the coefficients of determination (R^2) for all the treated temperatures, as shown in Table 5.3. The half-order model produces the highest mean R^2 values over all the four treatment temperatures for third-instar Mediterranean fruit fly (Table 5.3). After the reaction order has been determined, the values of k and c in Eq. 5.4 can then be determined for each temperature (see Table 5.4) using linear regression (Eq. 5.4). Normally, the k values change with temperature, whereas c values remain constant at around 1.0.

The experimentally determined kinetic data for third-instar Mediterranean fruit flies with curves predicted using a half-order of reaction are shown in Fig. 5.4. The slope of the curve for each temperature corresponds to the value of k in Eq. 5.4 and Table 5.4. It is clear from Fig. 5.5 that the value of k increases with temperature.

The change of k in Eqs 5.2 and 5.4 with temperature often follows an Arrhenius relationship, as follows (Stumbo, 1973):

$$k = k_{ref}e^{\frac{-Ea}{R}\left(\frac{1}{T} - \frac{1}{T_{ref}}\right)} \tag{5.5}$$

Table 5.3. Coefficients of determination (R^2) from kinetic order (n) models for thermal mortality of third-instar Mediterranean fruit fly at four different temperatures (from Gazit et al., 2004).

Temp. (°C)	R^2 for different kinetic orders (n)				
	0.0	0.5	1.0	1.5	2.0
46	0.945	0.992	0.842	0.562	0.472
48	0.924	0.990	0.831	0.608	0.557
50	0.889	0.981	0.932	0.663	0.545
52	0.789	0.893	0.957	0.954	0.869
Mean	0.870	0.964	0.891	0.697	0.611

Table 5.4. Thermal death kinetic parameters for the half-order kinetic model $(N/N_0)^{0.5} = -kt + c$ and standard errors (SE) of the means for third-instar Mediterranean fruit fly at four temperatures (from Gazit *et al.*, 2004).

Temp. (T, °C)	$k \pm$ SE	$c \pm$ SE	Log t	Log k	$1/T * 10^3$ (K)
46	0.0165 ± 0.0007	1.0327 ± 0.0383	1.778	−2.084	3.133
48	0.0674 ± 0.0035	1.0361 ± 0.0427	1.176	−1.472	3.114
50	0.2545 ± 0.0177	0.9755 ± 0.0609	0.602	−0.895	3.095
52	1.0417 ± 0.2082	0.9370 ± 0.1698	0.000	−0.283	3.076

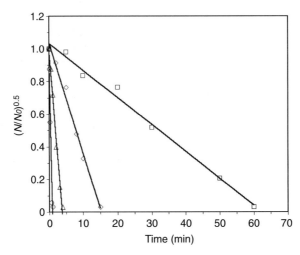

Fig. 5.4. Thermal mortality curves of third-instar Mediterranean fruit fly at four different temperatures using a half-order kinetic model. □ 46°C; ◇ 48°C; △ 50°C; ○ 52°C (from Gazit *et al.*, 2004).

where T is the absolute temperature (K), k_{ref} is the reaction rate constant at the reference temperature T_{ref}, E_a is the activation energy (J/mol) and R is the universal gas constant (8.314 J/mol K). Taking the logarithm on both sides of Eq. 5.5 yields the following equation:

$$\log k = \log k_{ref} - \frac{E_a}{R}\left(\frac{1}{T} - \frac{1}{T_{ref}}\right) \tag{5.6}$$

That is, plotting log k against the reciprocal of absolute temperature results in a straight line. This is illustrated in Fig. 5.5 for third-instar Mediterranean fruit flies. The activation energy for thermal inactivation of insects is calculated, from the slope times the value of R, to be 593.9 kJ/mol (Gazit *et al.*, 2004). The activation energy for thermal inactivation of pests results in a slope of a reaction rate *versus* the temperature. The higher the activation energy, the more sensitive pests are to temperature changes.

Based on fundamental thermal death kinetics, a half-order reaction model has been used successfully to describe the larval thermal death kinetics of

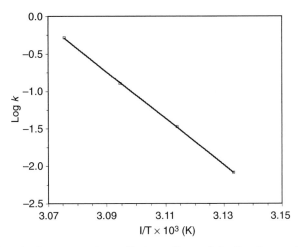

Fig. 5.5. Arrhenius plot for temperature effects on thermal death rate constant for third-instar Mediterranean fruit fly. The straight line (log k = 95.098–31.016*1000/T) was obtained by linear regression for third-instars (R^2 = 0.999) (from Gazit *et al.*, 2004).

codling moth (Wang *et al.*, 2002a), navel orangeworm (Wang *et al.*, 2002b), Indianmeal moth (Johnson *et al.*, 2003), Mediterranean fruit fly (Gazit *et al.*, 2004), Mexican fruit fly (Hallman *et al.*, 2005) and red flour beetle (Johnson *et al.*, 2004). The thermal death kinetic data showed that the navel orangeworm was the most heat-resistant insect at temperatures > 50°C.

Logarithmic Model

The logarithmic model has been used extensively in thermobacteriology, which deals with the thermal death kinetics of pathogenic and spoilage bacteria in food products. The most important concepts in this model are the D and z values determined from thermal death time (TDT) curves for thermally inactivating pathogens (Stumbo, 1973). The TDT curve can be developed by plotting the minimum exposure time required at each temperature to achieve 100% kill of a given bacterial population on a semi-log scale (see Fig. 5.6a). Those concepts are based on the observation that the mortality of microorganisms at a given temperature follows a logarithmic constant rate:

$$\log N = \log N_0 - \frac{t}{D} \qquad (5.7)$$

where t is the treatment time (min). The susceptibility of bacteria to heat at a specific temperature is characterized by the value of D, the time (min) required to obtain one log reduction (tenfold reduction) in a bacterial population (see Fig. 5.6a). The value of D is also obtained from the –1/slope of the regression equation of the log survivor curve against time. Holdsworth (1997) proved that Eq. 5.7 represents a first order of reaction. The D value can be related to the

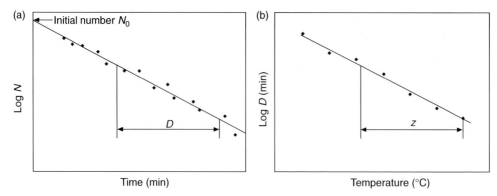

Fig. 5.6. Semi-logarithmic survival curve (a) to obtain D value and thermal death time curve (b) to obtain z value.

rate-constant k in Eq. 5.4 for the first order of reaction ($n = 1$) by the following relationship:

$$D = \frac{2.303}{k} \tag{5.8}$$

where D varies with temperature. When the D value is plotted against the temperature, a linear relationship is observed as a TDT curve (see Fig. 5.6b).

The thermal susceptibility of bacteria to temperature change is characterized by a z value (the degree of temperature increase to result in one log reduction in D value), such that:

$$z = \frac{T_2 - T_1}{\log D_{T_1} - \log D_{T_2}} \tag{5.9}$$

where D_T represents the value of D measured at temperature T, and T_1 and T_2 are two different temperatures. The z value is also obtained by the -1/slope of the regression equation of the log D value against temperature.

King *et al.* (1979) concluded that there was no apparent difference between death rate curves for unicellular microorganisms and multicellular insect pests. The theory and concepts evolved from the TDT curves and classical thermal kinetic model for describing thermal mortality of bacteria may be readily used to study insect thermal mortality. The activation energy for thermal kill of insects can also be related to the z value according to the following relationship (Tang *et al.*, 2000):

$$E_a = \frac{2.303 R T_{min} T_{max}}{z} \tag{5.10}$$

where T_{min} and T_{max} are the minimum and maximum temperatures (K) of a test range, respectively.

Normally, after obtaining the average data from the replicates for four to five exposure times at each of four different temperatures, the logarithmic thermal death kinetic model can be established. The negative inverse of the slope of the TDT curve can be used to determine the z value, which can be used to estimate the activation energy needed for thermal death of the test insects according to Eq. 5.10.

The log times required for 100% mortality of 300 third-instar Mediterranean fruit flies obtained by Gazit *et al.* (2004) are listed in Table 5.5 and used to plot TDT curves at four temperatures (see Fig. 5.7). The z value derived from the TDT curves is 3.6°C, resulting in activation energy of 551.9 kJ/mol as calculated using Eq. 5.10. TDT curves that define minimum times over a temperature range are also determined with HBS for other insects (see Fig. 5.8, Table 5.5).

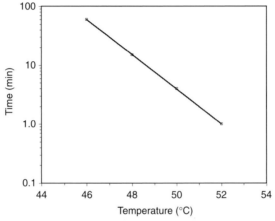

Fig. 5.7. Thermal death–time curves log t = 15.365–0.295 and T (R^2 = 0.999) for third-instar Mediterranean fruit fly (n = 300) at a heating rate of 15°C/min where t = time (min) T = temperature (°C) (from Gazit *et al.*, 2004).

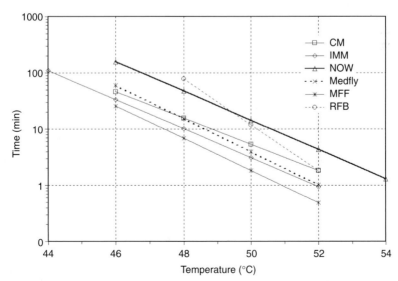

Fig. 5.8. Minimum time–temperature combinations for complete kill of 600 fifth-instars of codling moth (CM), Indianmeal moth (IMM) and navel orangeworm (NOW), and of 600 third-instars of Mediterranean fruit fly (Medfly), Mexican fruit fly (MFF) and red flour beetle (RFB).

Table 5.5. Comparison of lethal times obtained by experiments, half-order kinetic models (Eq. 5.4), probit analysis, complementary log-log (CLL) and logit models for third-instar Mediterranean fruit fly at four temperatures.

Temp (°C)	Observed time for 100% mortality of 300 insects ($\approx LT_{99.67}$) (min)	Half-order kinetic model			Probit analysis			CLL			Logit		
		LT_{95}	$LT_{99.67}$	$LT_{99.9968}$ (Probit-9)	LT_{95}	$LT_{99.67}$	$LT_{99.9968}$ (Probit-9)	LT_{95}	$LT_{99.67}$	$LT_{99.9968}$ (Probit-9)	LT_{95}	$LT_{99.67}$	$LT_{99.9968}$ (Probit-9)
46	60	49.0	59.1	62.2	45.9	74.0	106.7	39.4	66.1	97.1	41.0	63.2	89.0
48	15	12.1	14.5	15.3	11.5	18.3	26.5	9.7	15.5	22.7	10.0	15.1	21.5
50	4	3.0	3.6	3.8	2.6	4.3	6.4	2.3	3.9	6.0	2.4	3.9	5.7
52	1	0.7	0.8	0.9	0.4	0.5	0.7	0.5	0.8	1.3	0.5	0.8	1.3

Modified First-order Model

In many cases, the first-order reaction model is not adequate to describe the thermal death of microorganisms. In order to improve the curve fitting of the first-order reaction model for bacterial spores, an empirical kinetic model is developed as a modification to the first-order reaction relationship by adding an exponential constant a (Alderton and Snell, 1970). This equation was used by Jang (1986, 1991), Moss and Jang (1991) and Moss and Chan (1993) for insects:

$$\left(\log S\right)^a = -kt + c \tag{5.11}$$

This equation is fitted to experimental data in two steps: (i) obtain a best value for the exponential constant a using multiple linear regression for data at all temperatures; and (ii) determine values of c and k by linear regression with $(\log S)^a$ as the dependent variable and t as an independent variable for each temperature. A drawback of this method is the lack of a physical meaning for parameter a.

Lethal Times (LT)

A large amount of the reported data related to thermal susceptibility of insect pests is in the form of LT_{95} (min) or LT_{99} (min), as shown in Table 5.6. Lethal time (LT_y, min) is defined as the time needed to kill $y\%$ (referred to as mortality rate) of insects at a given temperature. Survival rate (S, %) and mortality rate (M, %) are related by the following relationship (Table 5.7):

$$S = \frac{N}{N_0} \times 100 = 1 - M \tag{5.12}$$

Based on the experimental values of N and N_0, the LT can be estimated directly. For example, LT_{95} results in one survivor out of 20 treated insects. Kinetic models that describe the survival rate of an insect population based on treatment time can be readily used to estimate LT values. Based on the kinetic model in Eq. 5.4, the LT value can be inversely derived:

$$LT = -\frac{\ln S - c}{k} \quad (n = 1)$$
$$LT = -\frac{S^{1-n} - c}{k} \quad (n \neq 1) \tag{5.13}$$

Similar relationships to Eq. 5.13 can be developed from other kinetic models to estimate LT_{95}, LT_{99}, $LT_{99.67}$ and $LT_{99.9968}$ (Probit-9) for each treatment temperature by inputting 0.05, 0.01, 0.0033 and 0.000032 for N/N_0, respectively. The estimated $LT_{99.67}$ – which is one survivor out of 300 treated insects – closely corresponds to the observed exposures that produce a complete kill of 300 insects (see Table 5.5). Lethal times increase with increasing insect sample populations (e.g. from 50 to 31,250), but the difference between $LT_{99.9968}$ and LT_{95} becomes small at higher temperatures.

Table 5.6. Different model predictions for thermal kill of insects at different population levels.

Insect(s)	Temp. (°C)	Model(s)	Predictions	Source
Lesser grain borer	70–80	Probit	LT_{50}, $LT_{99.9}$	Evans (1981)
Lesser grain borer	45–55	Logit; Probit; CLL	LT_{99}, $LT_{99.9}$	Beckett and Morton (2003)
Lesser grain borer	42–53	Probit	LT_{50}, LT_{99}, $LT_{99.9}$	Beckett and Morton (2003)
Oriental fruit fly	43–48	Empirical	LT_{90}, $LT_{99.997}$	Jang (1991)
Codling moth	45–51	Probit	LT_{50}	Yokoyama *et al.* (1991)
Hessian fly	47.5–80	Log likelihood regression	LT_{50}, LT_{95}, LT_{99}, $LT_{99.99}$	Sokhansanj *et al.* (1992)
Apple moth	39–47	CLL	LT_{99}	Jones *et al.* (1995)
Mexican fruit fly	43–45	Empirical; CLL; logit	LT_{99}, $LT_{99.9}$, $LT_{99.9968}$	Thomas and Mangan (1997)
Codling moth	43–49	CLL	LT_{99}	Jones and Waddell (1997)
Cook island fruit fly	43–48	CLL	LT_{99}	Waddell *et al.* (1997)
Queensland fruit fly	42–48	CLL	LT_{99}	Waddell *et al.* (2000)
Codling moth	46–52	Half-order kinetic	LT_{95}, LT_{99}, $LT_{99.83}$, $LT_{99.9968}$	Wang *et al.* (2002a)
Navel orangeworm	46–54	Empirical; half-order kinetic	LT_{95}, LT_{99}, $LT_{99.83}$, $LT_{99.9968}$	Wang *et al.* (2002b)
Indianmeal moth	44–52	Empirical; half-order kinetic	LT_{95}, LT_{99}	Johnson *et al.* (2003)
Medfly egg and third-instars	46–52	Half-order kinetic	LT_{95}, LT_{99}, $LT_{99.67}$, $LT_{99.9968}$	Gazit *et al.* (2004)
Red flour beetle	48–52	Half-order kinetic	LT_{95}, LT_{99}	Johnson *et al.* (2004)

Table 5.7. Lethal time (LT) levels as a function of mortality (*M*), survival ratio (*S*), insect population (*N*, survival number; N_0, total number) and probit levels.

Lethal time level (LT)	*M* (%)	*S*	N/N_0	Probit level
LT_{95}	95	0.05	1/20	6.645
LT_{99}	99	0.01	1/100	7.326
$LT_{99.83}$	99.83	0.0017	1/600	7.929
$LT_{99.9968}$	99.9968	0.000032	1/31250	9.000

Therefore, adding a short holding time at high temperatures dramatically increases the mortality of insects.

Probit Analysis

First proposed by Baker (1939) in relation to tephritid fruit flies, probit analysis is another method commonly used to describe changes in insect populations according to treatment times (Russell *et al.*, 1977; Throne *et al.*, 1995). This method is based on the assumption that the frequency of individual deaths in an insect population under a constant temperature follows the standard normal distribution over the logarithm of time (Finney, 1971). Under this assumption, the plot of insect mortality against logarithm of time at a given temperature resembles a sigmoidal curve, as shown in Fig. 5.9.

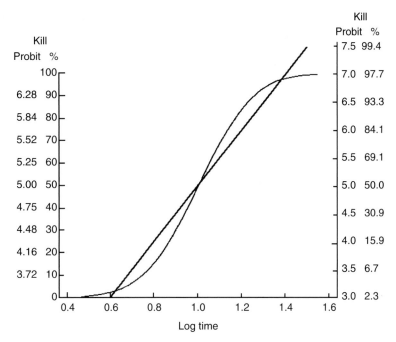

Fig. 5.9. Effect of probit transformation in which the normal sigmoid curve is transformed into a straight line when using a probit scale instead of percentages (from Finney, 1971).

The probit of the mortality *M* is defined as the abscissa, which corresponds to a probability *P* in a normal distribution with a mean value of 5 and variance (σ^2) of 1. The probit of *P* is *Y* (Finney, 1971), thus:

$$P = \int_{-\infty}^{Y-5} \frac{e^{-\mu^2/2}}{\sqrt{2\pi}} d\mu \qquad (5.14)$$

where μ is the mean of the normal distribution. The relation between the probit of exposed insect mortality and dose (time) is the linear equation ($Y = 5 + (t-\mu)/\sigma$). The effect of transformation from percentages to probits is demonstrated in Fig. 5.9. The relationship between the log time and probit of the responses becomes a straight line to obtain the slope and interception in place of μ and σ.

For probit analysis, the percentage responses (insect mortality) observed for each dose (log time at a temperature) should first be calculated and converted to probits using Table 5.7 or other tables in the literature such as those in Finney (1971). The probits are then plotted against the dose (log *t*) to obtain a linear relationship by regression. The estimated probits are used to recalculate the corresponding percentages and precision of the estimated insect mortality.

Converting probit-transformed data back to the original units is not straightforward, but can be conducted using tables and mathematical computer programs (Throne *at al.*, 1995). Although it is useful to confirm the effectiveness of quarantine treatments, it may not be appropriate for the development of new thermal methods if the information is not easily extrapolated for the purpose of predicting mortality in new test conditions (Tang *et al.*, 2000).

Complementary Log–Log Transformation (CLL)

CLL is widely used in the field of entomology for insect mortality predictions because it is included in common statistical software such as SAS and Minitab (Jones *et al.*, 1995; Thomas and Mangan, 1997; Waddell *et al.*, 1997). The CLL is the inverse of the cumulative distribution function and ensures predicted probabilities between 0 and 1 proportionally. The relationship between insect mortality and time at a given temperature is assumed as:

$$M = 1 - e^{-e^{at+b}} \tag{5.15}$$

After taking logarithms of Eq. 5.15 for two times to obtain CLL, Eq. 5.15 becomes:

$$\log\left(-\log(1-M)\right) = -kt + c \tag{5.16}$$

Equation 5.16 shows that plotting the CLL of the survival rate (1–M) against time results in a linear line. When this model is used, the values of k and c are obtained at each temperature based on linear regression of the converted mortality *versus* time. Similar to the modified first-order model demonstrated by Eq. 5.11, Eq. 5.16 can be used inversely to predict the LT needed to kill the required insect samples.

Logit Transformation

The logit regression is the inverse of the cumulative logistic distribution function. The relationship of mortality against time is assumed as:

$$M = \frac{e^{at+b}}{1 + e^{at+b}} \tag{5.17}$$

Reorganizing Eq. 5.17 leads to:

$$\frac{M}{1-M} = e^{at+b} \tag{5.18}$$

After taking logarithms of Eq. 5.18, we have:

$$\log\left(\frac{M}{1-M}\right) = -kt + c \tag{5.19}$$

Based on experimental data, plotting logit transformation of insect mortality, $\log\left(\dfrac{M}{1-M}\right)$, against the exposure time results in a linear relationship.

When the linear regression is taken, the constants k and c are obtained. The developed model (Eq. 5.19) can be inversely applied to predict the LT for different insect populations as shown in Table 5.5 (Thomas and Mangan, 1997; Beckett and Morton, 2003). Using an iterative reweighted least square algorithm, logit regression methods estimate parameters in the model, so it is optimized.

Degree–Minute Model

A commonly used method to account for the cumulative lethal effect of insect infestation treatment is the so-called degree–minute (DM) model (Shellie and Mangan, 1994; Nyanjage et al., 1998). This model suggests that the cumulated temperature (in degrees) beyond a threshold value times the duration of exposure (in minutes) is a critical factor that yields a certain level of insect mortality, regardless of the treatment process. A general DM model can be expressed with the following equation:

$$DM = \int_0^t \left[T(t) - T_s \right] dt \qquad (5.20)$$

where $T(t)$ is the temperature in fruit (°C) as a function of time t, and T_s is the threshold temperature (°C). For thermal treatments where insects are fully exposed to a constant temperature, Eq. 5.20 is reduced to:

$$DM = \left[T(t) - T_s \right] t \qquad (5.21)$$

Researchers have not based their selection of the threshold temperature on any solid theoretical basis. For example, 30°C was arbitrarily selected for Mexican fruit fly larvae, Anastrepha ludens (Loew) [Diptera: Tephritidae] (Shellie and Mangan, 1994), 42°C for Queensland fruit fly eggs, Bactrocera tryoni (Froggatt) [Diptera: Tephritidae] (Waddell et al., 2000) and 48°C for warehouse beetle larvae, Trogoderma variabile Ballion [Coleoptera; Dermestidae] (Wright et al., 2002).

In addition, the DM model treats each degree beyond the threshold temperature as equally important when calculating cumulative temperature–time effects on insect mortality. That is, two degrees for 1 min beyond the threshold temperature has the same affect as one degree for 2 min beyond the threshold temperature. Table 5.8 shows the differences in lethal effects at the same temperature–time combinations between the DM model – where we selected 42°C as the threshold temperature for codling moth larvae – and the kinetic model for four different constant temperature treatments. The equivalent total lethal time M_{ref} (min) in Table 5.8 determines the cumulative lethal effect for any given combination of temperature–time at a reference temperature, which is defined by Eq. 5.22 (Hansen et al., 2004; Wang et al., 2004).

It is clear from Table 5.8 that the DM model does not reflect the true effect of temperatures, because lethality increased only 2.0 times when the temperature increased from 48°C to 54°C. In reality, the lethal effect increases

Table 5.8. Comparisons of temperature effects on cumulative mortality of codling moth larvae between the degree–minute and kinetic models (from Hansen *et al.*, 2004).

Temp. (°C) + holding time (min)	Degree–minute model (°C min)	M_{ref} (temp) from the kinetic model (min)
48 + 2	12	2.0
50 + 2	16	6.3
52 + 2	20	20.0
54 + 2	24	63.3

exponentially with increasing temperature. As shown in Fig. 5.8, the treatment time to achieve the same lethal effect (complete kill of 600 insects) dramatically decreased with increased temperature. The cumulated lethal time calculated based on a half-order thermal death kinetic model for codling moth (Wang *et al.*, 2002a) was 31.6 times larger at 54°C than at 48°C for the same 2 min (see Table 5.8). The increased lethal time with increasing temperature based on the half-order thermal death kinetic model reflects the real temperature effect on insect mortality.

5.4 Model Comparisons

Selection of the appropriate model depends on researcher preference and intended use. Thomas and Mangan (1997) reviewed several models for Mexican fruit flies, and recommended a thermal death kinetic model for estimation of quarantine treatment levels when developing new thermal treatment methods.

The best fit between mortality data and exposure time depends on the target insect, temperature range and insect population level. Converting transformed data back to original units is important in selecting which transformation results in a model that best fits the data. Throne *et al.* (1995) reported that the log-probit and log-logit models had the best fit to the parasitoid wasp *Bracon hebetor* (Say) [Hymenoptera: Braconidae] data at the LT_{99} level (see Fig. 5.10) when compared to CLL. They suggested that model selections also depend on minimizing the confidence limits on that lethal level and time.

Based on experimental data from navel orangeworms, half-order kinetic models are more successfully used to predict the required time at different lethal levels as compared with modified first-order models (Wang *et al.*, 2002b). Table 5.9 shows that the predicted $LT_{99.83}$ by the half-order fundamental kinetic model is close to the experimental results, and that < 20% extra time is needed to increase the efficacy of a heat treatment from 99% mortality to 99.9968% (Probit-9) when fifth-instar navel orangeworm are exposed to a constant temperature.

The lethal times predicted by the half-order kinetic model are longer than those obtained by the modified first-order model for fewer Probit-9 treatments.

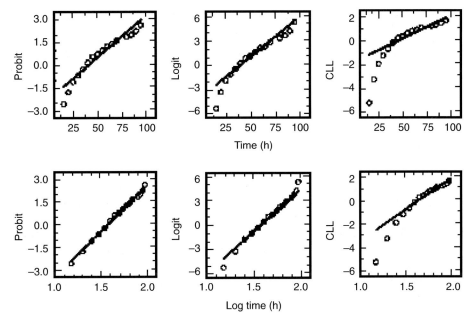

Fig. 5.10. Observed (open circles) and predicted (line) mortality of the parasitoid wasp transformed by probit, logit and complementary log–log (CLL) over time and log time (from Throne *et al.*, 1995).

Table 5.9. Comparison of lethal times (min) obtained by experiments, modified first-order (Eq. 5.11) and half-order kinetic models (Eq. 5.4) for fifth-instar navel orangeworms at five temperatures (from Wang *et al.*, 2002b).

Temp. (°C)	Observed 100% mortality (min)	Modified first-order model				Half-order kinetic model			
		LT_{95}	LT_{99}	$LT_{99.83}$	$LT_{99.9968}$ (Probit-9)	LT_{95}	LT_{99}	$LT_{99.83}$	$LT_{99.9968}$ (Probit-9)
46	140	106.6	124.9	140.4	165.9	120.0	137.6	146.1	151.1
48	50	36.0	42.3	47.6	56.3	40.9	46.8	49.6	51.3
50	15	11.5	13.3	14.8	17.4	13.5	15.3	16.2	16.8
52	6	3.7	4.5	5.2	6.3	4.3	5.0	5.4	5.6
54	1	0.7	0.8	0.9	1.1	0.8	0.6	1.0	1.0

For Probit-9 quarantine treatments, however, the treatment times predicted using the modified first-order model are longer than those obtained by the half-order fundamental kinetic model. The time difference between the two methods is < 10% except for 46°C, where the larger treatment time interval causes the error. Generally, the fundamental kinetic model (Eq. 5.4) is simple to obtain and use with the acceptable precision when compared with the modified first-order model in Eq. 5.11.

The half-order kinetic model is further compared with probit, CLL and logit models based on the experimental data for third-instar Mediterranean fruit fly at four temperatures (see Table 5.5). The predicted time at $LT_{99.67}$ by the half-order kinetic model agrees well with the experimental values. The lethal time at Probit-9 level by the half-order kinetic model is slightly underestimated compared with the other three models, especially at 46 and 48°C. For predicted lethal times at 50 and 52°C, the half-order kinetic model agrees well with the CLL and logit models.

5.5 Model Applications

Once the z, k and E_a values are determined for a target insect, the cumulative temperature–time effect of a thermal treatment with a known temperature–time history on reduction of the insect population can then be predicted with good accuracy by using the half-order kinetic model.

Time–temperature histories at the commodity core and surface are different from the ideal step-temperature profile. Total insect mortality is dependent on the cumulative thermal exposure during the course of treatment over the entire section of the commodity. The cumulative lethal effect imparted by the treatment to each part of a fruit can be estimated if the actual temperature–time history of that location is recorded during the treatment.

In developing an effective heat treatment, we use the temperature history of the least heated part of a fruit to calculate cumulative heat effect. For first-order kinetics, we use the cumulated lethal time model to determine the cumulative lethal effect for any given combination of temperature and time (Tang *et al.*, 2000; Wang *et al.*, 2002a) in terms of equivalent total lethal time M_{ref} (min) at a reference temperature, T_{ref} (°C):

$$M_{ref} = \int_0^t 10^{\frac{T(t)-T_{ref}}{z}} dt \qquad (5.22)$$

where $T(t)$ is the recorded core temperature as a function of time t (in min) and z is the temperature difference required for a tenfold change in the thermal death time curve (°C).

For example, in our early study, we used an average z value of 4°C for codling moth larvae and 48°C as the reference temperature based on heating block studies (Wang *et al.*, 2002a). However, one can use other lethal temperatures (e.g. 50 or 54°C). The equivalent total lethal time M_{ref} at one temperature, T_1, can be translated to a lethal time of another temperature, T_2, with the same mortality by the following equation (Wang *et al.*, 2004):

$$M_{ref}(T_1) = f \cdot M_{ref}(T_2) \qquad (5.23)$$

where f is the conversion factor that can be calculated by the following equation for a given value of z:

$$f = 10^{-\frac{T_1-T_2}{z}} \qquad (5.24)$$

Thus, the cumulated lethal time at 50, 52, 54 and 55°C can be translated to that at 48°C by the following relationship with the same lethality effect:

$$M_{ref}(Temp) = f \cdot M_{ref}(48) \tag{5.25}$$

and

$$f = 10^{-\frac{Temp-48}{z}} \tag{5.26}$$

Based on f values calculated from Eq. 5.26 for codling moths ($z = 4$°C), 1 min full exposure at 48°C has the same lethal effect as 0.316, 0.100, 0.0316 and 0.0178 min exposures at 50, 52, 54 and 55°C, respectively.

Experimentally determined mortality and calculated equivalent exposure times at 48°C are plotted in Fig. 5.11 as an example of infested cherries with third-instar codling moth, together with the predicted insect mortality by the heating block study (Wang *et al.*, 2002a). The estimated lethal time based on the temperature–time history of the fruit core can be used to define a boundary for the least mortality. However, the probit analysis might not provide an accurate treatment guideline when a different temperature–time history in fruits is used.

The kinetic model is an improvement over probit analysis for estimation of treatment parameters (Robertson *et al.*, 1994). Thus, a treatment efficacy based on cumulated lethal times can be accurately estimated for validating the developed protocol. Because internal temperatures are measured, the heat may be from different sources, including forced hot air, water baths and electromagnetic energy. This approach can also be used for different types of

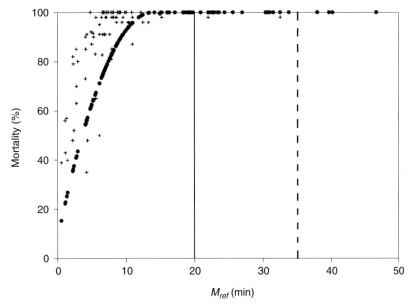

Fig. 5.11. Relationship between corrected cumulative times (M_{ref}) set at 48°C and percentage mortality of codling moth larvae, where + is experimentally determined insect mortality and ●●● is predicted mortality based on cherry core temperature from the heating block system (from Hansen *et al.*, 2004).

commodities. Furthermore, because the cumulated lethal times at a reference temperature include the treatment temperature profile, the procedure compensates for slow heating due to fruit size. This would be particularly helpful in identifying treatments against internal pests in commodities that have a wide range of host sizes, such as codling moths in apples or fruit flies in mangoes (Hansen *et al.*, 2004).

5.6 Closing Remarks

Thermal death kinetic models are effective tools in the development of thermal treatments for insect infestations of commodities in a more systematic and efficient manner compared with traditional trial-and-error approaches. The thermal death kinetic models not only allow prediction of insect mortality with treatment times when subjected to steady-state isothermal heating, but can also be incorporated in other models to predict insect thermal mortality under transit and non-isothermal conditions.

Fundamental death kinetic models can be used to derive other useful information, such as LT values for desired mortality levels, TDT curves and the activation energy for thermal kill of a specific insect pest. Such information allows reliable comparison of heat tolerance among different insect pests and at different life stages, and their sensitivity to temperature changes. While still useful, empirical models are case-dependent and cannot be extended beyond the test conditions from which they are developed.

In developing fundamental kinetics models, it is critical that test methods eliminate the influence of factors such as non-isothermal test conditions. The reality of non-isothermal conditions in commodities during hot air or water treatments can be taken care of by applying heat transfer theory via numeric simulation, along with intrinsic thermal kinetic models for treatment of insect pests.

5.7 References

Alderton, G. and Snell, N. (1970) Chemical states of bacterial spores: heat resistance and its kinetics at intermediate water activity. *Applied Microbiology* 19, 565–572.

Baker, A.C. (1939) The basis for treatment of products where fruit flies are involved as a condition for entry into the USA. *USDA Circular* 551, 1–7.

Beckett, S.J. and Morton, R. (2003) The mortality of three species of *Psocoptera*, *Liposcelis bostrychophila* Badonnel, *Liposcelis decolor* Pearman and *Liposcelis paeta* Pearman, at moderately elevated temperatures. *Journal of Stored Products Research* 39, 103–115.

Evans, D.E. (1981) The influence of some biological and physical factors on the heat tolerance relationships for *Rhyzopertha dominica* (F.) and *Sitophilus oryzae* (L.) (Coleoptera: Bostrychidae and Curculionidae). *Journal of Stored Products Research* 17, 65–72.

Evans, D.E. (1986) The influence of rate of heating on the mortality of *Rhyzopertha dominica* (L.) (Coleoptera: Bostrychidae).

Journal of Stored Products Research 23, 73–77.

Feng, X., Hansen, J.D., Biasi, B. and Mitcham, E.J. (2004) Use of hot water treatment to control codling moths in harvested California 'Bing' sweet cherries. *Postharvest Biology and Technology* 31, 41–49.

Finney, D.J. (1971) *Probit Analysis*, 3rd edn. Cambridge University Press, London.

Follett, P.A. and Sanxter, S.S. (2001) Hot water immersion to ensure quarantine security for *Cryptophlebia* spp. (Lepidoptera: Tortricidae) in lychee and longan exported from Hawaii. *Journal of Economic Entomology* 94, 1292–1295.

Gazit, Y., Rossler, Y., Wang, S., Tang, J. and Lurie, S. (2004) Thermal death kinetics of egg and third-instar Mediterranean fruit fly *Ceratitis capitata* (Wiedemann) (Diptera: Tephritidae). *Journal of Economic Entomology* 97, 1540–1546.

Hallman, G.J. (1990) Vapor heat treatment of carambolas infested with Caribbean fruit fly (Diptera: Tephritidae). *Journal of Economic Entomology* 83, 2340–2342.

Hallman, G.J. (1996) Mortality of third instar Caribbean fruit fly (Diptera: Tephritidae) reared in diet or grapefruit and immersed in heated water or grapefruit juice. *Florida Entomologist* 79, 168–172.

Hallman, G.J., Wang, S. and Tang, J. (2005) Reaction orders of thermal mortality of third-instar Mexican fruit fly *Anastrepha ludens* (Loew) (Diptera: Tephritidae). *Journal of Economic Entomology* 98, 1905–1910.

Hansen, J.D. and Heidt, M.L. (2006) Codling moth survival under heated anaerobic conditions. *Journal of the Kansas Entomological Society* 79, 207–209.

Hansen, J.D. and Sharp, J.L. (1998) Thermal death studies of third-instar Caribbean fruit fly (Diptera: Tephritidae). *Journal of Economic Entomology* 91, 968–973.

Hansen, J.D. and Sharp, J.L. (2000) Thermal death of third instars of the Caribbean fruit fly (Diptera: Tephritidae) treated in different substrates. *Journal of Entomological Science* 35, 196–204.

Hansen, J.D., Armstrong, J.W., Hu, B.K.S. and Brown, S.A. (1990) Thermal death of oriental fruit fly (Diptera: Tephritidae) third instars in developing quarantine treatments for papayas. *Journal of Economic Entomology* 83, 160–167.

Hansen, J.D., Wang, S. and Tang, J. (2004) A cumulated lethal time model to evaluate efficacy of heat treatments for codling moth *Cydia pomonella* (L.) (Lepidoptera: Tortricidae) in cherries. *Postharvest Biology and Technology* 33, 309–317.

Hayes, C.F., Chingon, H.T.G., Nitta, F.A. and Wang, W.J. (1984) Temperature control as an alternative to ethylene dibromide fumigation for the control of fruit flies (Diptera: Tephritidae) in papaya. *Journal of Economic Entomology* 77, 683–686.

Hayes, C.F., Chingon, H.T.C., Nitta, F.A. and Leung, A.M.T. (1987) Calculation of survival from double hot water immersion treatment for papayas infested with oriental fruit flies (Diptera: Tephritidae). *Journal of Economic Entomology* 80, 887–890.

Heather, N.W., Corcoran, R.J. and Kopittke, R.A. (1997) Hot air disinfestation of Australian 'Kensington' mangoes against two fruit flies (Diptera: Tephritidae). *Postharvest Biology and Technology* 10, 99–105.

Holdsworth, S.D. (1997) *Thermal Processing of Packaged Foods*. Blackie Academic & Professional, London.

Ikediala, J.N., Tang, J., Neven, L.G. and Drake, S.R. (1999) Quarantine treatment of cherries using 915 MHz microwaves: temperature mapping, codling moth mortality and fruit quality. *Postharvest Biology and Technology* 16, 127–137.

Ikediala, J.N., Tang, J. and Wig, T. (2000) A heating block system for studying thermal death kinetics of insect pests. *Transactions of the ASAE* 43, 351–358.

Jang, E.B. (1986) Kinetics of thermal death in eggs and first instars of three species of fruit flies (Diptera: Tephritidae). *Journal of Economic Entomology* 79, 700–705.

Jang, E.B. (1991) Thermal death kinetics and heat tolerance in early and late third instars of the oriental fruit fly (Diptera: Tephritidae). *Journal of Economic Entomology* 84, 1298–1303.

Jang, E.B., Nagata, J.T., Chan, H.T. and Laidlaw, W.G. (1999) Thermal death kinetics in eggs and larvae of *Bactrocera latifrons* (Diptera: Tephritidae) and comparative thermotolerance to three other tephritid fruit fly species in Hawaii. *Journal of Economic Entomology* 92, 684–690.

Johnson, J.A., Wang, S. and Tang, J. (2003) Thermal death kinetics of fifth instar *Plodia interpunctella* (Lepidoptera: Pyralidae). *Journal of Economic Entomology* 96, 519–524.

Johnson, J.A., Valero, K.A., Wang, S. and Tang, J. (2004) Thermal death kinetics of red flour beetle, *Tribolium castaneum* (Coleoptera: Tenebrionidae). *Journal of Economic Entomology* 97, 1868–1873.

Jones, V.M. and Waddell, B.C. (1997) Effect of hot water on mortality of *Cydia pomonella* (Lepidoptera: Tortricidae). *Journal of Economic Entomology* 90, 1357–1359.

Jones, V.M., Waddell, B.C. and Maindonald, J.H. (1995) Comparative mortality responses of three tortricid (Lepidoptera) species to hot water. *Journal of Economic Entomology* 88, 1356–1360.

King, A.D., Bayne, H.G. and Alderton, G. (1979) Non-logarithmic death rate calculations for *Byssochlamys fulva* and other microorganisms. *Applied Environment Microbiology* 37, 596–600.

Lurie, S., Jemric, T., Weksler, A., Akiva, R. and Gazit, Y. (2004) Heat treatment of 'Oroblanco' citrus fruit to control insect infestation. *Postharvest Biology and Technology* 34, 321–329.

Mangan, R.L., Shellie, K.C. and Ingle, S.J. (1998) High temperature forced-air treatments with fixed time and temperature for 'Dancy' tangerines, 'Valencia' oranges, and 'Rio Star' grapefruit. *Journal of Economic Entomology* 91, 934–939.

Mitcham, E.J., Veltman, R.H., Feng, X., de Castro, E., Johnson, J.A., Simpson, T.L., Biasi, W.V., Wang, S. and Tang, J. (2004) Application of radio frequency treatments to control insects in in-shell walnuts. *Postharvest Biology and Technology* 33 (1), 93–101.

Moss, J.I. and Chan, H.T. (1993) Thermal death kinetics of Caribbean fruit fly (Diptera: Tephritidae) embryos. *Journal of Economic Entomology* 86, 1162–1166.

Moss, J.I. and Jang, E.B. (1991) Effects of age and metabolic stress on heat tolerance of Mediterranean fruit fly (Diptera: Tephritidae) eggs. *Journal of Economic Entomology* 84, 537–541.

Neven, L.G. (1994) Combined heat treatments and cold storage effects on mortality of fifth-instar codling moth (Lepidoptera: Tortricidae). *Journal of Economic Entomology* 87, 1262–1265.

Neven, L.G. (1998a) Effects of heating rate on the mortality of fifth-instar codling moth (Lepidoptera: Tortricidae). *Journal of Economic Entomology* 91, 297–301.

Neven, L.G. (1998b) Respiratory response of fifth-instar codling moth (Lepidoptera: Tortricidae) to rapidly changing temperatures. *Journal of Economic Entomology* 91, 302–308.

Neven, L.G. and Mitcham, E.J. (1996) CATTS (Controlled Atmosphere/Temperature Treatment System): a novel tool for the development of quarantine treatments. *American Entomologist* 42, 56–59.

Neven, L.G. and Rehfield, L.M. (1995) Comparison of pre-storage heat treatments on fifth-instar codling moth (Lepidoptera: Tortricidae) mortality. *Journal of Economic Entomology* 88, 1371–1375.

Neven, L.G., Rehfield, L.M. and Shellie, K.C. (1996) Moist and vapour forced air treatment of apples and pears: effects on the mortality of fifth instar codling moth (Lepidoptera: Tortricidae). *Journal of Economic Entomology* 89, 700–704.

Nyanjage, M.O., Wainwright, H. and Bishop, C.F.H. (1998) The effects of hot water treatments in combination with cooling and/or storage on the physiology and disease of mango fruits. *Journal of Horticultural Science and Biotechnology* 73, 589–597.

Paull, R.E. (1994) Response of tropical horticultural commodities to insect disinfestation treatments. *HortScience* 29, 988–996.

Robertson, J.L., Preisler, H.K., Frampton, E.R. and Armstrong, J.W. (1994) Statistical analyses to estimate efficacy of disinfestation treatments. In: Sharp, J.L. and Hallman, G.J. (eds). *Quarantine Treatments for Pests of Food Plants*. Westview Press, Inc., Boulder, Colorado.

Russell, R.M., Robertson, J.L. and Savin, N.E. (1977) POLO: a new computer program for probit analysis. *Bulletin of the Entomological Society of America* 23, 209–213.

Sharp, J.L., Ouye, M.T., Ingle, S.J. and Hart, W.G. (1989) Hot water quarantine treatment for mangoes from Mexico infested with Mexican fruit fly and West Indian fruit fly (Diptera: Tephritidae). *Journal of Economic Entomology* 82, 1657–1662.

Shellie, K.C. and Mangan, R.L. (1994) Postharvest quality of Valencia orange after exposure to hot, moist, forced-air for fruit-fly disinfestation. *HortScience* 29, 1524–1527.

Shellie, K.C., Neven, L.G. and Drake, S.R. (2001) Assessing 'Bing' sweet cherry tolerance to a heated controlled atmosphere for insect pest control. *HortTechnology* 11, 308–311.

Soderstrom, E.L., Brandl, D.G. and Mackey, B.E. (1996) High temperature alone and combined with controlled atmospheres for control of diapausing codling moth (Lepidoptera: Tortricidae) in walnuts. *Journal of Economic Entomology* 89, 144–147.

Sokhansanj, S., Venkatesam, V.S. and Wood, H.C. (1992) Thermal kill of wheat midge and Hessian fly. *Postharvest Biology and Technology* 2, 65–71.

Stumbo, C.R. (1973) *Thermobacteriology in Food Processing*. Academic Press, Inc., New York.

Tang, J., Ikediala, J.N., Wang, S., Hansen, J.D. and Cavalieri, R.P. (2000) High-temperature, short-time thermal quarantine methods. *Postharvest Biology and Technology* 21, 129–145.

Thomas, D.B. and Mangan, R.L. (1997) Modelling thermal death in the Mexican fruit fly (Diptera: Tephritidae). *Journal of Economic Entomology* 90, 527–534.

Thomas, D.B. and Shellie, K.C. (2000) Heating rate and induced thermotolerance in larvae of Mexican fruit fly, a quarantine pest of citrus and mangoes. *Journal of Economic Entomology* 93, 1373–1379.

Throne, J.E., Weaver, D.K. and Baker, J.E. (1995) Probit analysis: assessing goodness of fit based on back-transformation and residuals. *Journal of Economic Entomology* 88, 1513–1516.

Waddell, B.C., Clare, G.K. and Maindonald, J.H. (1997) Comparative mortality responses of two Cook Island fruit fly (Diptera: Tephritidae) species to hot water immersion. *Journal of Economic Entomology* 90, 1351–1356.

Waddell, B.C., Jones, V.M., Petry, R.J., Sales, F., Paulaud, D., Maindonald, J.H. and Laidlaw, W.G. (2000) Thermal conditioning in *Bactrocera tryoni* eggs (Diptera: Tephritidae) following hot-water immersion. *Postharvest Biology and Technology* 21, 113–128.

Wang S. and Tang, J. (2004) Radio frequency heating: a potential method of postharvest pest control in nuts and dry products. *Journal of Zhejiang University Science* 5, 1169–1174.

Wang, S., Ikediala, J.N., Tang, J., Hansen, J.D., Mitcham, E., Mao, R. and Swanson, B. (2001a) Radio frequency treatments to control codling moth in in-shell walnuts. *Postharvest Biology Technology* 22, 29–38.

Wang, S., Tang, J. and Cavalieri, R.P. (2001b) Modeling fruit internal heating rates for hot air and hot water treatments. *Postharvest Biology and Technology* 22, 257–270.

Wang, S., Ikediala, J.N., Tang, J. and Hansen, J.D. (2002a) Thermal death kinetics and heating rate effects for fifth-instar *Cydia pomonella* (L.) (Lepidoptera: Tortricicae). *Journal of Stored Products Research* 38, 441–453.

Wang, S., Tang, J., Johnson, J.A. and Hansen, J.D. (2002b) Thermal death kinetics of fifth-instar *Amyelois transitella* (Walker) (Lepidoptera: Pyralidae) larvae. *Journal of Stored Products Research* 38, 427–440.

Wang, S., Tang, J., Johnson, J.A., Mitcham,

E., Hansen, J.D., Cavalieri, R.P., Bower, J. and Biasi, B. (2002c) Process protocols based on radio frequency energy to control field and storage pests in in-shell walnuts. *Postharvest Biology and Technology* 26, 265–273.

Wang, S., Yin, X., Tang, J. and Hansen, J.D. (2004) Thermal resistance of different life stages of codling moth (Lepidoptera: Tortricidae). *Journal of Stored Products Research* 40, 565–574.

Wang, S., Johnson, J.A., Tang, J. and Yin, X. (2005) Heating condition effects on thermal resistance of fifth-instar navel orange-worm (Lepidoptera: Pyralidae). *Journal of Stored Products Research* 41, 469–478.

Wright, E.J., Sinclair, E.A. and Annis, P.C. (2002) Laboratory determination of the requirements for control of *Trogoderma variable* (Coleoptera: Dermestidae) by heat. *Journal of Stored Products Research* 38, 147–155.

Yokoyama, V.Y. and Miller, G.T. (1987) High temperature for control of oriental fruit fly (Lepidoptera: Tortricidae) in stone fruits. *Journal of Economic Entomology* 80, 641–645.

Yokoyama, V.Y., Miller, G.T. and Dowell, R.V. (1991) Response of codling moth (Lepidoptera: Tortricidae) to high temperature, a potential quarantine treatment for exported commodities. *Journal of Economic Entomology* 84, 528–531.

6 Biology and Thermal Death Kinetics of Selected Insects

J. Tang,[1]* S. Wang[1]** and J.A. Johnson[2]

[1]*Department of Biological Systems Engineering, Washington State University, Pullman, Washington, USA; e-mails: *jtang@wsu.edu; **shaojin_wang@wsu.edu; [2]USDA-ARS Horticultural Crops Research Laboratory, Parlier, California, USA; e-mail: jjohnson@fresno.ars.usda.gov*

6.1 Introduction

A major problem in the storage and marketing of agricultural products is infestation with postharvest insect pests. Common storage pests such as the cosmopolitan species Indianmeal moth (*Plodia interpunctella* (Hübner) [Lepidoptera: Pyralidae]) and red flour beetle (*Tribolium castaneum* (Herbst) [Coleoptera: Tenebrionidae]) cause considerable economic loss to the food industry in the form of direct damage to the product, control costs and customer returns of the infested product. A number of economically important field pests, e.g. the codling moth (*Cydia pomonella* (L.) [Lepidoptera: Tortricidae]), navel orangeworm (*Amyelois transitella* (Walker) [Lepidoptera: Pyralidae]) and various tephritid fruit fly species may also be present in marketed products and become the subject of serious phytosanitary and quarantine issues for exported commodities.

Due to regulatory requirements and consumer expectations, the presence of a single living insect in marketed commodities is generally not acceptable, prompting the use of postharvest treatments to insure insect-free products. The application of heat may be an effective disinfestation method for many products, but development of effective heat treatments requires a thorough understanding of the temperature response of the target insects and treated commodity. This chapter presents a brief introduction to the general biology of several economically important insect pests and provides relevant thermal death kinetic information. It also includes information regarding the effects of life stages and pre-treatment conditions on the thermal resistance of some of those insects.

6.2 Biology and Economic Impact of Target Species

Codling Moth

Codling moth is a serious pest of many tree fruits within the rose family. Known to most as the familiar worm in the apple, it also attacks pears, crabapples, quince, hawthorn, walnuts and stone fruits. Originally from Asia Minor, it spread along with cultivated apples throughout Europe and later to the Americas. It is now common in most areas where apples are grown, with a few notable exceptions (see Fig. 6.1).

Adult codling moths (see Fig. 6.2a) have a wingspan of about 15–20 mm and are around 8–10 mm long at rest. Their bodies are grey, marked with lighter grey wavy lines across the wings, and a distinctive copper patch at the tip of the wings. These colours provide them with excellent camouflage when resting on tree bark. Adults are active at dusk when temperatures are > 15°C. At this time they mate, with females laying between 30 and 100 disc-shaped eggs, attaching them singly to leaves, spurs, fruits or nuts.

The caterpillars that emerge from the eggs seek out and bore into fruit or nuts, where they continue to feed and develop. There are five instars; at maturity, larvae (see Fig. 6.2b) usually leave the fruit or nuts to search for suitable pupation sites in tree trunks, soil or debris. There they spin a silken cocoon, pupate and later emerge as adults. The number of generations occurring each year depends largely on temperature, with three full generations and a partial fourth common in milder climates like California, while only two generations are found in colder regions.

Codling moths overwinter as specialized larvae that enter a type of dormancy known as diapause. The larvae enter diapause at maturity, surviving

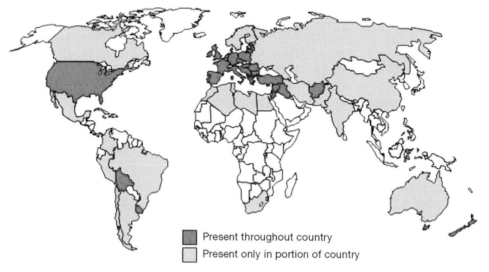

Present throughout country
Present only in portion of country

Fig. 6.1. Worldwide distribution of codling moth, *Cydia pomonella* (modified from the Codling Moth Information Support System, http://ippc.orst.edu/CodlingMoth/).

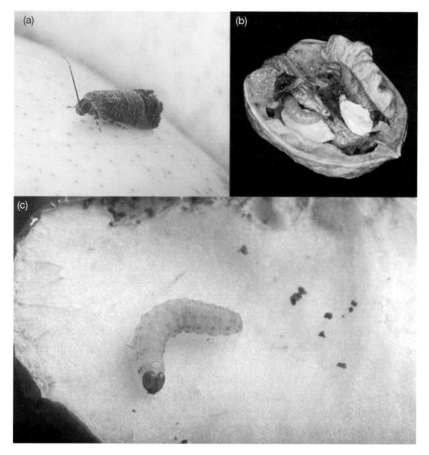

Fig. 6.2. Codling moth, *Cydia pomonella*: (a) adult (Photo by Scott Bauer, ARS); (b) larvae infesting a walnut (Photo by Charles Curtis, ARS); (c) larva infesting an apple (Photo by Peggy Greb, ARS).

the winter in cocoons in protected sites. In walnut orchards, diapausing larvae are occasionally found within the nuts, but in fruit orchards diapausing larvae leave the fruit before harvest. Because the diapausing stage of many insects is often the most difficult to kill with postharvest control methods, particularly fumigants, it is often of special interest to researchers. However, walnuts are the only product in which diapausing codling moth is found.

Codling moths can ruin 10–90% of the fruit and nut orchards they infest. On tree fruit, the codling moth causes damage as the feeding larva burrow through to the core of the fruit, leaving behind tunnels filled with reddish-brown faeces known as frass. This extensive feeding causes interior breakdown of the fruit, and may cause it to drop early in the season. If left uncontrolled, codling moth can infest 20–90% of a fruit orchard (Ohlendorf, 1999).

When infesting walnuts, codling moth larvae enter the nut and feed directly on the kernels. Again, nuts damaged early in the season may drop from the tree; nuts damaged later remain on the tree but are filled with dark, crumbly frass, rendering them inedible. If left uncontrolled in walnut orchards, codling moth may infest 10–15% of the nuts (Ohlendorf, 1999).

Because codling moth can render products unmarketable, it is considered a major pest. Apple, pear and walnut growers are able to keep damage to a minimum through an aggressive management programme that may include sanitation, insecticidal sprays and non-chemical techniques such as mating confusion, viral pesticide applications and release of natural enemies. To properly time management practices, populations are often monitored with pheromone traps. In the USA, apple, pear and walnut growers spend roughly 10% of their total cultural operations budget on codling moth control measures (Caprile *et al.*, 2001; Buchner *et al.*, 2002; Ingels *et al.*, 2003).

Although widely spread throughout temperate regions, codling moth has a limited distribution in East Asia and Africa. Consequently, Japan, South Korea, Taiwan, China and Western Australia place restrictions on importation of host plants into their countries. Japan, in particular, requires strict quarantine treatments of methyl bromide alone or methyl bromide in combination with cold treatment for all possible hosts. Such restrictions add considerably to the cost of sending products to foreign markets.

Navel Orangeworm

Navel orangeworm is a moth believed to be of neotropical origin. First described in Mexico, it was later discovered in Arizona on navel oranges, giving it its common name (Wade, 1961). Introduced into California in the 1940s, it became a serious pest of walnuts, almonds, pistachios and figs. Considered to be a scavenger, it may be found on a variety of damaged, dropped or mummified stone and citrus fruits, but rarely becomes a pest on these commodities. Primarily a field pest of nut crops and figs because it attacks the marketable commodity, it is of great phytosanitary concern.

Adult navel orangeworms (see Fig. 6.3a) are silver grey, marked with black irregular lines and with a wingspan of about 20 mm. The paired labial palps form a prominent 'snout' at the front of the head, which allows them to be distinguished easily from most other orchard moths. Two days following emergence, adult females can lay an average of about 80 flat, oval eggs. Eggs are first cream-coloured, then turn red about 24 h after oviposition (see Fig. 6.3c). Eggs are laid singly on fruits or nuts that remain on the tree after harvest ('mummies'), or on developing fruits or nuts.

Navel orangeworm is usually found on the split hulls of blighted, codling moth-infested or otherwise damaged walnuts. Newly hatched larvae feed directly on the fruit or nutmeats (see Fig. 6.3b, d). In figs, larvae enter through the ostiole and feed internally, while in almonds larvae may feed on the hull as well as the kernel. Often, several larvae are found within the same nut or fig. Like codling moth, adult navel orangeworm do not feed on fruits or nuts.

In the central valley of California, there are normally three complete generations of navel orangeworm each year, with a partial fourth generation in the warmer southern region (University of California, 2002). Although there is

Fig. 6.3. Navel orangeworm, *Amyelois transitella*: (a) adult (Photo by Judy Johnson, ARS); (b) severely infested walnut (Photo by Judy Johnson, ARS); (c) eggs (Photo by Charles Curtis, ARS); (d) larva on almond (Photo by Charles Curtis, ARS).

no well-defined diapause stage, navel orangeworm populations tend to overwinter as larvae within mummified nuts or fruit, either on the tree or the ground. Because mummies are the main source of adults in the spring, removal by knocking them from the trees and disking them into the soil has long been recommended as a control strategy.

Additional treatment methods include early harvest, pesticide sprays and natural control through parasitoids. Pesticide applications target the exposed egg stage. From 7–14% of the cultural costs of maintaining a commercial tree nut orchard in California go towards navel orangeworm control (Buchner *et al.*, 2002; Viveros *et al.*, 2003; Beede *et al.*, 2004).

Nuts infested with navel orangeworm are inedible, so are rejected by those responsible for the cleaning, processing and marketing of final nut products (see Fig. 6.3b). Kernel damage by navel orangeworm also increases the likelihood of mould growth and aflatoxin development, which can cause rejection of entire shipments. With adequate pest management practices, growers can reduce infestation to low levels, generally <4%. However, because the presence of live navel orangeworm in marketed products is not acceptable, processors are required to disinfest products before shipment, generally by fumigating with phosphine or methyl bromide.

Indianmeal Moth

Indianmeal moth is one of the most common and serious pests of stored products. It feeds on a wide variety of dry vegetable matter, including grains, nuts, dry beans, dried fruits, dried flowers, bird seed, dry pet foods, processed foods such as cereals, and confections. It does better in coarse meals than in finely ground flours, but can survive in processing dust and debris. Generally believed to be of European origin, it is now cosmopolitan in its distribution. Often cited as the leading cause of customer product returns, it is also a very common pest of home pantries.

Adult moths are about 8 mm long at rest, with a wingspan of about 15 mm (see Fig. 6.4a). The outer tip of the forewings is coppery brown, with darker markings; the base of the forewings is grey. Indianmeal moth can be distinguished easily from other stored product moths, which are more uniformly coloured. The number of eggs laid by each female moth varies from 100 to 300, and is usually highest on grain and nut products.

The white, spherical eggs are < 1 mm in diameter and are deposited singly or in groups on the product surface. Newly emerged larvae, about 1 mm long,

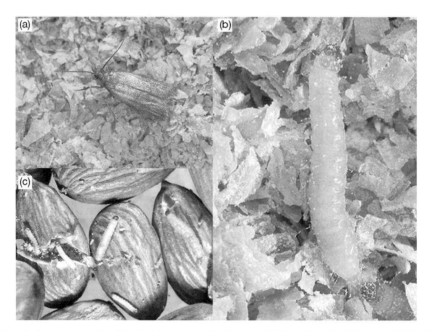

Fig. 6.4. Indianmeal moth, *Plodia interpunctella*: (a) adult (Photo by Scott Bauer, ARS); (b) larva (Photo by Scott Bauer, ARS); (c) infested almonds (Photo by Judy Johnson, ARS).

immediately start feeding on the product. Because these larvae are capable of entering packaging through cracks and holes as small as 0.4 mm (Tsuji, 1998), they are considered to be highly invasive of bulk-stored and packaged products.

Fully grown Indianmeal larvae are about 12 mm long and may be white, light yellow, pink or greenish (see Fig. 6.4b). As the larva crawls, it leaves a very fine, almost invisible, white silk thread as a trail marker. Heavy infestations in storage may be detected by the appearance of a sheet of silk on the product surface. Damage is due largely to contamination of the product with silk, faeces, cast skins, the insects themselves and product loss through larval feeding (see Fig. 6.4c).

Adults do not feed on the product. Larvae pupate within a silken cocoon directly in the product or nearby. In temperate climates, Indianmeal moths pass the winter as fully grown diapausing larvae in cocoons constructed within product or in cracks, under boards, between sheets of paper or in other dark – preferably dry – places. Although it is not usually found in the field, Indianmeal moth is capable of survival outdoors in warmer climates such as the central valley of California. In colder areas, Indianmeal moth is usually limited to heated buildings.

For both foreign and domestic markets, there is no tolerance for live insects in agricultural products. Consequently, control of postharvest infestations of Indianmeal moth is a high priority with processors. Because Indianmeal moth is capable of breeding in processing debris and dust, control in storage and processing facilities begins with sanitation. Other pest management strategies include use of pheromone traps to identify developing populations, space treatments with pesticides, reducing population growth through aeration and release of parasitoids. Processors also rely heavily on the use of chemical fumigants to disinfest products before shipment.

Red Flour Beetle

Red flour beetle, like the Indianmeal moth, is a common and serious pest of stored products. It is very similar to the confused flour beetle, *Tribolium confusum* Jacquelin du Val. Both beetles feed on a large variety of products, including nuts, dried fruits, legumes, spices and processed foods, but flours and other milled products are favoured. Generally, they do not attack sound, unbroken grain, instead feeding on broken kernels, particularly the germ. They are also known to feed on other stored product insects and their eggs. Believed by some to be Indo-Australian in origin, the red flour beetle is cosmopolitan in distribution. Although somewhat limited to warmer regions, it is capable of surviving in heated buildings in cold climates.

Adult red flour beetles are a shiny reddish-brown, flat, and only about 3 mm in length (see Fig. 6.5a). They are very similar in appearance to confused flour beetles, differing only slightly in the structure of the antennae and thorax. Adults of both species are very active, but only the red flour beetle is known to fly. Adults are long-lived; some have survived for 3 years, with females laying eggs for more than 1 year.

Fig. 6.5. Red flour beetle, *Tribolium castaneum*: (a) adult; (b) eggs; (c) larvae; (d) pupae.

Each female may produce 400–500 white, spherical eggs (see Fig. 6.5b), depositing them directly in the food material. Larvae (see Fig. 6.5c) are cream-coloured, about 6 mm in length at maturity, with a short, forked projection from the tip of the abdomen. Pupae are cream-coloured, and found naked within the product (see Fig. 6.5d). Unlike stored product moths, adult beetles as well as larvae live in and feed on the products they infest. Consequently, any life stage may be found in the product at any time. Development time from egg to adult is dependent on environmental factors – particularly temperature, with a minimum of about 25 days (Arbogast, 1991).

Red flour beetles are serious pests in a variety of situations, including farm-stored grains, flour mills, food processing plants and bulk-stored dried fruits and nuts. While not particularly good at penetrating through packaging material, like the Indianmeal moth they are adept at invading through small cracks and gaps in package seals. Direct damage to products is caused by feeding larvae and adults, but a more serious problem is contamination with cast skins, dead insects and faeces. In addition, large beetle populations produce quinones, which cause objectionable odours and reduce product quality.

Management of red flour beetle is similar to that of Indianmeal moth, and begins with aggressive prevention strategies that include sanitation and exclusion. Monitoring for beetle populations through pheromone or food-baited traps is used to make treatment decisions. Entire flour mills or food processing plants may be fumigated or heat-treated to kill resident populations.

Insecticidal crack and crevice sprays, as well as dusts made of diatomaceous earth, are also used to reduce populations. Stored grain may be aerated to reduce grain temperatures and slow pest population growth.

Although it is difficult to estimate the economic impact of storage pests such as the red flour beetle and Indianmeal moth, these management practices are a significant cost for farmers and food processors. While product loss due to feeding by red flour beetle populations may be minimal, far more product is lost through contamination and the odours created by beetle infestations. Farm animals may refuse infested grains or do poorly on them. Of particular concern is the loss of consumer confidence when infested products are found.

Tephritid Fruit Flies

Fruit flies of the family Tephritidae are among the most economically damaging of insect pests, with many causing severe phytosanitary problems in international trade (White and Elson-Harris, 1992). Worldwide, about 70 tephritid species are considered to be important pests of agricultural products, with many others ranked as minor or potential pests. The most economically important belong to five genera: *Bactrocera*, *Anastrepha*, *Ceratitis*, *Dacus* and *Rhagoletis*. The most significant genus, *Bactrocera*, is found primarily in the Old World tropics, and contains about 40 pest species, including the Oriental fruit fly (*B. dorsalis*), melon fly (*B. cucurbitae*) and olive fruit fly (*B. oleae*). The genus *Anastrepha* is most prevalent in the Neotropics, with about 15 important pest species, including the Mexican fruit fly, *A. ludens* (see Fig. 6.6b).

About 17 pest species belong to the genus *Rhagoletis*, which occur primarily in the Neotropics, North America and Eurasia, and these include the western cherry fruit fly, *R. indifferens*, and the apple maggot, *R. pomonella*. *Ceratitis* and *Dacus* species are found mostly in Africa, with each genus having about ten important pest species. These include the pumpkin fly, *Dacus bivittatus*, and the notorious Mediterranean fruit fly, *Ceratitis capitata* (see Fig. 6.6a).

Most pest tephritids feed as larvae on fruits of various types. The most serious pest species are polyphagous, having a wide host range. The Mediterranean fruit fly, or Medfly, has been found on more than 300 host plants. Most fruit fly species, however, are oligophagous, attacking a few closely related hosts; many other species are monophagous, feeding on a single host species. Adults feed on plant exudates (such as those from oviposition wounds), nectar, honeydew, rotting fruit and other food sources.

Tephritids often engage in elaborate courtship and mating behaviours, and sex pheromones have been identified from both sexes for most pest species, with the notable exception of *Anastrepha*. Generally, females position their elongated eggs just under the surface of the host fruit and will often mark the fruit with a pheromone meant to deter other females from ovipositing on the same fruit. The legless larvae (maggots, Fig. 6.6c) hatch several days later, depending upon species and temperature, and go through three larval instars. Larvae lack definite, sclerotized head capsules, appearing headless. Most fruit-feeding species leave the fruit in order to pupate within the soil. Many temperate

Fig. 6.6. Tephritid fruit flies: (a) Mediterranean fruit fly adult , *Ceratitis capitata* (Photo by Scott Bauer, ARS); (b) Mexican fruit fly adult, *Anastrepha ludens* (Photo by Jack Dykinga, ARS); (c) papaya infested by fruit fly larvae (Photo by Scott Bauer, ARS); (d) cherries infested with fruit fly larvae (Photo by Peggy Greb, ARS).

species may overwinter as diapausing pupae. Tropical species normally have numerous generations per year; some temperate species have only one.

Most damage from pest fruit flies occurs when feeding maggots reduce the host fruit to a juicy, inedible, rotting lump (see Fig. 6.6d). Ovipositing females may cause some disfigurement to fruit as well. Because tephritid larvae feed internally within the fruit, they are very difficult to control with conventional pesticides, resulting in high infestation levels in areas where fruit flies are endemic. Many of the more serious fruit fly pests, such as Mediterranean fruit fly, oriental fruit fly and olive fruit fly, can reach infestation levels of nearly 100% if left unchecked. Worldwide distribution of the Mediterranean fruit fly is shown in Fig. 6.7.

Because of the potential severity of these pests, countries where the flies do not occur restrict import or require stringent quarantine treatments of fruit from infested areas. Industry experts estimate that in the state of Hawaii, where Mediterranean fruit fly, Oriental fruit fly, melon fly and Malaysian fruit fly are all established, nearly US$300 million is lost each year from locally grown produce markets. This figure does not include potentially high-value export markets. Estimates for potential direct and indirect losses to agriculture should the Mediterranean fruit fly become established in California run as high as US$1.4 billion annually (Kaplan, 2004).

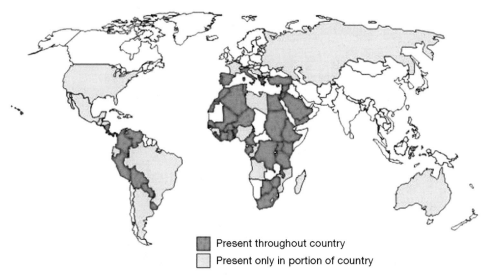

Present throughout country
Present only in portion of country

Fig. 6.7. Worldwide distribution of Mediterranean fruit fly, *Ceratitis capitata* (modified from OEPP/EPPO, 2005).

In areas where tephritid fruit flies are endemic, control in the field varies, depending largely upon the crop and pest species. Because the larvae are difficult to control within the fruit, control strategies tend to target adults, focusing on mating and feeding behaviours. The application of chemical pesticides must be carefully coordinated with monitoring data to kill adult females before eggs are laid.

Bait sprays combine a proteinaceous liquid feeding attractant with a toxicant to attract and kill female flies, and may be more effective than pesticides without baits. Spray programmes are combined with cultural control methods such as orchard sanitation and early harvest. Monitoring methods that use feeding attractants or sex pheromones to attract adult flies to traps are sometimes modified for mass-trapping systems. The oriental fruit fly was successfully eradicated from Okinawa, Japan, by the distribution of fibreboard squares soaked in a male attractant and toxicant, thereby reducing the number of viable males and eventually eliminating the entire population (Fruit Fly Eradication Project Office, 2006).

Attempts to eradicate introduced fruit flies often involve the sterile insect technique in which male flies sterilized with ionizing radiation are released in huge numbers. Wild females produce unviable eggs after mating with the sterile males. This method is usually employed with area-wide bait spray programmes to reduce fruit fly populations. Eradication programmes are normally used only where fruit flies have recently been introduced and are not well established. However, Japan successfully eradicated the established melon fly from the islands of Okinawa using the sterile insect technique (Fruit Fly Eradication Project Office, 2006).

While eradication may not be feasible in areas where fruit fly populations are well established, some of the techniques used in eradication programmes may be used in area-wide integrated pest management programmes to reduce fruit fly populations to more acceptable levels. The state of Hawaii, USA, has had success with this approach, combining field sanitation, bait sprays timed with careful monitoring and periodic releases of sterile males to reduce population outbreaks (Kaplan, 2004). A similar area-wide programme involving Israel, Jordan and the Palestinian Authority targets Mediterranean fruit fly, and has resulted in agricultural exports from the region increasing from US$1 million in 1998 to US$50 million by 2004 (FAO, 2005).

Produce from areas infested with fruit flies is subject to quarantines from countries where the flies do not occur (see Table 6.1). To access these markets, the produce must either undergo a disinfestation treatment or be certified as pest-free. There is a large body of research dealing with quarantine treatments for fruit flies, employing a variety of methods.

The USA Department of Agriculture (USDA)-Animal and Plant Health Inspection Service (APHIS) treatment manual (USDA-APHIS, 2004) lists methyl bromide fumigation, various forms of heat treatment, cold treatment and irradiation as approved treatments for fruit flies. Although the Montreal Protocol ban on methyl bromide allows an exemption for quarantine treatments, cost increases and reduced availability for the fumigant have generated renewed interest in improving existing alternative treatments. APHIS has recently approved generic irradiation doses for broad classes of insects, including fruit flies, making it easier to use this method as a quarantine treatment. Generic heat treatment levels for fruit flies have also been suggested (Follett and Neven, 2006).

6.3 Thermal Death Data

When considering thermal treatments to control the above insects, it is desirable to know how they respond to heat under ideal isothermal conditions. The following sections provide relevant fundamental kinetic data for those insects.

Fundamental Kinetic Model

The heating block systems described in Chapter 5, this volume, were used to obtain thermal mortality data for different life stages of codling moth, navel orangeworm, Indianmeal moth, red flour beetle, Mediterranean fruit fly and Mexican fruit fly (Wang *et al.*, 2002a, b; Johnson *et al.*, 2003, 2004; Gazit *et al.*, 2004; Hallman *et al.*, 2005). Those data fit well to a 0.5th order reaction model:

$$\frac{dS}{dt} = -kS^{0.5} \tag{6.1}$$

or:

Table 6.1. Tephritid fruit flies of quarantine importance.

Scientific name: common name	Major economic hosts	Distribution
Anastrepha fraterculus: South American fruit fly	Guava, mango, apple, citrus, peach, plum	Mexico, southern Texas, widely distributed in Central and South America
Anastrepha ludens: Mexican fruit fly	*Sargentia greggii* (wild host), citrus, mango	Mexico, Texas, widely distributed in Central and South America
Anastrepha obliqua: West Indian fruit fly	*Spondias* spp. (wild hosts), mango	Widely distributed in South and Central America
Anastrepha suspensa: Caribbean fruit fly	Bush cherry, guava	Widely distributed in the Caribbean
Ceratitis capitata: Mediterranean fruit fly	Wide variety of fruits, including coffee, apple, avocado, citrus, fig, mango, pear	Wide distribution (see Fig. 6.7)
Ceratitis cosyra: Mango fruit fly	Mango, avocado, citrus, peach	Widely distributed in Africa
Ceratitis quinaria: Five-spotted fruit fly	Apricot, citrus, guava, peach	Widely distributed in Africa
Ceratitis rosa: Natal fruit fly	Wide variety of fruits, including pome and stone fruit, citrus, guava, grape, mango	Widely distributed in Africa
Bactrocera cucumis: Cucumber fly	Cucurbit, tomato	Australia
Bactrocera cucurbitae: Melon fly	Cucurbit	Wide distribution
Bactrocera dorsalis: Oriental fruit fly	Wide variety of fruits, including apple, plum, banana, guava, mango, orange, peach, tomato	Wide distribution
Bactrocera minax: Chinese citrus fly	Citrus	Bhutan, China, India
Bactrocera tryoni: Queensland fruit fly	Wide variety of fruits, including citrus, guava, apple, mango, peach, plum, pear	Australia
Bactrocera tsuneonis: Japanese orange fly	Citrus	China, Japan, Taiwan, Vietnam
Bactrocera zonata: Peach fruit fly	Guava, mango, peach, apricot, fig, citrus	Widely distributed in Asia
Dacus ciliatus: Ethiopian fruit fly	Cucurbit	Widely distributed in Asia and Africa
Rhagoletis completa: Walnut husk fly	Walnut, occasionally peach	Mexico, USA, Italy, Switzerland
Rhagoletis cingulata: Eastern cherry fruit fly	Cherry	Canada, Mexico, USA
Rhagoletis fausta: Black cherry fruit fly	Cherry	Canada, USA
Rhagoletis indifferens: Western cherry fruit fly	Cherry	Canada, Mexico, USA, Switzerland
Rhagoletis mendax: Blueberry maggot	Blueberry, huckleberry	Canada, USA
Rhagoletis pomonella: Apple maggot	Apples, occasionally other tree fruit	Canada, Mexico, USA, Costa Rica, Colombia
Rhagoletis ribicola: Dark currant fly	Gooseberry, red currant, other *Ribes* spp.	Canada, USA

$$S^{0.5} = -kt + c \tag{6.2}$$

where S is the ratio of the surviving numbers of insects (N) to the initial number of the population (N_0). The thermal death rate constants were obtained by linear regression at each temperature. Changes of k in Eq. 6.2 for the above-mentioned insects are described by the Arrhenius relationship:

$$\log k = \log k_0 - \frac{E_a}{2.303R} \frac{1}{T} \tag{6.3}$$

Parameters for the most heat-resistant life stage of several insects are listed in Table 6.2, with their Arrhenius relationships listed in Table 6.3.

Thermal Death Time (TDT) Curves

As described in Chapter 5, this volume, TDT curves provide information on the minimum time required to kill completely a certain percentage of a targeted

Table 6.2. Thermal death constants for a 0.5th order reaction model of the last larval instars of six insects at different temperatures.

| Insect | Temp. (°C) | 0.5th order kinetic model | | |
		k	c	R^2
Codling moth	46	0.0189	1.0555	0.954
	48	0.0691	0.9584	0.965
	50	0.2016	0.9401	0.857
	52	0.5056	0.9488	0.957
Indianmeal moth	44	0.0078	1.069	0.96
	46	0.0313	0.963	0.96
	48	0.0999	0.983	0.99
	50	0.3595	1.011	0.96
	52	0.9406	1.046	0.94
Navel	46	0.0070	1.0633	0.964
orangeworm	48	0.0209	1.0778	0.960
	50	0.0662	1.1155	0.900
	52	0.1662	0.9323	0.938
	54	0.8881	0.9202	0.961
Mediterranean	46	0.0165	1.0327	0.992
fruit fly	48	0.0674	1.0361	0.990
	50	0.2545	0.9755	0.981
	52	1.0417	0.9370	0.893
Mexican	44	0.0100	1.0152	0.990
fruit fly	46	0.0412	1.1160	0.935
	48	0.1690	1.1127	0.945
	50	0.4607	0.9631	0.928
Red flour	48	0.0122	1.0365	0.992
beetle	50	0.1042	1.0509	0.977
	52	0.5182	0.9075	0.898

insect population at different temperatures (Tang *et al.*, 2000). For example, complete kill of 600 fifth-instar navel orangeworm larvae requires minimum exposure times of 140, 50, 15, 6 and 1 min at 46, 48, 50, 52 and 54°C, respectively (Wang *et al.*, 2002b).

Different insect species and life stages have different responses to heat, but over a temperature range 46–54°C, the minimum exposure time–temperature combination required to kill insects completely (n = 600) follows a semi-logarithmic relationship: $\log t = a - bT$, where t is the time (min), T is temperature (°C), and a and b are positive constants that depend upon insect species and life stages (see Table 6.3).

The z value that indicates the temperature difference resulting in one log reduction of the mortality rate was obtained from the reciprocal of b. The z values are 4.2, 3.9 and 2.5°C for codling moth, navel orangeworm and red flour beetle, respectively. The much smaller z value for red flour beetle than that for the other five insects indicates a stronger response to increases in temperature (Wang *et al.*, 2002a, b; Johnson *et al.*, 2004).

Lethal Times (LT)

Established thermal death kinetic models can be used to predict lethal times (LT) to reach 95%, 99%, 99.83% or 99.9968% insect mortality following the procedures described in Chapter 5, this volume. LT values for two field pests, codling moth and navel orangeworm, are listed in Table 6.4, along with the minimum exposures needed for 100% mortality (n = 600; Wang *et al.*, 2002a, b). The duration of the exposure required to kill 599 insects in a 600-insect sample corresponds to $LT_{99.83}$. Therefore, the treatment time to achieve 100% kill of 600 insects should be slightly greater than $LT_{99.83}$. According to Table 6.4, the observed minimum exposures for complete kill of 600 insects are in general agreement with $LT_{99.83}$ predicted from the 0.5th order reaction model. The discrepancies are mainly the result of the time intervals (> 2 min) used in the experimentation.

Table 6.3. Summary of thermal death time (TDT) and Arrhenius relationships for the last larval instars of the six insects listed in Table 6.2 (n = 600) (from Wang *et al.*, 2002a, b; Johnson *et al.*, 2003, 2004; Gazit *et al.*, 2004; Hallman *et al.*, 2005).

Insect	TDT[a]	Arrhenius[b]
Codling moth	$\log t = 12.41 - 0.23\ T\ (R^2 = 0.996)$	$\log k = 75.22 - 24.64*1000/T\ (R2 = 0.995)$
Indianmeal moth	$\log t = 13.39 - 0.26\ T\ (R^2 = 0.998)$	$\log k = 82.22 - 26.83*1000/T\ (R^2 = 0.998)$
Navel orangeworm	$\log t = 14.19 - 0.26\ T\ (R^2 = 0.987)$	$\log k = 82.21 - 26.93*1000/T\ (R^2 = 0.989)$
Mediterranean fruit fly	$\log t = 15.37 - 0.30\ T\ (R^2 = 0.999)$	$\log k = 95.10 - 31.02*1000/T\ (R^2 = 0.999)$
Mexican fruit fly	$\log t = 14.55 - 0.29\ T\ (R^2 = 0.997)$	$\log k = 88.30 - 28.72*1000/T\ (R^2 = 0.995)$
Red flour beetle	$\log t = 21.46 - 0.41\ T\ (R^2 = 0.993)$	$\log k = 130.21 - 42.52*1000/T\ (R^2 = 0.994)$

[a] $\log t\ (\text{min}) = a - bT\ (°C)$.

[b] $\log k = \log k_0 - \dfrac{E_a}{2.303RT(K)}$.

Table 6.4. Comparison of lethal times (LT, min) obtained by experiments and 0.5th order kinetic models for two field insects: fifth-instar codling moth and navel orangeworm at different temperatures and at a heating rate of 18°C/min (from Wang *et al.*, 2002a, b).

Insect	Temp. (°C)	Observed time for 100% mortality of 600 insects (min; ~> $LT_{99.83}$)	Prediction with 0.5th order kinetic model			
			LT_{95}	LT_{99}	$LT_{99.83}$	$LT_{99.9968}$ (Probit-9)
Codling moth	46	50	44.0	50.6	53.7	55.6
	48	15	10.6	12.4	13.3	13.8
	50	5	3.6	4.2	4.5	4.6
	52	2	1.4	1.7	1.8	1.9
Navel	46	140	120.0	137.6	146.1	151.1
orangeworm	48	50	40.9	46.8	49.6	51.3
	50	15	13.5	15.3	16.2	16.8
	52	6	4.3	5.0	5.4	5.6
	54	1	0.8	0.6	1.0	1.0

The exposure times required to achieve 100% mortality of fifth-instar codling moth using a heating block system were confirmed with efficacy tests in which infested cherries were treated with hot water (Feng *et al.*, 2004; Hansen *et al.*, 2004) and infested walnuts treated with radio frequency (RF) heating (Wang *et al.*, 2001b). The results from heating block studies for codling moth also agree with those obtained by Yokoyama *et al.* (1991), who placed test insects in 15 ml glass vials and submerged them in a hot water bath. In that study, 100% mortality of 300–550 larvae was achieved after exposures of > 55, 10 and 4 min at 46, 48 and 50°C, respectively.

Table 6.5 lists the LT calculated from the 0.5th order kinetic model for two fruit flies (Gazit *et al.*, 2004; Hallman *et al.*, 2005). The calculated values for

Table 6.5. Comparison of lethal times (LT, min) obtained by the 0.5th order kinetic model for two fruit flies at four different temperatures (from Gazit *et al.*, 2004; Hallman *et al.*, 2005).

Insect	Temp. (°C)	Observed minimum time for 100% mortality of 300 insects (min; > $LT_{99.67}$)	Lethal time (min)			
			LT_{95}	LT_{99}	$LT_{99.67}$	$LT_{99.9968}$ (Probit-9)
Mediterranean fruit	46	25	18.8	21.7	22.7	24.0
fly (eggs)	48	5.0	3.7	4.3	4.5	4.7
	50	3.0	2.1	2.5	2.6	2.7
	52	0.67	0.44	0.56	0.60	0.65
Mediterranean fruit	46	60	49.0	56.5	59.1	62.2
fly (third-instars)	48	15	12.1	13.9	14.5	15.3
	50	4.0	3.0	3.4	3.6	3.8
	52	1.0	0.7	0.8	0.8	0.9
Mexican fruit fly	44	100	79.2	91.5	97.4	101.0
(third-instars)	46	25	21.7	24.7	26.1	27.0
	48	6.0	5.3	6.0	6.3	6.6
	50	2.0	1.6	1.9	2.0	2.1

$LT_{99.67}$ (299 out of 300 insects, or 99.67% mortality) are in general agreement with the observed minimum exposure times needed for complete kill of 300 insects, with some discrepancies due to the selection of exposure times used in the experiments.

In both Tables 6.4 and 6.5, the difference between $LT_{99.9968}$ and $LT_{99.83}$ or $LT_{99.67}$ is very small, and diminishes with increasing temperature. For practical purposes, the predicted kinetic model shows that the experimental exposures that cause complete mortality in a population of 300–600 insects would be not much different to those required for Probit-9 quarantine treatments.

The established thermal death 0.5th order kinetic model predicts the LT to reach 95% and 99% mortality for two storage pests, Indianmeal moth and red flour beetle, as shown in Table 6.6 (Johnson *et al.*, 2003, 2004). Red flour beetle larvae are quite resistant to 48°C, with an estimated LT_{95} and LT_{99} of 67 and 77 min, respectively, the highest LT values for all the insects discussed in this section.

However, lengthy exposures at lower treatment temperatures may not be appropriate for large-scale commercial applications where rapid, continuous treatment is required to process large product volumes. In those cases, treatment temperatures of > 50°C should be used. Such product temperatures are readily obtained in walnuts and other tree nuts using RF energy (Wang *et al.*, 2001a, 2002c). As indicated by the LT values, navel orangeworm is the most heat-resistant insect, at 50°C, among all discussed insects (see Tables 6.4, 6.5 and 6.6).

6.4 Influence of Life Stages and Species on Thermal Mortality

It is obvious from the above data that the heat sensitivity of different insect species and developmental stages varies considerably due to various physiological and environmental factors. While developing a heat treatment for

Table 6.6. Lethal times (min) obtained by the 0.5th order kinetic model and observed exposure to achieve 100% mortality for two storage pests at the last larval instar (Johnson *et al.*, 2003, 2004)

Insect	Temp. (°C)	Observed minimum time for 100% mortality in 600 insects (min; > $LT_{99.83}$)	0.5th order kinetic model			
			LT_{95}	95% CI	LT_{99}	95% CI
Indianmeal moth	44	120	105.9	93.7–118.1	121.4	106.8–136.0
	46	30	23.6	21.5–25.7	27.5	24.9–30.1
	48	10	7.6	6.7–8.4	8.8	7.8–9.8
	50	3	2.2	1.6–2.7	2.5	1.9–3.1
	52	1	0.9	0.7–1.3	1.0	0.7–1.3
Red flour beetle	48	85	66.6	61.2–72.0	76.8	70.4–83.2
	50	12	7.9	6.9–8.9	9.1	8.0–10.2
	52	2	1.3	0.8–1.8	1.6	1.1–2.1

any given commodity, it is necessary to determine the most heat-resistant species and life stage among the pests of concern for that commodity. Regulatory organizations such as the Ministry of Agriculture, Forestry and Fisheries (MAFF) in Japan often require that the most resistant life stage of a target pest be used in the development of quarantine treatments (Hansen *et al.*, 2004). In this section, we will discuss relative heat tolerances of different life stages for four different insects based on information obtained using the Washington State University heat block system.

Most Heat-resistant Life Stages

Codling moth

Differences in the thermal resistance among five immature life stages (white-ring eggs, black-head eggs, third-instars, fifth-instars and diapausing larvae) were determined by Wang *et al.* (2004). The temperature and exposure combinations chosen for the study cut across the previously determined TDT curve so that a range of mortality levels would result. As shown in Table 6.7 and Fig. 6.8, complete kill of all five life stages was achieved with both 5 min at 50°C or 2 min at 52°C.

In sublethal treatments, fifth-instar diapausing larvae were the most heat-resistant life stage ($P < 0.05$). No significant difference was observed in mortality between white-ring and black-head eggs for all treatments except for 5 min at 48°C or 2 min at 0°C (Wang *et al.*, 2004). For the above-mentioned temperature–time combinations, white-ring eggs were more heat resistant than black-head eggs. In addition, third-instars were found to be the least heat-

Table 6.7. Thermal mortality (corrected, %) of different life stages of codling moths after heating at 15°C/min (three replicates) in a heating block system as a function of temperature-time combinations and the corresponding equivalent lethal time at 50°C (Wang *et al.*, 2004).

Temperature + holding time (°C + min)	Equivalent treatment times (min) at 50°C	White-ring eggs	Black-head eggs	Third-instars	Fifth-instars	Diapausing fifth-instars
48 + 2a	0.67	60.2 ± 17.3a*	72.9 ± 5.3a	17.8 ± 5.1b	21.8 ± 16.7b	0.9 ± 0.3c
48 + 5	1.62	65.0 ± 4.1c	87.3 ± 3.4b	96.2 ± 2.4a	70.2 ± 11.8c	1.0 ± 0.9d
48 + 10	3.20	91.9 ± 4.1c	91.4 ± 2.5c	100a	97.4 ± 0.5b	1.1 ± 0.0d
50 + 2	2.12	90.1 ± 5.2c	95.6 ± 2.2ab	97.4 ± 1.1a	92.3 ± 0.1bc	2.7 ± 1.0d
50 + 3	3.12	98.1 ± 0.6b	95.6 ± 1.4b	99.6 ± 0.8a	96.9 ± 1.1b	1.6 ± 1.4c
50 + 5	5.12	100a	100a	100a	100a	100a
52 + 0.5	1.95	92.2 ± 4.7a	92.6 ± 1.7a	93.1 ± 1.7a	70.6 ± 17.2b	0.7 ± 0.3c
52 + 1	3.53	97.9 ± 0.4ab	98.2 ± 1.6a	98.5 ± 1.4a	90.1 ± 8.2b	1.9 ± 0.5c
52 + 2	6.70	100a	100a	100a	100a	100a

* Different letters within a row indicate that means are significantly different ($P < 0.05$).

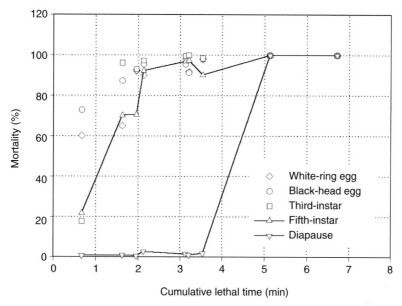

Fig. 6.8. Relationship of the mean percentage mortality over three replicates of different life stages of codling moth (*Cydia pomonella*) for different equivalent treatment times (M_{ref}) at 50°C (from Wang *et al.*, 2004).

resistant life stage for treatments of 5 min at 48°C, 10 min at 48°C or 3 min at 50°C + 3 min ($P < 0.05$).

But for short, sublethal treatment temperatures at 50–52°C, fifth-instars were more heat-resistant or at least had a similar thermotolerance to eggs and third-instars. Although the mortality of non-diapausing fifth-instars was consistently higher than that of diapausing larvae at sublethal levels, the temperature–time combination for complete kill of diapausing larvae was the same as that of non-diapausing larvae. Therefore, when developing treatment protocols adequate for all stages at 50 and 52°C, fifth-instars were selected as the target stage for efficacy studies because they were easier to work with and rear than diapausing larvae.

To compare the effects of total thermal exposure, including ramping and holding, all treatment conditions can be converted to equivalent treatment times at 50°C using the following equation (Eq. 5.22 in Chapter 5, this volume):

$$M_{ref} = \int_0^t 10^{\frac{T(t)-50}{z}} dt \tag{6.4}$$

where the z value for codling moth was determined to be 4°C (Wang *et al.*, 2002a).

Among the calculated equivalent treatment times in Table 6.7 and Fig. 6.8, the treatments at 50°C for more than 5 min resulted in complete codling moth mortality regardless of life stage (Wang *et al.*, 2004). For the treatment conditions that did not cause 100% mortality, diapausing larvae were the most heat-resistant life stage, followed by fifth-instars. Thus, the equivalent lethal time M_{ref} concept can be used as an effective tool to compare the relative thermotolerances of different life stages.

Navel orangeworm

The life stages of navel orangeworm most commonly found on dehydrated walnuts are larvae and pupae (navel orangeworm has no well-defined diapause stage). The thermal resistances of third-instars, fifth-instars and pupae were compared using a heating rate of 18°C/min (Wang *et al.*, 2002b; Table 6.8). Third-instars were less heat tolerant than the other life stages at all temperatures ($P < 0.05$). Although mortalities for fifth-instars were consistently less than those for pupae, the difference was significant only at 52°C ($P < 0.05$). The high temperatures ($> 52°C$) and short treatment times developed with fifth-instars as the targeted life stage should provide effective control for third-instars and pupae.

Mediterranean fruit fly

TDT curves were used by Gazit *et al.* (2004) to compare thermotolerances for egg and third-instar Mediterranean fruit flies with a sample size of 300 individuals (see Table 6.5). These TDT data show that longer exposure times were needed to achieve complete kill of third-instars than eggs at each temperature, especially at lower treatment temperatures (46 and 48°C). That is, the third-instars were more heat resistant than eggs.

Red flour beetle

The thermal resistances of five red flour beetle life stages were compared by Johnson *et al.* (2004); percentage mortality data are shown in Table 6.9 and Fig. 6.9. Overall, older larvae were more heat-tolerant than other life stages,

Table 6.8. Mortality (mean ± SD, %) of navel orangeworms ($n = 600$) at three life stages after heating at 18°C/min (three replicates) in a heating block system (from Wang *et al.*, 2002c).

Temperature + holding time (°C + min)	Third-instars	Fifth-instars	Pupae
48 + 25a	87.4 ± 3.0a*	26.5 ± 1.5b	27.9 ± 2.1b
50 + 10	97.6 ± 0.6a	38.0 ± 8.4b	59.3 ± 6.3b
52 + 4	99.8 ± 0.2a	73.5 ± 8.0b	98.2 ± 0.8a

* Different letters within row indicate that means are significantly different ($P < 0.05$).

Table 6.9. Mortality of five life stages of red flour beetle ($n = 600$) exposed to high temperatures (from Johnson *et al.*, 2004).[a]

Temperature + exposure time (°C + min)	Eggs	Younger larvae	Older larvae	Pupae	Adults
48 + 20	100.0a	96.2 ± 1.2ab	24.7 ± 2.6c	72.9 ± 19.7b	78.6 ± 5.8b
48 + 30	100.0a	99.0 ± 0.6a	30.8 ± 4.9c	99.0 ±0.6a	86.8 ± 8.7b
48 + 40	100.0a	97.6 ± 2.4a	57.1 ± 12.0b	100.0a	96.9 ± 2.0a
50 + 4	94.0 ± 6.0a	97.3 ± 1.1a	30.0 ± 11.9c	53.2 ± 4.2cb	62.2 ± 10.2b
50 + 8	100.0a	99.5 ± 0.5a	74.9 ± 9.0b	99.4 ± 0.5a	97.0 ± 1.3a
50 + 12	100.0a	100.0a	97.8 ± 0.9b	100.0a	100.0 a
52 + 0.5	99.2 ± 0.8a	92.3 ± 2.2a	27.5 ± 15.4b	46.1 ± 6.8b	52.3 ± 12.4b
52 + 1	98.9 ± 1.1a	96.5 ± 2.8a	78.2 ± 9.8a	73.1 ± 13.8a	81.4 ± 7.5a
52 + 2	99.2 ± 0.8a	92.9 ± 5.0a	100.0a	99.4 ± 0.6a	96.3 ± 1.3a

[a] Values are means ± SE of four replicates; within rows, values followed by a different letter are significantly different (LSD mean separation, $P < 0.05$).

particularly at lower treatment temperatures. In particular, mortality of older larvae was significantly less at 48 and 50°C compared with that of the other life stages ($P < 0.05$; Johnson *et al.*, 2004).

Eggs and younger larvae at any temperature–time combinations were usually the least thermotolerant, with no difference observed between these two life stages. Pupae and adults were more tolerant than eggs and younger larvae at 48°C for 20 min and 52°C for 0.5 min. In summary, the relative heat tolerance of red flour beetle stages is as follows: older larvae > pupae and adults > eggs and younger larvae.

Comparing Thermotolerance among Insects

Because several insects may infest the same commodity, one must select the most thermotolerant insect species as the targeted pest when developing phytosanitary or quarantine heat treatments. Since the late larval instar was the most thermotolerant life stage for the four insect pests discussed above, this life stage was selected for comparing the thermotolerances of the four insect species. The TDT curves shown in Fig. 5.10 (Chapter 5, this volume), all obtained with the heating block method, compare the thermal resistance of fifth-instar codling moth (Wang *et al.*, 2002a), Indianmeal moth (Johnson *et al.*, 2003), navel orangeworm (Wang *et al.*, 2002b), third-instar Mediterranean fruit fly (Gazit *et al.*, 2004), Mexican fruit fly (Hallman *et al.*, 2005) and red flour beetle (Johnson *et al.*, 2004).

Except for red flour beetle at 48°C, navel orangeworm was the most thermotolerant and Mexican fruit fly the least thermotolerant. Based on the slope of the TDT curves, the third-instar red flour beetle was more sensitive to temperature changes than the other five larvae. Common in thermal control of microorganisms in food research, it is hoped that use of a consistent

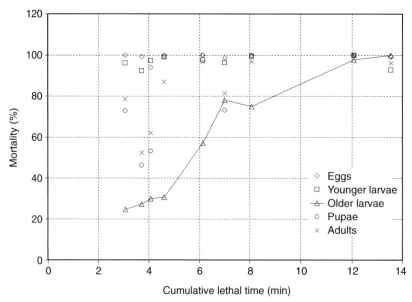

Fig. 6.9. Relationship of the mean percentage mortality over three replicates of different life stages of red flour beetle (*Tribolium castaneum*) for different equivalent treatment times (M_{ref}) at 50°C.

experimental method (the heating block method) to compare the intrinsic thermotolerance across different species may lead to a common understanding of the fundamental mechanisms that kill insect pests.

6.5 Activation Energies for Thermal Kill of Insect Pests

The activation energy (E_a) for thermal inactivation of pests can be obtained from the temperature-dependent characteristics of insect death rates described in Chapter 5, this volume. Insects with high activation energy levels are more sensitive to changes in temperature. Activation energies for the selected insects calculated from the slope of the Arrhenius relationships listed in Table 6.3 range from 209 to 958 kJ/mol for insects (see Table 6.10).

The E_a values for fifth-instar codling moth (Wang *et al.*, 2002a), fifth-instar Indianmeal moth (Johnson *et al.*, 2003), fifth-instar navel orangeworm (Wang *et al.*, 2002b) and third-instar Medfly (Gazit *et al.*, 2004) obtained from the same heating block system are of the same order of magnitude, which is also similar to that of Caribbean fruit fly eggs (Moss and Chan, 1993) and Queensland fruit fly eggs (Waddell *et al.*, 2000).

The E_a values of four species of fruit flies in Hawaii (Jang, 1986, 1991) were higher than those of the Medfly in Israel (Gazit *et al.*, 2004), probably due to differences in heating methods, heating rates or environments. In general, insects are more susceptible to increased temperatures than commodities because the activation energy for thermal kill of insects is slightly greater than

Table 6.10. Comparisons of insect and microorganism thermal kill activation energies with those of food quality changes due to heat treatments (from Wang *et al.*, 2002a).

Insect/Material	Temperature range (°C)	Activation energy E_a (kJ/mol)	Source
Mediterranean fruit fly			
Eggs	45–47	784	Jang, 1986
	46–52	491	Gazit *et al.*, 2004
First-instar	45–48	656	Jang, 1986
Third-instar[a]	46–52	552	Gazit *et al.*, 2004
Melon fly			
Eggs	43–46	518	Jang, 1986
First-instar	45–48	650	Jang, 1986
Oriental fruit fly			
Eggs	43–46	958	Jang, 1986
First-, early and late third-instar	43–48	209–401	Jang, 1986, 1991
Caribbean fruit fly (eggs)	37–42	440	Moss and Chan, 1993
	43–50	445	Moss and Chan, 1993
Queensland fruit fly (eggs)	42–48	538[b]	Waddell *et al.*, 2000
Codling moth (fifth-instar)[a]	46–52	473	Wang *et al.*, 2002a
Indianmeal moth (fifth-instar)[a]	44–52	514	Johnson *et al.*, 2003
Navel orangeworm (fifth-instar)[a]	46–54	510–520	Wang *et al.*, 2002b
Red flour beetle (third-instar)[a]	48–52	814	Johnson *et al.*, 2004
Quality (texture – softening or firmness, colour, flavour, etc.)	50–70	42–126	Lund, 1977; Rao and Lund, 1986
Microorganisms (spores)	100–130	222–502	Lund, 1977

[a] Parameter obtained at 15–18°C/min heating rate.
[b] Estimated by the authors from the reported data.

that for product quality changes such as texture softening or thermal inactivation of pathogenic microbial spores (see Table 6.10).

6.6 Preconditioning Effects on Thermotolerance of Pests

Enhanced thermotolerance to lethal temperatures following pre-treatment exposure to non-lethal elevated temperatures has been reported for a number of insects, including fruit flies (Beckett and Evans, 1997), flesh flies (Yocum and Denlinger, 1992) and lightbrown apple moths (Lester and Greenwood, 1997). For example, third-instar Caribbean fruit flies reared at 30°C are significantly more heat resistant than those reared at 20°C (Hallman, 1994). Exposure of target insects to elevated harvest, processing or storage temperatures (≥ 30°C) prior to heat treatments may alter the efficacy of the treatments, requiring that treatment times be extended or target temperatures slightly raised to obtain the desired level of control.

When treatment times required at 46°C to achieve 99% mortality (LT$_{99}$) for *Bactrocera tryoni* eggs preconditioned at sublethal temperatures (34–42°C) were compared with non-preconditioned eggs (Waddell *et al.*, 2000; Fig. 6.10), a significant amount of extra time was required to achieve the same level of mortality when the insects were pre-conditioned at 34–40°C. The increased heat tolerance diminished when preconditioned at 42°C.

Similar observations were made with codling moths. Yin *et al.* (2006a) preconditioned fifth-instar codling moth by exposing them to 35°C for 120, 360 and 1,080 min before determining the time required to achieve 100% mortality at three lethal temperatures (48, 50 and 52°C). The results revealed significant increases of thermal tolerance in almost all samples, as shown in Fig. 6.11. The minimum exposure times required to completely kill 300 fifth-instar codling moth preconditioned at 35°C for 360 min were 30, 7 and 3 min at 48, 50 and 52°C, respectively, compared with 15, 5 and 2 min at those same temperatures without preconditioning.

The enhanced thermotolerance in insects exposed to sublethal temperatures is possibly caused by the expression of heat-shock proteins (Thomas and Shellie, 2000; Yin *et al.*, 2006b), which will be discussed in detail in Chapter 11, this volume. However, the induced thermotolerance during the sublethal temperature exposure may be reversed by conditioning the insects at room temperature. For example, codling moth larvae preconditioned at 35°C

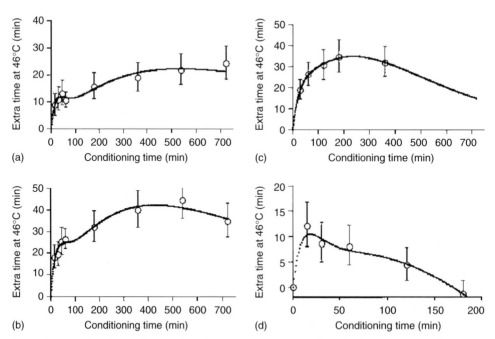

Fig. 6.10. Relationship between static temperature conditioning of *Bactrocera tryoni* eggs at 34°C (a), 38°C (b), 40°C (c) and 42°C (d) and the amount of extra treatment time required at 46°C to achieve 99% mortality compared with unconditioned eggs (from Waddell *et al.*, 2000).

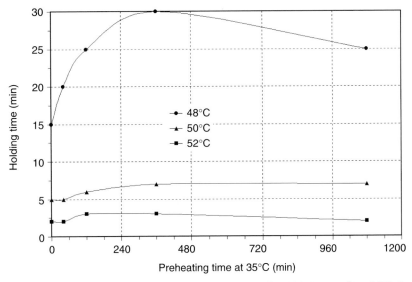

Fig. 6.11. Pre-treatment effect on treatment time required for 100% mortality of fifth-instar codling moth at three temperatures (three replicates for 300 larvae; from Yin *et al.*, 2006a).

for 6 h completely lost any enhanced thermotolerance when held for 2 h at 22°C (see Fig. 6.12). It may be possible, therefore, to use a 2 h recovery period at 20°C before a heat treatment to reduce thermotolerance of fifth-instar codling moth that have experienced abnormally warm sublethal storage conditions.

6.7 Effect of Heating Rates in Thermal Treatments

Similar to the induction of heat tolerance in insects through exposure to sublethal temperatures, slow heating rates may also result in heat tolerance. Lester and Greenwood (1997) reported that a low heating rate and consequent long exposure of insects to warm, but not lethal temperatures (< 42°C) may condition insects such that the heat treatment becomes less effective. Thomas and Shellie (2000) reported that the exposure times needed to achieve 99% mortality of Mexican fruit fly at 44°C were 62 and 42 min when heated at 0.175 and 1.400°C/min, respectively. Evans (1986) and Neven (1998a, b) reported significant effects of heating rates between 0.13 and 0.20°C/min on insect metabolism and physiological tolerance to heat treatment.

Consequently, a longer holding time is required at a final temperature in order to achieve the same mortality as that achieved with a faster heating rate. For conventional heating, heating rates in the interior of commodities range between 0.05 and 2.00°C/min depending on heating methods, type and size of commodity and final treatment temperature (Wang *et al.*, 2001b). In addition, the heating rate at the interior of a commodity decreases with time under a

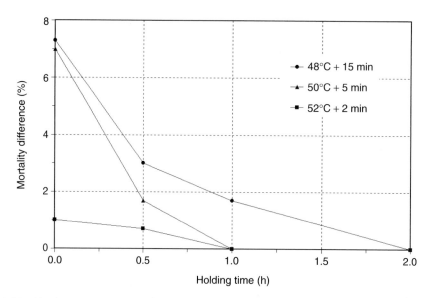

Fig. 6.12. Mortality differences of fifth-instar codling moth between unconditioned samples and samples conditioned at 35°C for 6 h as a function of holding at 22°C using three different temperature-time combinations (Yin *et al.*, 2006a)

constant treatment condition due to decreasing temperature gradients between the heating medium and fruit. As a result, conventional heat treatments typically require lengthy exposures to achieve the desired security level against insects. Most insects have adequate time to adapt to heat by increasing their thermal resistance (Waddell *et al.*, 2000).

Enhanced thermal resistance due to slow heating can be avoided by using fast heating methods. Radio frequency (RF) or microwave (MW) heating can raise fruit and nut temperatures at 1–20°C/min (Wang *et al.*, 2001a, 2002c; Birla *et al.*, 2004). Studies have shown that the relatively short ramp times resulting from heating rates between 1 and 18°C/min are not adequate to allow fifth-instar codling moth or fifth-instar navel orangeworm to increase heat tolerance (Wang *et al.*, 2002a, 2005).

6.8 Closing Remarks

This chapter first describes the general biology of several economically important insect pests in the international trade of agricultural commodities. It also presents fundamental thermal death kinetic data for those insects. Thermal death kinetic data obtained from reliable experimental methods reveal useful information about the lethal effects of heat to various insects at different life stages and as influenced by pre-treatment conditions.

Thermal death kinetic data that truly reflect the intrinsic heat tolerance of insects as described in this chapter are, however, rare and limited to only a small group of species. More research is needed to obtain thermal death kinetic data that are not confounded with the influence of non-isothermal test conditions.

Care must be taken when using thermal kinetic data based on laboratory experiments to develop treatment protocols for insect pests. Unanticipated problems may occur when one extrapolates kinetic information based on small experimental populations to large insect populations used to validate commercial treatments. In addition, the heat tolerance of laboratory-reared insects that have been used to develop kinetics models may be different from those that infest commodities. Unexpected pre-treatment conditions may also significantly increase the thermal resistance of insects infesting agricultural commodities. Readers may refer to Chapters 9 and 13, this volume, for additional precautions needed when developing treatment protocols.

6.9 References

Arbogast, R.T. (1991) Beetles: Coleoptera. In: Gorham, J.R. (ed.) *Ecology and Management of Food Industry Pests.* FDA Technical Bulletin 4, Association of Official Analytical Chemists, Arlington, Virginia, pp. 131–176.

Beckett, S.J. and Evans, D.E. (1997) The effects of thermal acclimation on immature mortality in the Queensland fruit fly *Bactrocera tryoni* and the light brown apple moth *Epiphyas postivttana* at a lethal temperature. *Entomologia Experimentalis et Applicata* 82, 45–51.

Beede, R.H., Kallsen, C.E., Freeman, M.A., Ferguson, L., Holtz, B.A., Klonsky, K.M. and De Moura, R.L. (2004) *Sample Costs to Establish and Produce Pistachios: San Joaquin Valley, Low Volume Irrigation.* University of California Cooperative Extension, Davis, California.

Birla, S.L., Wang, S. and Tang, J. (2004) Improving heating uniformity of fresh fruit in radio frequency treatments for pest control. *Postharvest Biology Technology* 33, 205–217.

Buchner, R.P., Edstrom, J.P., Hasey, J.K., Krueger, W.H., Olson, W.H., Reil, W.O., Klonsky, K.M. and De Moura, R.L. (2002) *Sample Costs to Establish a Walnut Orchard and Produce Walnuts: English.* University of California Cooperative Extension, Davis, California.

Caprile, J.L., Grant, J.A., Holtz, B.A., Kelley, K.M., Mitcham, E.J., Klonsky, K.M. and

De Moura, R.L. (2001) *Sample Costs to Establish an Apple Orchard and Produce Apples: Granny Smith Variety.* University of California Cooperative Extension, Davis, California.

Evans, D.E. (1986) The influence of rate of heating on the mortality of *Rhyzopertha dominica* (L.) (Coleoptera: Bostrychidae). *Journal of Stored Product Research* 23, 73–77.

FAO (2005) *Area-wide Integrated Pest Management.* Food and Agriculture Organization of the United Nations, Rome (http://www.fao.org/ag/magazine/0506sp1.htm).

Feng, X., Hansen, J.D., Biasi, B. and Mitcham, E.J. (2004) Use of hot water treatment to control codling moths in harvested California 'Bing' sweet cherries. *Postharvest Biology and Technology* 31, 41–49.

Follett, P.A. and Neven, L.G. (2006) Current trends in quarantine entomology. *Annual Review of Entomology* 51, 359–385.

Fruit Fly Eradication Project Office (2006) Okinawa Prefectural Government (OPF-FEPO) http://www.pref.okinawa.jp/mibae/en/index.html.

Gazit, Y., Rossler, Y., Wang, S., Tang, J. and Lurie, S. (2004) Thermal death kinetics of egg and third-instar Mediterranean fruit fly *Ceratitis capitata* (Wiedemann) (Diptera: Tephritidae). *Journal of Economic Entomology* 97, 1540–1546.

Hallman, G.J. (1994) Mortality of third-instar Caribbean fruit fly (Diptera: Tephritidae) reared at three temperatures and exposed to hot water immersion or cold storage. *Journal of Economic Entomology* 87, 405–408.

Hallman, G., Wang, S. and Tang, J. (2005) Reaction orders of thermal mortality of third-instar Mexican fruit fly *Anastrepha ludens* (Loew) (Diptera: Tephritidae). *Journal of Economic Entomology* 98, 1905–1910.

Hansen, J., Wang, S. and Tang, J. (2004) A cumulated lethal time model to evaluate efficacy of heat treatments for codling moth *Cydia pomonella* (L.) (Lepidoptera: Tortricidae) in cherries. *Postharvest Biology and Technology* 33, 309–317.

Ingels, C.A., Klonsky, K.M. and De Moura, R.L. (2003) *Sample Costs to Produce Pears: Green Bartlett Variety.* University of California Cooperative Extension, Davis, California.

Jang, E.B. (1986) Kinetics of thermal death in eggs and first instars of three species of fruit flies (Diptera: Tephritidae). *Journal of Economic Entomology* 79, 700–705.

Jang, E.B. (1991) Thermal death kinetics and heat tolerance in early and late third instars of the oriental fruit fly (Diptera: Tephritidae). *Journal of Economic Entomology* 84, 1298–1303.

Johnson, J.A., Wang, S. and Tang, J. (2003) Thermal death kinetics of fifth instar *Plodia interpunctella* (Lepidoptera: Pyralidae). *Journal of Economic Entomology* 96, 519–524.

Johnson, J.A., Valero, K.A., Wang, S. and Tang, J. (2004) Thermal death kinetics of red flour beetle, *Tribolium castaneum* (Coleoptera: Tenebrionidae). *Journal of Economic Entomology* 97, 1868–1873.

Kaplan, J.K. (2004) Fruit flies flee paradise. *Agricultural Research* February, 4–9.

Lester, P.J. and Greenwood, D.R. (1997) Pretreatment induced thermotolerance in lightbrown apple moth (Lepidoptera: Tortricidae) and associated induction of heat shock protein synthesis. *Journal of Economic Entomology* 90, 199–204.

Lund, D.B. (1977) Design of thermal processes for maximizing nutrient retention. *Food Technology* 31, 71–78.

Moss, J.I. and Chan, H.T. (1993) Thermal death kinetics of Caribbean fruit fly (Diptera: Tephritidae) embryos. *Journal of Economic Entomology* 86, 1162–1166.

Neven, L.G. (1998a) Effects of heating rate on the mortality of fifth-instar codling moth (Lepidoptera: Tortricidae). *Journal of Economic Entomology* 91, 297–301.

Neven, L.G. (1998b) Respiratory response of fifth-instar codling moth (Lepidoptera: Tortricidae) to rapidly changing temperatures. *Journal of Economic Entomology* 91, 302–308.

OEPP/EPPO (2005) *Distribution Maps of Quarantine Pests for Europe. Ceratitis captitata.* A2 no 105 http://www.eppo.org/QUARANTINE/listA2.htm).

Ohlendorf, B. (1999) *Integrated Pest Management for Apples and Pears* 2nd edition, University of California Division of Agriculture Natural Resource Publication 3340, Oakland, California.

Rao, M.A. and Lund, D.B. (1986) Kinetics of thermal softening of foods – a review. *Journal of Food Processing and Preservation* 10, 311–329.

Tang, J., Ikediala, J.N., Wang, S., Hansen, J.D. and Cavalieri, R.P. (2000) High-temperature short-time thermal quarantine methods. *Postharvest Biology and Technology* 21, 129–145.

Thomas, D.B. and Shellie, K.C. (2000) Heating rate and induced thermotolerance in larvae of Mexican fruit fly, a quarantine pest of citrus and mangoes. *Journal of Economic Entomology* 93, 1373–1379.

Tsuji, H. (1998) Experimental invasion of a food container by first-instar larvae of the Indian meal moth, *Plodia interpunctella* Hübner, through pinholes. *Medical Entomology and Zoology* 49, 99–104.

University of California (2002) *UC IPM Pest Management Guidelines: Almond.* University of California Agriculture and Natural Resources, Publication 3431, Davis, California.

USDA-APHIS (2004) *Treatment Manual.* United State Department of Agriculture Animal and Plant Health Inspection

Service http://www.aphis.usda.gov/ppq/ manuals/port/ Treatment_Chapters.htm.

Viveros, M.A., Freeman, M.A., Klonsky, K.M. and De Moura, R.L. (2003) *Sample Costs to Establish an Almond Orchard and Produce Almonds: San Joaquin Valley South, Flood Irrigation.* University of California Cooperative Extension, Davis, California.

Waddell, B.C., Jones, V.M., Petry, R.J., Sales, F., Paulaud, D., Maindonald, J.H. and Laidlaw, W.G. (2000) Thermal conditioning in *Bactrocera tryoni* eggs (Diptera: Tephritidae) following hot-water immersion. *Postharvest Biology and Technology* 21, 113–128.

Wade, W.H. (1961) Biology of the navel orangeworm, *Paramyelois transitella* (Walker), on almonds and walnuts in northern California. *Hilgardia* 31, 129–171.

Wang, S., Ikediala, J.N., Tang, J., Hansen, J.D., Mitcham, E., Mao, R. and Swanson, B. (2001a) Radio frequency treatments to control codling moth in in-shell walnuts. *Postharvest Biology Technology* 22, 29–38.

Wang, S., Tang, J. and Cavalieri, R.P. (2001b) Modeling fruit internal heating rates for hot air and hot water treatments. *Postharvest Biology and Technology* 22, 257–270.

Wang, S., Ikediala, J.N., Tang, J. and Hansen, J.D. (2002a) Thermal death kinetics and heating rate effects for fifth-instar *Cydia pomonella* (L.) (Lepidoptera: Tortricicae). *Journal of Stored Products Research* 38, 441–453.

Wang, S., Tang, J., Johnson, J.A. and Hansen, J.D. (2002b) Thermal death kinetics of fifth-instar *Amyelois transitella* (Walker) (Lepidoptera: Pyralidae) larvae. *Journal of Stored Products Research* 38, 427–440.

Wang, S., Tang, J., Johnson, J.A., Mitcham, E., Hansen, J.D., Cavalieri, R.P., Bower, J. and Biasi, B. (2002c) Process protocols based on radio frequency energy to control field and storage pests in in-shell walnuts. *Postharvest Biology and Technology* 26, 265–273.

Wang, S., Yin, X., Tang, J. and Hansen J. (2004) Thermal resistance of different life stages of codling moth (Lepidoptera: Tortricidae). *Journal of Stored Products Research* 40, 565–574.

Wang, S., Johnson, J.A., Tang, J. and Yin, X. (2005) Heating condition effects on thermal resistance of fifth-instar navel orangeworm (Lepidoptera: Pyralidae). *Journal of Stored Products Research* 41, 469–478.

White, I.M. and Elson-Harris, M.M. (1992) *Fruit Flies of Economic Significance: their Identification and Bionomics.* CAB International, Wallingford, UK.

Yin, X., Wang, S., Tang, J. and Hansen, J. (2006a) Thermal resistance of fifth-instar codling moth (Lepidoptera: Tortricidae) as affected by pretreatment conditioning. *Journal of Stored Products Research* 42, 75–85.

Yin, X., Wang, S., Tang, J., Lurie, S. and Hansen, J.D. (2006b) Thermal conditioning of fifth-instar *Cydia pomonella* (Lepidoptera: Tortricidae) affects HSP 70 accumulation and insect mortality. *Physiological Entomology* 31, 241–247.

Yocum, G.D. and Denlinger, D.L. (1992) Prolonged thermotolerance in the fresh fly, *Sarcophaga crassipalpis*, does not require continuous expression or persistence of the 72 kDa heat shock protein. *Journal of Insect Physiology* 38, 603–609.

Yokoyama, V.Y., Miller, G.T. and Dowell, R.V. (1991) Response of codling moth (Lepidoptera: Tortricidae) to high temperature, a potential quarantine treatment for exported commodities. *Journal of Economic Entomology* 84, 528–531.

7

Thermal Control of Fungi in the Reduction of Postharvest Decay

E. FALLIK* AND S. LURIE**

*Department of Postharvest Science, ARO, The Volcani Center, Israel;
e-mails: *efallik@agri.gov.il; **slurie43@agri.gov.il*

7.1 Introduction

Advances in agronomic, processing, preservation, packaging, shipping and marketing technologies on a global scale have enabled the fresh fruit and vegetable industry to supply consumers with a wide range of high-quality produce year-round. However, economic losses caused by postharvest pathogens can be high, and the avoidable losses between the farm gate and consumer should be reduced.

Fresh fruits and vegetables, because of the added cost of harvesting and handling, increase several times in value as they are moved from the field to the consumer. To decrease our dependency on chemical control in ensuring quality of harvested produce, alternative methods of postharvest decay control are being developed. Thermal treatments can be used for postharvest insect control for perishable commodities such as fresh fruits, vegetables, bulbs and cut flowers (Hallman, 2000). These heat treatments have also been very effective in controlling fungi that are the main causes of postharvest decay development (Barkai-Golan, 2001; Vicente *et al.*, 2002).

Pre-storage heat treatments to control decay development during the storage and marketing period are often applied for a relatively short time (minutes), because the target decay-causing agents are found on the surface or in the first few cell layers under the skin of fresh produce (Barkai-Golan, 2001). Heat treatments against pathogens may be applied to freshly harvested produce in several ways: by vapour heat, hot dry air, hot water dips (Lurie, 1998) or a short, hot water rinsing and brushing (Fallik *et al.*, 2001a; Fallik, 2004).

Various heat treatments have been extensively studied to control many postharvest fungi and bacteria (Couey, 1989; Lurie, 1998; Ferguson *et al.*, 2000). However, agricultural commodities vary greatly in size and shape, and thus respond differently to applied heat treatment. Inappropriate heat treatment can lead to ripening acceleration or heat damage (McDonald *et al.*, 1999). This chapter summarizes the information on thermal control of fungi, *in vitro* and *in vivo*, since the last review by Lurie (1998).

7.2 Responses of Fungi to Thermal Heat: *in vitro* Studies

Fungi are vary considerably in their sensitivity to pre-storage treatments of high temperatures (Barkai-Golan, 2001). Pathogen kill is not always proportional to the temperature–time product of the treatment, although reports have indicated a linear relationship between the logarithm of the decimal reduction time and the temperature of the heat treatment. The vegetative cells and conidia of most fungi are inactivated when exposed to 60°C for 5–10 min *in vitro* (Civello *et al.*, 1997). Spore germination and germ tube elongation were found to be more sensitive to heat treatments than dormant spores, which are unaffected by hot water (D'Hallewin *et al.*, 1997).

The sensitivity of fungal spores to heat treatments does not necessarily depend upon spore size, shape or inoculum age. For example, spores of *Colletotrichum gloeosporioides* were found more heat-sensitive than *Dothiorella dominicana* in mango fruits, while *Botrytis cinerea* was found to be more susceptible to hot water treatment than *Alternaria alternata* in sweet bell pepper, based on spore germination and germ-tube elongation of these fungi (Barkai-Golan, 2001). Exposing fungal spores of *A. alternata* and *Fusarium solani* to 60°C for about 15 s *in vitro* resulted in 48 and 42% reduction in spore germination, respectively (see Fig. 7.1) (Fallik *et al.*, 2000).

Studies showed *Monilinia fructicola* to be more sensitive than *Penicillium expansum* to high temperature (60°C for 20 s) (Karabulut *et al.*, 2002). The viability of five pathogens was decreased by treatment with hot water when tested *in vitro*. *Polyscytalum pustulans* was the most sensitive and *Rhizoctonia solani* the least sensitive. Arthospores of *Geotrichum citriaurantii*, the cause of sour rot in citrus, were inactivated in water at 5°C lower than spores of *Penicillium digitatum*, the cause of green mould on citrus. Nevertheless, hot water treatments that controlled green mould effectively on lemons failed to control sour rot (Smilanick *et al.*, 2003).

A minimum exposure period of 20 s at 56°C was required to inhibit *P. digitatum* spore germination *in vitro* (Porat *et al.*, 2000a). No surviving spores of *B. cinerea* were observed after 15 min at 45°C (Marquenie *et al.*, 2002). Spores of *B. cinerea* exposed to a range of temperatures up to 49.5°C for 30 s survived, but were killed at the lethal heat dose of 46.3°C for 2 min (Lichter *et al.*, 2003). *Monilinia fructigena* was even more sensitive, requiring 3 min at 45°C for complete spore inactivation (Marquenie *et al.*, 2002).

A conidial suspension of *Phaeomoniella chlamydospora*, two species of *Phaeoacremonium* – *P. inflatipes* and *P. aleophilum* – and plugs of agar with mycelia of these fungi were placed in glass vials and incubated in hot water for 15–120 min. Conidia were sensitive to hot water treatment after 15 and 30

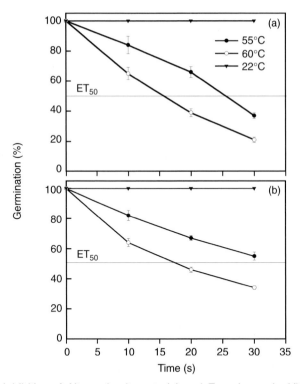

Fig. 7.1. *In vitro* inhibition of *Alternaria alternata* (a) and *Fusarium solani* (b) with increasing times of exposure. Data recorded after 24 h incubation at 20°C following treatment. Bars indicate standard error (SE) for each point (from Fallik *et al.*, 2000).

min, respectively. Nevertheless, the mycelia of *P. inflatipes* from agar plugs grew on potato dextrose agar at 22°C after 120 min incubation at 51°C (Whiting *et al.*, 2001).

Black rot of pineapple (*Chalara paradoxa* (De Seyen.) *Sacc.*) is a common postharvest problem in many countries. *In vitro* studies on heat-treated (50°C for 3 min) spores of the pathogen showed a mean colony count of 11 ± 0.7 following 48 h incubation at 28°C, while plates with heat-treated spores at 54 and 58°C showed a mean colony count of 1 ± 1.0 and 1 ± 0.7, respectively (Wijeratnam *et al.*, 2005).

A combination of 10% ethanol and hot water immersion improved the inhibition rate of *B. cinerea* spore germination. At 55°C for 30 s, 18% of the water-treated spores germinated, while none of the ethanol-treated spores did. No germination occurred after immersion in water or 10% ethanol at 60°C for 30 s. At 30°C, more than 80% of the spores germinated after a 60 s immersion in water or ethanol at 10%. After a 60 s immersion in water at 50°C, 15% of the spores germinated, while none germinated after immersion in water at 55 or 60°C. After a 60 s immersion in 10% ethanol at 50, 55 or 60°C, none of the spores germinated (Karabulut *et al.*, 2004a).

Spores of *Rhizopus stolonifer*, *Aspergillus niger*, *B. cinerea* and *A. alternata* were exposed to solutions containing up to 30% (v/v) ethanol at 25–50°C for 30 s, and their survival was then determined by germination on semi-solid media. Sub-inhibitory ethanol concentrations at ambient temperatures became inhibitory when heated at temperatures much lower than those that caused thermal destruction of the spores by water alone.

At 40°C, the estimated ethanol concentrations that inhibited the germination of 50% (LD_{50}) of the spores of *B. cinerea, A. alternata, A. niger* and *R. stolonifer* were 9.7, 13.5, 19.6 and 20.6%, respectively. *Botrytis cinerea* and *A. alternata* were less resistant to the combination than either *A. niger* or *R. stolonifer* (Gabler *et al.*, 2004). The combined treatment of UV-C (4.1 kJ/m^2) and heat treatment (45°C, 3 h in air) also reduced fungal infections and delayed *in vitro* germination of *B. cinerea* conidia (Pan *et al.*, 2004).

In vitro evaluation has shown that spores are much more susceptible to heat treatments than mycelia. However, the efficacy of any heat treatment to kill fungal development depends upon the genetic background of the fungus and its life cycle before infection. Better fungal eradication can be achieved by combined treatments of heat and ethanol or UV illumination.

7.3 Methods of Thermal Treatment

Differences between Various Methods of Heating

Contamination of fresh produce with pathogenic agents may occur at any point during production, harvesting, packing, processing, distribution or marketing. The different methods of treatment to reduce or eliminate decay development include hot air, which requires a longer treatment time than hot water. Hot air is applied either as a vapour or in a forced stream. A relatively new method involves a machine that subjects produce to a short, hot water rinsing and brushing treatment (Fallik *et al.*, 2001a; Fallik, 2004). Hot water is a more effective heat transfer medium than air and, when properly circulated through a load of fruit, establishes a uniform temperature profile more quickly than either vapour or dry heat (Lurie, 1998).

Hot Air (Forced Air and Vapour Heat): Long-period Treatments

Hot air treatments are carried out in a high-airflow controlled temperature (HAFCT) fruit treatment facility (Dentener *et al.*, 1996). In dry heat treatment, heat is applied without adding steam or water. As a general rule, dry-heat processes use higher temperatures and longer exposure times than steam-based processes, but the time–temperature requirements depend on the properties and size of the objects being treated.

The basic design of the treatment chamber is simple. If proper precautions are taken to exclude hazardous materials, emissions from the dry heat system are minimal. Condensation on fruit surfaces or in the treatment chamber is prevented by keeping the dew point temperature 2–3°C below the dry-bulb temperature throughout the duration of the test. This precise control of temperature and relative humidity is advantageous because it prevents condensation inside the treatment area and on the fruit surface, thus preventing fruit desiccation and scalding (Gaffney and Armstrong, 1990; Sharp *et al.*, 1991).

The influence of postharvest heat conditioning at 38°C for 24, 48 or 72 h on ripe Gialla cactus pear (*Opuntia ficus-indica* (L.) Miller) fruit was investigated during 21 days of storage at 6°C and 90–95% relative humidity (RH) followed by 7 days at 20°C and 70–75% RH (simulated marketing). Treatment for 24–72 h was effective in reducing decay, with conditioning for 23 h being the most effective (Schirra *et al.*, 1997a, b). Postharvest heat treatment of cactus pear at 37°C for 30 h under saturated humidity postharvest reduced decay during 45 days at 6°C plus 4 additional days at 20°C (Schirra *et al.*, 1999).

Fully red strawberries (*Fragria × ananassa* Duch. cv. Selva) were treated for 1–5 h at temperatures ranging from 39 to 50°C. After treatments, fruits were placed at 0°C overnight and then held at 20°C for 3 days. Most of the heat treatments improved strawberry shelf life, with the best results obtained for fruit heated at 42 and 48°C for 3 h. Heat treatments prevented fungal development and decreased the number of damaged fruits (Civello *et al.*, 1997).

Northeaster strawberries were heated with moist hot air at 36, 39, 42, 45, 50 and 55°C for 0, 20, 40, 60, 80 or 100 min. Fruit were injured when exposed to temperatures 50°C or higher, or durations of 60 min or longer, at 45°C. Treatment at 45°C for 40 min or at 42°C for 60–100 min resulted in the least decay incidence after 5 days at 0°C, 3 days at 10°C and 1 day at 20°C (Wang, 2000).

Strawberries cv. Selva were heat-treated in an air oven (45°C, 3 h) and then stored at 0°C for 0, 7 or 14 days. After 7 days at 0°C followed by 72 h at 20°C, the percentage of decayed fruit was lower in heat-treated than in control fruits. The treatment decreased the initial bacterial population, but did not reduce the amount of fungi initially present. This difference was still significant after 48 h at 20°C.

In the case of fungi, heat-treated fruit that were stored for 7 or 14 days at 0°C and then transferred to 20°C for 48 h showed a lower colony-forming units (CFU) value than did controls (Vicente *et al.*, 2002). Keeping strawberries (*Fragaria × ananassa* cv Seascape) at 45°C for 3 h reduced decay development caused by *B. cinerea* and *R. stolonifer* (Pan *et al.*, 2004).

Hot air (40, 50, 60, 70, 80 and 90°C for 1–10 min at 1-min intervals) affected the severity of rot caused by *R. stolonifer* on tomatoes. All temperatures tested reduced the severity of decay at various times within the 10-day storage period. The disease rating was highest in untreated fruits, less in fruits treated at 90°C and least in those treated at 40°C. The period of exposure to hot air before storage was important. There was also significant interaction between temperature and time of treatment. The best temperature–time treatments for decay control were 60°C for 6 min or 70°C for 1 to 5 min (Aborisade and Ojo, 2002).

A range of vapour heat temperatures (50–55°C) and treatment time intervals (12–32 min) were initially evaluated for their effects on *B. cinerea* in artificially inoculated table grapes. Disease development was monitored during storage at 20°C for 9 days after treatment. Berries inoculated after treatment were more susceptible to disease than the controls. All heat treatments applied after inoculation significantly reduced the infection compared with the controls. The most promising treatments were 52.5°C for 21 or 24 min and 55°C for 18 or 21 min, which reduced infection levels by 72–95% on the 9th day compared with controls. These treatments were also found to give effective disease control in inoculated and naturally infected grape bunches stored at 0.5°C after treatment (Lydakis and Aked, 2003).

Curing is considered a moderate heat treatment. The effectiveness of curing oranges and lemons at 33°C for 65 h followed by storage under ambient and cold-storage conditions was investigated. This treatment effectively reduced the incidence of *P. digitatum* (Pers) Sacc. and *P. italicum* Wehmer decay on inoculated and naturally infected oranges and lemons stored at 20°C for 7 days. However, it failed to control green and blue mould infections on fruits placed in long-term cold storage, except for green mould on oranges, which was effectively controlled (Plaza *et al.*, 2004).

Combinations of hot air and other treatments that control fungi development have been reported. The viability of *P. expansum* Link conidia in sporulating culture declined rapidly when exposed to 38°C for 4 days and, when conidia were exposed to 38°C prior to inoculation of apple fruit (*Malus* × *domestica* Borkh.), the resulting lesions were smaller than those on fruit inoculated with non-heated conidia. However, when apples were treated with calcium or with calcium plus a biological control agent and then heated, the reduction of decay was greater than with heat alone (Conway *et al.*, 1999).

Linear heating rates of 4, 6, 8, 10 and 12°C/h to treatment temperatures of 44 and 46°C were used to treat eight cultivars of apples (Red Delicious, Golden Delicious, Granny Smith, Fuji, Gala, Jonagold, Braeburn and Cameo) and two cultivars of winter pears (d'Anjou and Bosc). Decay organisms in heated fruit were suppressed compared with untreated fruits (Neven *et al.*, 2000).

Golden Delicious apples were wound-inoculated with conidial suspensions of either *C. acutatum* or *P. expansum*, then treated with heat (38°C) for 4 days, antagonists or a combination of both. Heat, or heat in combination with either antagonist, eliminated decay caused by *P. expansum*. Either heat or the antagonists alone reduced decay caused by *C. acutatum*, but a combination of the two was required to completely eliminate the decay caused by this pathogen (Conway *et al.*, 2004). Fruit treatment with hot air (at 38°C) for 4 days had eradicative but no residual activity against blue mould caused by *P. expansum* on apple.

1-Methylcyclopropene (1-MCP) treatment increased bitter rot and blue mould decays, which were effectively controlled by a combination of the antagonist and heat treatments. *Colletotrichum acutatum* was found to be a weaker pathogen than *P. expansum*, and bitter rot caused by *C. acutatum*, even on the control treatments, developed only after 4 months in controlled atmosphere (CA) storage followed by 2 weeks' incubation at 24°C. In contrast, non-treated fruit inoculated with *P. expansum* were completely decayed after 2 months in CA. The antagonist controlled bitter rot more effectively than blue mould, while heat treatment more effectively controlled blue mould.

The use of 1-MCP on harvested fruit to inhibit maturation can predispose fruit to decay, but the alternatives to synthetic fungicides are capable of preventing this increase in decay (Janisiewicz *et al.*, 2003). Before storage in air at 0°C, apples were treated with either 1-MCP for 17 h at 20°C, 38°C air for 4 days, 1-MCP plus heat or left untreated. Some sets of untreated fruit were stored in CA of 1.5 kPa O_2 and 2.5 kPa CO_2 at 0°C, while other sets were removed from cold storage in air after 2.5 or 5 months, warmed to 20°C and treated with 1-MCP for 17 h. Pre-storage 1-MCP, heat, 1-MCP plus heat treatments and CA storage decreased decay severity caused by wound-inoculated *P. expansum* Link, *B. cinerea* Pers.: Fr. and *C. acutatum* (Saftner *et al.*, 2003).

Strawberries cv. Selva (75 or 90% superficial red colour) were packaged with films having different permeability properties, heat-treated in an air oven (45°C, 3 h), stored at 0°C for 0, 7 or 14 days and then transferred to 20°C for 2 days. The application of heat treatment alone reduced fungal decay. When the treatment was performed on fruit in films that allowed the retention of a portion of the CO_2 produced during heating, the reduction of decay was enhanced (Vicente *et al.*, 2003).

The effect of forced moist (100% relative humidity, RH) or dry (50%) hot air (48.5°C or 50°C for 4 h) with or without thiabendazole on fungal development of Maradol papaya (*Carica papaya* L.) fruit stored at 5°C or 20°C for up to 42 days was investigated. Dry hot air treatment, especially in combination with thiabendazole, decreased growth of inoculated *C. gloeosporioides* (Perez-Carrillo and Yahia, 2004).

Despite the use of heat treatment to reduce decay development, it can sometimes increase it. Papaya fruit (*Carica papaya* L. cv. Waimanalo Solo) at colour-break ripeness were either not heated (controls) or forced-air heated to centre temperatures of 47.5, 48.5 or 49.5°C, and held at these temperatures for 20, 60, 120 or 180 min. Both control and heat-treated fruit had relatively high levels of decay. Heat treatment increased the incidence of body decay but did not affect the incidence of stem-end rots (Lay-Yee *et al.*, 1998).

Treating Tarocco blood oranges (*Citrus sinensis* Linn. Obsek) at 37°C for 48 h under saturated humidity remarkably promoted the development of secondary fungal infections such as *Phytophthora* rots (Schirra *et al.*, 2002). Moist hot air at 48.5 or 50°C for 4 h caused fruit injury in papaya, while dry, hot treatment decreased fungal development (Perez-Carrillo and Yahia, 2004).

The application of hot air treatments is for several hours to several days, with temperatures of 38–50°C. A combination of hot air treatment with modified atmosphere packaging (MAP), CA, 1-MCP or fungicide improves the efficacy of this treatment in controlling microorganisms.

Hot Water Dips: Medium-period Treatments

Most hot water dip treatment facilities that are in commercial use are either part of a batch or continuous system. In a batch system, produce is loaded onto a platform inside a metal or plastic basket, which is then lowered into a hot water immersion tank where the fruit or vegetables remain at the prescribed temperature for a certain length of time before being taken out, usually by means of an overhead hoist. In the continuous system fruit are dipped, either loosely or in a plastic basket, on a conveyor belt that moves slowly from one end of a hot water tank to the other. The belt speed is set to ensure the fruit are submerged for a pre-determined length of time (Tsang *et al.*, 1995).

Fruits, vegetables and flowers can benefit from hot water dips. Peaches and nectarines infected with *M. fructicola* were immersed in hot water for 1, 2, 4 or 8 min at temperatures of 46 or 50°C to control decay. Immersion of fruit in water at 46 or 50°C for 2.5 min reduced the incidence of decayed fruit from 82.8 to 59.3 and 38.8%, respectively (Margosan *et al.*, 1997).

Mature-green (*Prunus salicina* Lindl. cv. Friar) plums were treated in water at 40°C for 40 min, 45°C for 35 min, 50°C for 30 min or 55°C for 25 min, and then stored at 0°C for 35 days plus 9 days of ripening at 20–25°C. Decay symptoms were retarded in fruits treated at 45 and 50°C and only those fruit were acceptable for the market (Abu-Kpawoh *et al.*, 2002).

Hot water treatments (53°C, 2 min) of organically farmed apples were able to control Gloeosporium disease, the most severe fungal disease of organically produced apples during storage. A reduction of Gloeosporium rot below 10% after 5–6 months' storage was achieved with hot water treatments after harvest, irrespective of storage conditions (cold or CA storage) (Trierweiler *et al.*, 2003).

Similar results were obtained for organic strawberries. Postharvest hot water dips of organically grown strawberries at 55 and 60°C for 30 s significantly reduced decay incidence to 3.4 and 2.7%, respectively, while that in the control was 28.5%. However, in another experiment, the efficacy of hot water treatment at 60°C was significantly higher than that of hot water treatment at 55°C (Karabulut *et al.*, 2004a).

Dipping red ginger flower (*Alpinia purpurata*) at 49 and 50°C for 12–15 min extended postharvest vase life by killing most pests and saprophytic moulds (Jaroenkit and Paull, 2003). Dipping cactus fruit (*Opuntia ficus indica*) for 3 min at 52°C significantly reduced fungal development and improved visual quality of the fruit after prolonged storage (Rodriguez *et al.*, 2005).

Pineapples inoculated with 10^4 spores/ml of *C. paradoxa*, followed by a hot water dip treatment at 54°C for 3 min, were free of disease when stored at 10°C for 21 days followed by 48 h at an ambient temperature (28 ± 2°C). Inoculated dip-treated fruit held at 28 ± 2°C for 6 days also remained healthy (Wijeratnam *et al.*, 2005). After hot water treatment (70, 80 and 90°C for 5 and 10 min) to suppress fungal growth on harvested chestnuts, the nuts were enclosed in plastic bags and incubated at 25 ± 1°C for 20 days. Hot water at 70 and 80°C did not control fungal growth. However, treatment at 90°C for 10 min reduced the infection level by 95% (Panagou *et al.*, 2005).

The quality characteristics of Olinda and Campbell oranges (Valencia late cultivar) were evaluated after exposure to a fruit core temperature of 44°C for 100 min or 46°C for 50 min, subsequent storage at 6°C for 2 weeks and an additional week of marketing simulation at 20°C. While neither heat treatment affected decay incidence in Olinda oranges, significantly less decay was found in heat-treated Campbell fruit compared with control fruit both during storage and shelf life (Schirra *et al.*, 2005). A hot water dip for 2 min at 52–53°C inhibited the development of decay in lemons inoculated with *P. digitatum* (Nafussi *et al.*, 2001).

Dipping rockmelons in 50, 55, 60 and 65°C water for 1 min inhibited disease development caused by *Fusarium* spp., *Alternaria* spp. and *Colletotrichum* spp. at all temperatures in comparison with the control (22°C). A 1-min exposure to 60°C was the optimum temperature, resulting in 97% of the fruit being disease-free after 3 weeks' storage at 5°C (simulated export conditions) compared with only 7% when dipped in 22°C water (McDonald *et al.*, 2005).

Monilinia fructigena was reduced by up to 83% by a hot water treatment at 53°C and 3 min dipping. With this treatment, Gloeosporium rot was also reduced by up to 92% but the incidence of decay with *Nectria galligena* increased with the 1 min treatment at all temperatures. For the reduction of postharvest decay, a treatment of 53°C for 3 min is recommended. For cultivars with a high sensitivity to skin disorders, dips for 2 or 3 min at the highest cultivar-specific temperature showing no symptoms are recommended (Maxin *et al.*, 2005).

Mature green tomato fruit (*Lycopersicon esculentum* Mill. cv. Sunbeam) were treated in water for 1 h at 27 (ambient), 39, 42, 45 or 48°C. After 14 days storage and shelf-life simulation, treatment at 42°C reduced decay by 60%, whereas the other water temperatures were less effective (McDonald *et al.*, 1999). Dipping bell pepper at 53°C for 4 min prior to storage was effective in reducing decay after 14 and 28 days storage at 8°C (Gonzalez-Aguilar *et al.*, 1999, 2000).

Combined treatments have shown synergistic effects to improve the control of decay development. The potential of using hot water (2.5 min at 45°C), 2% sodium carbonate or 2% sodium bicarbonate solutions alone or combined with hot water for control of *P. digitatum* (green mould) was investigated on commercially ripe clementines. Carbonate and bicarbonate solutions effectively controlled green mould during storage but hot water did not (Larrigaudiere *et al.*, 2002).

Applying a biocontrol agent following hot water treatment (45°C for 2 min) was as effective as the fungicidal treatment, which gave 100% control of both green and blue moulds on artificially inoculated Valencia and Shamouti orange cultivars (Obagwu and Korsten, 2003). The efficacy of hot water, biological control and CA, alone and in combination, in controlling grey mould on harvested strawberry fruit was tested. All fruit were wound-inoculated with *B. cinerea* Pers.:Fr and subsequently dipped in hot water at 63°C for 12 s. Fruit treated with the combination of heat, bio-control and CA had significantly less decay than those from all other treatments (Wszelaki and Mitcham, 2003).

Effects of hot water treatment and Modified Atmosphere Packaging (MAP) on the quality of tomatoes were studied. Prior to packaging with low-density polyethylene (LDPE) film, tomatoes were immersed in hot water (42.5°C) for 30 min. The use of a combination of hot water dips and MAP reduced decay during 2 weeks at 10°C and then for 3 days at 20°C (Suparlan-Itoh, 2003).

Immersion of naturally infected, freshly harvested table grapes for 30 s in 30% ethanol at 24°C reduced decay by approximately 50% after 35 days of storage at 1°C. The addition of ethanol significantly improved the efficacy of a hot water treatment applied to grapes that had been inoculated with *B. cinerea* 2 h prior to immersion in heated solutions. Immersion of inoculated, freshly harvested table grapes for 3 min at 30, 40 or 50°C in 10% ethanol reduced decay to 20.7, 6.7 and 0.1 berries/kg, respectively after 30 days of storage at 1°C, while decay after immersion in water at these temperatures was 35.9, 17.6 and 1.7 berries/kg, respectively (Karabulut *et al.*, 2004c).

The incidence of grey mould (caused by *B. cinerea*) among grape berries that were untreated or immersed for 1 min in ethanol (35 % v/v) at 25 or 50°C was 78.7, 26.2 and 3.4 berries/kg, respectively, after 1 month of storage at 0.5°C and 2 days at 25°C. Heated ethanol was effective up to 24 h after inoculation, but less effective when berry pedicels were removed before inoculation (Gabler *et al.*, 2005).

The most effective treatment for controlling decay development on sweet cherry inoculated by *P. expansum* and *B. cinerea* was immersion in 10% ethanol at 60°C for 3 min. Ethanol treatments at 20, 30, 40 or 50% and water treatments at 55 or 60°C significantly reduced natural fungal populations on the surfaces of fruit in all of the experiments. The addition of 10% ethanol to water significantly increased the efficacy of water in reducing the fungal populations at elevated temperatures (Karabulut *et al.*, 2004b). Also, a combination of calcium (1%) and hot water dips (50°C for 2 min) of sweet cherry (*Prunus avium* L.) significantly reduced fruit susceptibility to decay (Vangdal *et al.*, 2005).

Bananas (*Musa acuminata*) are a perishable fruit susceptible to anthracrose, crown rot and blossom end rot at the postharvest stage. Disease severity in banana fruit was significantly reduced by hot water treatment (50 ± 2°C for 5 min) and fungicide application (prochloraz, 250 ppm), and the latter treatment also reduced disease incidence. Fruit stored at low temperatures (10, 14 and 18°C) exhibited similar disease severity levels throughout the period of investigation and at levels much lower than those observed in fruit held at room temperature.

On the 25th day of storage, the highest disease severity (61.8%) occurred in the untreated fruit at room temperature, whereas the fruit treated with fungicide and hot water showed remarkably small areas covered by disease (< 3.4%). *Collechotrichum musae*, responsible for the most important postharvest disease known as anthracnose, was the most abundant pathogen isolated (Hassan *et al.*, 2004). Optimum temperature and exposure times for postharvest hot water treatment of banana were determined to be 50°C and 3 min, respectively. However, combining hot water treatment with the bacterial antagonist (*Burkholderia cepacia*) also gave more effective control of anthracnose, crown rot and blossom end rot than using the two treatments individually (De Costa and Erabadupitiya, 2005).

Hot water treatment (52°C for 10 min) combined with the fungicide bavistin (0.1%) was found to be the best in controlling the incidence of anthracnose and stem-end rot in Kesar mango fruit. The untreated (control) fruits were found to have been infected with *C. gloeosporioides* and *Diploidia natalensis* (Waskar, 2005). A combination of several heat treatments was also evaluated on mango. The efficacy of heat treatments to control *C. gloeosporioides* Penz. and anthracnose disease on mango was evaluated by using hot water at 55°C (HWD55), vapour heat (VHT) and hot water at 38°C for 5 min, combined with vapour heat (HWD38 + VHT).

Conidial germination was inhibited by the application of HWT55 for 5 min. After 36 h of incubation, the heat-treated conidia produced only 1% appressoria compared with the level of appressoria of untreated conidia, which developed normally. Disease incidence on mango inoculated with heat-treated conidia at 55°C for 5 min was reduced by 93%. One day after inoculation, at a depth of 1 mm, fungal colonization was reduced by 80%. Disease incidence on fruit treated with HWT55, VHT and HWT38 + VHT was reduced by 0.24, 0.26 and 0.14%, respectively. Disease development on heat-treated fruit was also delayed compared with those that had been untreated (Sopee and Sangchote, 2005).

Hot Water Rinsing and Brushing: Very Short-period Treatments

Hot water rinsing and brushing technology, which was reviewed recently by Fallik (2004), is the shortest physical treatment that can reduce decay development of fresh produce after harvest. This treatment is no longer than 30 s at temperatures at 48–63°C. In general, this technology involves a stainless steel machine equipped with parallel brushes, a thermostatically controlled hot water tank (300–500 l) and a dryer.

Fruits or vegetables are rinsed with pressurized hot water at 48–63°C for 10–25 s depending upon produce type and cultivar, from nozzles that point down either vertically or at predetermined angles onto the produce, which rolls on brushes. At the end of the hot washing treatment, forced-air fans or hot forced air is used to dry the produce inside a tunnel for 1–2 min (Fallik, 2004).

Hot water rinsing and brushing of red and yellow sweet bell pepper cultivars at 55°C for about 12 s significantly reduced decay incidence while maintaining quality compared with both untreated control and most other commercial treatments (Fallik *et al.*, 1999). The optimal HWRB treatment to reduce decay development on Galia-type melon fruit caused by *A. alternata* and *Fusarium* spp. was 59°C for 15 s (see Fig. 7.2) (Fallik *et al.*, 2000). Treating freshly harvested pink tomatoes with HWRB at 52°C for 15 s, or dipping the fruit in 52°C for 1 min, significantly reduced decay development caused by *B. cinerea* after 3 weeks' storage at 2 or 12°C and an additional 5 days at 20°C (Ilic *et al.*, 2001).

Deciduous fruits were found to benefit from HWRB treatments as well. Hot water rinsing while brushing at 55°C for 15 s significantly reduced decay development in *P. expansum*-inoculated apple fruit after 4 weeks at 20°C, or in naturally infected (*P. expansum*) apples after prolonged storage of 4 months at 1°C plus 10 days at 20°C (Fallik *et al.*, 2001b).

In vivo studies of peaches and nectarines inoculated with *M. fructicola* followed by HWRB at 55 or 60°C for 20 s gave 70 and 80% decay inhibition, respectively, compared with the control. The inhibition percentage of *M.*

Fig. 7.2. The effect of hot water rinsing and brushing (HWRB) on Galia-type melon (a) at 59°C for 15 s on decay development compared with the control (b) after 15 days at 5°C plus 3 days at 20°C (Image courtesy of E. Fallik).

fructicola with HWRB was similar, if HWRB was applied shortly after inoculation or 24 h later. In contrast, the sensitivity of *P. expansum* spores inoculated into wounds increased when the fruits were treated with HWRB 24 h after the inoculation, compared with treatment just after inoculation (Karabulut *et al.*, 2002). The optimal HWRB treatment to reduce decay development while maintaining kumquat quality was found to be 55°C for 20 s (Ben-Yehoshua *et al.*, 2000). When organic citrus fruits were treated with HWRB at 56, 59 and 62°C for 20 s after artificial inoculation with *P. digitatum*, decay development in infected wounds was reduced to 20, 5 and < 1%, respectively, of that in untreated control fruits or fruits treated with tap water (see Fig. 7.3) (Porat *et al.*, 2000a).

A 20 s HWRB treatment at 59 or 62°C reduced decay in Star Ruby grapefruit that had been artificially inoculated with *P. digitatum* by 52 and 70%, respectively, compared with unwashed fruit. Tap water wash (~ 20°C) or HWRB at 53 or 56°C were ineffective (Porat *et al.*, 2000b). Green mould incidence caused by *P. digitatum* was reduced by HWRB from 97.9 and 98% on untreated Eureka lemons (*Citrus limon*) and Valencia oranges (*Citrus sinensis*) to 14.5 and 9.4%, respectively, by a brief 30 s treatment at 62.8°C after storage for 3 weeks at 12.8°C (Smilanick *et al.*, 2003).

Sour rot (*Geotrichum citri-aurantii*) incidence on lemons averaged 84.3% after all water treatments, and was not significantly reduced, although arthrospores of *G. citri-aurantii* died at lower water temperatures than spores of *P. digitatum* and *P. italicum* in *in vitro* tests (Smilanick *et al.*, 2003). When the storage response of cactus pear (*Opuntia ficus-indica* Miller (L.)) following HWRB (60–70°C, 10–30 s) was investigated after 4 weeks at 6 ± 1°C plus 1 week at 20 ± 1°C, the effectiveness of treatments to control decay was found to increase with temperature and treatment time. Based on the results of this

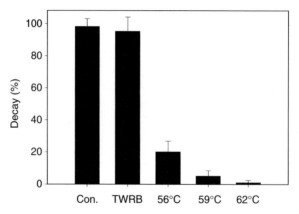

Fig. 7.3. Effects of hot water rinsing and brushing (HWRB) treatments (56, 59 and 62°C) on green mould decay development after artificial inoculation of Star Ruby red grapefruit. Surface wounds were inoculated with a *P. digitatum* spore suspension (10^4 spores/ml), and after 24 h the fruits were rinsed and brushed at three different temperatures for 20 s. Tap water rinsing and brushing (TWRB) served as the control (Con.). Decay was evaluated after 4 days of incubation at 24°C. Values are means ± SE of 60 wounds per treatment (from Porat *et al.*, 2000a).

study, treatments at 60°C for 30 s or 65°C for 20 s reduced decay incidence by 86–91% without compromising fruit quality (Dimitris *et al.*, 2005).

Depending on the cultivar, HWRB at 48–62°C for 15–20 s with 225 mg/ml prochloraz was found to be the most effective treatment for control of Alternaria rot in fruit with a high relative quiescent surface infection (Prusky *et al.*, 1999, 2002). Hot water rinsing and brushing at 55°C for 20 s significantly reduced chemical use (prochloraz) in controlling decay development caused mainly by *Penicillium* spp. in lychees after 5 weeks' storage and a marketing simulation period (Lichter *et al.*, 2000).

Table 7.1 provides an overview of the various heat treatments available as applied to a wide range of produce.

7.4 Conclusions

Pre-storage heat treatments comprise the largest area of research into postharvest decay control treatments. Nevertheless, there are many avenues that merit further exploration. Heat treatments, whether hot air, hot water dips or HWRB technology, are relatively simple techniques, economical, safe from rapid temperature changes and applicable to many potential postharvest problems. Heat treatments have been shown to combine well with modified atmosphere, biocontrol agents, ethanol, calcium and even low fungicidal dosages.

A better understanding of the physiology, pathology, biochemistry and molecular biology of heat-treated produce will enable the development of more

Table 7.1. The optimal temperature and time of exposure of heat treatments alone, or in combination with other treatments, on decay control in freshly harvested commodities.

Crop	Heat treatment regime	Optimal temperature (°C)/time	Fungus/pathogen	Reference(s)
Apple	HAT + Ca + biocontrol	38/4 days	*Penicillium expansum*	Conway *et al.* (1999)
Apple	HAT + bio-control	38/4 days	*Colletotrichum acutatum/P. italicum*	Conway *et al.* (2004)
Apple	HAT + biocontrol + CA	38/4 days	*C. acutatum/P. italicum*	Janisiewicz *et al.* (2003)
Apple	HAT + CA + 1-MCP	38/4 days	*C. acutatum/P. italicum/Botrytis cinerea*	Saftner *et al.* (2003)
Apple: organic	HWD	53/2 min	*Gloeosporium rot*	Trierweiler *et al.* (2003)
Apple: organic	HWD	53/3 min	*Gloeosporium rot*	Maxin *et al.* (2005)
Apple	HWRB	55/15 s	*P. expansum*	Fallik *et al.* (2001a, b)
Banana	HWD + Prochloraz[a]	50/5 min	*Collechotrichum musae*	Hassan *et al.* (2004)
Banana	HWD + biocontrol	50/3 min	Anthracrose, crown rot and blossom end rot	De Costa and Erabadupitiya (2005)
Cactus	HWD	52/3 min	–	Rodriguez *et al.* (2005)
Cactus pear	HAT	37/30 h	–[b]	Schirra *et al.* (1997a, b; 1999)
Chestnut	HWD	90/10 min	–	Panagou *et al.* (2005)
Citrus fruit	HWRB	62.8/30 s	*P. digitatum/P. italicum*	Smilanick *et al.* (2003)
Citrus fruit: organic	HWRB	62/30 s	*P. digitatum*	Porat *et al.* (2000a, b)
Clementine	HWD + SC + SBC	45/2.5 min	*P. digitatum*	Larrigaudiere *et al.* (2002)
Kumquat	HWRB	55/20 s	–	Ben-Yehoshua *et al.* (2000)
Lemon	HWD	52–53/2 min	*P. digitatum*	Nafussi *et al.* (2001)
Lemon and orange	HAT	33/65 h	*P. digitatum/P. italicum*	Plaza *et al.* (2004)
Lychee	HWRB	55/20 s	*P. digitatum*	
Mango	HWD + Bavistina	52/10 min	*Penicillium* spp.	Lichter *et al.* (2000)
Mango	HWD+HAT	38/5 min	*C. gloeosporioides/Diploidia natalensis*	Waskar (2005)
Mango	HWRB + Prochloraz[a]	48–62/20 s	*Alternaria alternata*	Sopee and Sangchote (2005)
Melon (Galia type)	HWRB	59/15 s		Prusky *et al.* (1999, 2002)
Orange	HWD	44/100 min, 50/50 min	*A. alternata/Fusarium* spp.	Fallik *et al.* (2000)
Orange	HWD + biocontrol	45/2 min	–	Schirra *et al.* (2005)
Papaya	HAT + TBZ[a]	48.5 or 50/4 h	*P. digitatum/P. italicum*	Obagwu and Korsten (2003)
			C. gloeosporioides	Perez-Carrillo and Yahia (2004)

Continued

Table 7.1. *Continued*

Crop	Heat treatment regime	Optimal temperature (°C)/time	Fungus/pathogen	Reference(s)
Peach and nectarine	HWD	46 or 50/2.5 min	*Monilinia fructicola*	Margosan *et al.* (1997)
Peach and nectarine	HWRB	60/20 s	*M. fructicola*	Karabulut *et al.* (2002)
Pepper (sweet)	HWD	53/4 min	–	Fallik *et al.* (1999)
Pepper (sweet)	HWRB	55/12 s	–	Gonzalez-Aguilar *et al.* (1999, 2000)
Pineapple	HWD	54/3 min	*Chlara paradoxa*	Wijeratnam *et al.* (2005)
Plum	HWD	45/35 min, 50/30 min	–	Abu-Kpawoh *et al.* (2002)
Red ginger (flower)	HWD	49–50/12–15 min	Saprophytic mould	Jaroenkit and Paull (2003)
Rockmelon	HWD	60/1 min	*Fusarium* spp., *Alternaria* spp. and *Colletotrichum* spp.	McDonald *et al.* (2005)
Strawberry	HAT	42 and 48/3 h	–	Civello *et al.* (1997)
Strawberry	HAT	45/40 min; 42/60–100 min	–	Wang *et al.* (2000)
Strawberry	HAT	45/3 h	–	Vicente *et al.* (2002)
Strawberry	HAT	45/3 h	*B. cinerea/Rhizopus stolonifer*	Pan *et al.* (2004)
Strawberry	HAT + MAP	45/3 h	–	Vicente *et al.* (2003)
Strawberry: organic	HWD	60/30 s,	–	Karabulut *et al.* (2004a, b)
Strawberry	HWD + CA + biocontrol	63/12 s	*B. cinerea*	Wszelaki and Mitcham (2003)
Sweet cherry	HWD + ethanol (10%)	60/3 min	*P. expansum/B. cinerea*	Karabulut *et al.* (2004b)
Sweet cherry	HWD + Ca (1%)	50/2 min	–	
Table grape	HAT	52.5/21–24 min; 55/18–21 min	*B. cinerea*	Lydakis and Aked (2003)
Table grape	HWD + ethanol (10%)	50/3 min,	*B. cinerea*	Karabulut *et al.* (2004a)
Table grape	HWD + ethanol (35%)	50/1 min	*B. cinerea*	Gabler *et al.* (2005)
Tomato	HAT	60/6 min, 70/1–5 min	*R. stolonifer*	Aborisade and Ojo (2002)
Tomato	HWD	42/1 h	–	McDonald *et al.* (1999)
Tomato	HWD + MAP	42.5/30 min	–	Suparlan-Itoh (2003)
Tomato	HWRB	52/15 s	*B. cinerea*	Ilic *et al.* (2001)

HAT, hot air treatment (dry or forced, or vapour); Ca, calcium; CA, controlled atmosphere; 1-MCP, 1-methylcyclopropene; HWD, hot water dip; HWRB, hot water rinsing and brushing; SC, sodium carbonate; SBC, sodium bicarbonate; MAP, modified atmosphere packaging.
[a] Fungicidal.
[b] Natural decay organisms.

precise and effective heat treatments in combination with other control agents, compounds and technologies. These simple technologies, which should also be explored on a broader range of freshly harvested commodities for minimally processed products or quarantine purposes, may successfully reduce our current extensive reliance on pesticides, and can therefore better protect the environment.

7.5 Acknowledgements

This is a contribution from the Agricultural Research Organization, the Volcani Centre, Bet Dagan, Israel, No. 452/05.

7.6 References

Aborisade, A.T. and Ojo, F.H. (2002) Effect of postharvest hot air treatment of tomatoes (*Lycopersicon esculentum* Mill) on storage life and decay caused by *Rhizopus stolonifer*. *Zeitschrift fur Pflanzenkrankheiten und Pflanzenschutz [Journal of Plant Diseases and Protection]* 109, 639–645.

Abu-Kpawoh, J.C., Xi, Y.F., Zhang, Y.Z. and Jin, Y.F. (2002) Polyamine accumulation following hot-water dips influences chilling injury and decay in 'Friar' plum fruit. *Journal of Food Science* 67, 2649–2653.

Barkai-Golan, R. (2001) *Postharvest Diseases of Fruits and Vegetables. Development and Control.* Elsevier, Netherlands, 418 pp.

Ben-Yehoshua, S., Peretz, J., Rodov, V., Nafussi, B., Yekutieli, O., Wiseblum, A. and Regev, R. (2000) Postharvest application of hot water treatment in citrus fruits: The road from the laboratory to the packing-house. *Acta Horticulturae* 518, 19–23.

Civello, P.M., Martinez, G.A., Chavas, A.R. and Anon, M.C. (1997) Heat treatments delay ripening and postharvest decay of strawberry fruit. *Journal of Agricultural Food Chemistry* 45, 4589–4594.

Conway, W.S., Janisiewicz, W.J., Klein, J.D. and Sams, C.E. (1999) Strategy for combining heat treatment, calcium infiltration, and biological control to reduce postharvest decay of 'Gala' apples. *HortScience* 34, 700–704.

Conway, W.S., Leverentz, B., Janisiewicz, W.J., Blodgett, A.B., Saftner, R.A. and Camp, M.J. (2004) Integrating heat treatment, biocontrol and sodium bicarbonate to reduce postharvest decay of apple caused by *Colletotrichum acutatum* and *Penicillium expansum*. *Postharvest Biology and Technology* 34, 11–20.

Couey, H.M. (1989) Heat treatment for control of postharvest diseases and insect pests of fruits. *HortScience* 24, 198–202.

De Costa, D.M. and Erabadupitiya, H.R.U.T. (2005) An integrated method to control postharvest diseases of banana using a member of the *Burkholderia cepacia* complex. *Postharvest Biology and Technology* 36, 31–39.

Dentener, P.R., Alexander, S.M., Lester, P.J., Petry, R.J., Maindonald, J.H. and McDonald, R.M. (1996) Hot air treatment for disinfestation of lightbrown apple moth and longtailed mealy bug on persimmons. *Postharvet Biology and Technology* 8, 143–452.

D'Hallewin, G., Dettori, A., Marceddu, S. and Schirra, M. (1997) Evoluzione dei processi infettivi di *Penicillium digitatum* Sacc. *In vivo e in vitro* dopo immersione in acqua calda. *Italus Hortus* 4, 23–26.

Dimitris, L., Pompodakis, N., Markellou, E. and Lionakis, S.M. (2005) Storage response of cactus pear fruit following hot water brushing. *Postharvest Biology and Technology* 38, 145–151.

Fallik, E. (2004) Prestorage hot water treatments (immersion, rinsing and brushing). *Postharvest Biology and Technology* 32, 125–134.

Fallik, E., Grinberg, S., Alkalai, S., Yekutieli, O., Wiseblum, A., Regev, R., Beres, H. and Bar-Lev, E. (1999) A unique rapid hot water treatment to improve storage quality of sweet pepper. *Postharvest Biology and Technology* 15, 25–32.

Fallik, E., Aharoni, Y., Copel, A., Rodov, R., Tuvia-Alkalai, S., Horev, B., Yekutieli, O., Wiseblum, A. and Regev, R. (2000) A short hot water rinse reduces postharvest losses of 'Galia' melon. *Plant Pathology* 49, 333–338.

Fallik, E., Tuvia-Alkalai, S., Copel, A., Wiseblum, A. and Regev, R. (2001a) A short water rinse with brushing reduces postharvest losses – 4 years of research on a new technology. *Acta Horticulturae* 553, 413–417.

Fallik, E., Tuvia-Alkalai, S., Feng, X. and Lurie, S. (2001b) Ripening characterization and decay development of stored apples after a short prestorage hot water rinsing and brushing. *Innovative Food Science and Emerging Technology* 2, 127–132.

Ferguson, I.B., Ben-Yehoshua, S., Mitcham, E.J., McDonald, R.E. and Lurie, S. (2000) Postharvest heat treatments: introduction and workshop summary. *Postharvest Biology and Technology* 21, 1–6.

Gabler, F.M., Mansour, M.F., Smilanick, J.L. and Mackey, B.E. (2004) Survival of spores of *Rhizopus stolonifer*, *Aspergillus niger*, *Botrytis cinerea* and *Alternaria alternata* after exposure to ethanol solutions at various temperatures. *Journal of Applied Microbiology* 96, 1354–1360.

Gabler, F.M., Smilanick, J.L., Ghosoph, J.M. and Margosan, D.A. (2005) Impact of postharvest hot water or ethanol treatment of table grapes on gray mould incidence, quality, and ethanol content. *Plant Disease* 89, 309–316.

Gaffney, J.J. and Armstrong, J.W. (1990) High-temperature forced air research facility for heating fruits for insect quarantine treatments. *Journal of Economical Entomology* 83, 1959–1964.

Gonzalez-Aguilar, G.A., Cruz, R., Baez, R. and Wang, C.Y. (1999) Storage quality of bell peppers pretreated with hot water and polyethylene packaging. *Journal of Food Quality* 22, 287–299.

Gonzalez-Aguilar, G.A., Gayosso, L., Cruz, R., Fortiz, J., Baez, R. and Wang, C.Y. (2000) Polyamines induced by hot water treatments reduce chilling injury and decay in pepper fruit. *Postharvest Biology and Technology* 18, 19–26.

Hallman, G.J. (2000) Factors affecting quarantine heat treatment efficacy. *Postharvest Biology and Technology* 21, 95–101.

Hassan, M.K., Shipton, W.A., Coventry, R. and Gardiner, C. (2004) Extension of banana shelf life. *Australasian Plant Pathology* 33, 305–308.

Ilic, Z., Polevaya, Y., Tuvia-Alkalai, S., Copel, A. and Fallik, E. (2001) A short prestorage hot water rinse and brushing reduces decay development in tomato, while maintaining its quality. *Tropical Agricultural Research Extension* 4, 1–6.

Janisiewicz, W.J., Leverentz, B., Conway, W.S., Saftner, R.A., Reed, A.N. and Camp, M.J. (2003) Control of bitter rot and blue mould of apples by integrating heat and antagonist treatments on 1-MCP treated fruit stored under controlled atmosphere conditions. *Postharvest Biology and Technology* 29, 129–143.

Jaroenkit, T. and Paull, R.E. (2003) Postharvest handling of Heliconia, red ginger, and bird-of-paradise. *HortTechnology* 13, 259–266.

Karabulut, O.A., Cohen, L., Wiess, B., Daus, A., Lurie, S. and Droby, S. (2002) Control of brown rot and blue mould of peach and nectarine by short hot water brushing and yeast antagonists. *Postharvest Biology and Technology* 24, 103–111.

Karabulut, O.A., Arslan, U. and Kuruoglu, G. (2004a) Control of postharvest diseases of organically grown strawberry with preharvest applications of some food additives and postharvest hot water dips. *Journal of Phytopathology* 152, 224–228.

Karabulut, O.A., Arslan, U., Kuruoglu, G. and Ozgenc, T. (2004b) Control of postharvest diseases of sweet cherry with ethanol and hot water. *Journal of Phytopathology* 152, 298–303.

Karabulut, O.A., Gabler, F.M., Mansour, M. and Smilanick, J.L. (2004c) Postharvest ethanol and hot water treatments of table grapes to control gray mould. *Postharvest Biology and Technology* 34, 169–177.

Larrigaudiere, C., Pons, J., Torres, R. and Usall, J. (2002) Storage performance of clementines treated with hot water, sodium carbonate and sodium bicarbonate dips. *Journal of Horticultural Science and Biotechnology* 77, 314–319.

Lay-Yee, M., Clare, G.K., Petry, R.J., Fullerton, R.A. and Gunson, A. (1998) Quality and disease incidence of 'Waimanalo Solo' papaya following forced-air heat treatments. *HortScience* 33, 878–880.

Lichter, A., Dvir, O., Rot, I., Akerman, M., Regev, R., Wiseblum, A., Fallik, E., Zauberman, G. and Fuchs, Y. (2000) Hot water brushing: an alternative method to SO$_2$ fumigation for color retention of lychee fruits. *Postharvest Biology and Technology* 18, 235–244.

Lichter, A., Zhou, H.W., Vaknin, M., Dvir, O., Zutchi, Y., Kaplunov, T. and Lurie, S. (2003) Survival and responses of *Botrytis cinerea* after exposure to ethanol and heat. *Journal of Phytopathology* 151, 553–563.

Lurie, S. (1998) Postharvest heat treatments of horticultural crops. *Horticultural Review* 22, 91–121.

Lydakis, D. and Aked, J. (2003) Vapour heat treatment of Sultanina table grapes. I: control of *Botrytis cinerea*. *Postharvest Biology and Technology* 27, 109–116.

Margosan, D.A., Smilanick, J.L., Simmons, G.F. and Henson, D.J. (1997) Combination of hot water and ethanol to control postharvest decay of peaches and nectarines. *Plant Disease* 81, 1405–1409.

Marquenie, D., Lammertyn, J., Geeraerd, A.H., Soontjens, C., Van Impe, J.F., Nicolai, B.M. and Michiels, C.W. (2002) Inactivation of conidia of *Botrytis cinerea* and *Monilinia fructigena* using UV-C and heat treatment. *International Journal of Food Microbiology* 74, 27–35.

Maxin, P., Huyskens-Keil, S., Klopp, K. and

Ebert, G. (2005) Control of postharvest decay in organic grown apples by hot water treatment. *Acta Horticulturae* 682, 2153–2157.

McDonald, K.L., McConchie, M.R., Bokshi, A. and Morris, S.C. (2005) Heat treatment: a natural way to inhibit postharvest diseased in rockmelon. *Acta Horticulturae* 682, 2029–2033.

McDonald, R.E., McCollum, T.G. and Baldwin, E.A. (1999) Temperature of water heat treatments influences tomato fruit quality following low-temperature storage. *Postharvest Biology and Technology* 16, 147–155.

Nafussi, B., Ben-Yehoshua, S., Rodov, V., Peretz, J., Ozer, B.K. and D'Hallewin, G. (2001) Mode of action of hot-water dip in reducing decay of lemon fruit. *Journal of Agricultural and Food Chemistry* 49, 107–113.

Neven, L.G., Drake, S.R. and Ferguson, H.J. (2000) Effects of the rate of heating on apple and pear fruit quality. *Journal of Food Quality* 23, 317–325.

Obagwu, J. and Korsten, L. (2003) Integrated control of citrus green and blue molds using *Bacillus subtilis* in combination with sodium bicarbonate or hot water. *Postharvest Biology and Technology* 28, 187–194.

Pan, J., Vicente, A.R., Martinez, G.A., Chaves, A.R. and Civello, P.M. (2004) Combined use of UV-C irradiation and heat treatment to improve postharvest life of strawberry fruit. *Journal of the Science of Food and Agriculture* 84, 1831–1838.

Panagou, E.Z., Vekiari, S.A., Sourris, P. and Mallidis, C. (2005) Efficacy of hot water, hypochlorite, organic acids and natamycin in the control of post-harvest fungal infection of chestnuts. *Journal of Horticultural Science and Biotechnology* 80, 61–64.

Perez-Carrillo, E. and Yahia, E.M. (2004) Effect of postharvest hot air and fungicide treatments on the quality of 'maradol' papaya (*Carica papaya* L.) *Journal of Food Quality* 27, 127–139.

Plaza, P., Usall, J., Torres, R., Abadias, M., Smilanick, J.L. and Vinas, I. (2004) The use of sodium carbonate to improve curing treatments against green and blue molds on citrus fruits. *Pest Management Science* 60, 815–821.

Porat, R., Daus, A., Weiss, B., Cohen, L., Fallik, E. and Droby, S. (2000a) Reduction of postharvest decay in organic citrus fruit by a short hot water brushing treatment. *Postharvest Biology and Technology* 18, 151–157.

Porat, R., Pavoncello, D., Peretz, Y., Weiss, B., Cohen, L., Ben-Yehoshua, S., Fallik, E., Droby, S. and Lurie, S. (2000b) Induction of resistance against *Penicillium digitatum* and chilling injury in Star Ruby grapefruit by a short hot water brushing treatment. *Journal of Horticultural Science and Biotechnology* 75, 428–432.

Prusky, D., Fuchs, Y., Kobiler, I., Roth, I., Weksler, A., Shalom, Y., Fallik, E., Zaurberman, G., Pesis, E., Akerman, M., Yekutieli, O., Wiseblum, A., Regev, R. and Artes, L. (1999) Effect of hot water brushing, prochloraz treatment and waxing on the incidence of black spot decay caused by *Alternaria alternata* in mango fruit. *Postharvest Biology and Technology* 15, 165–174.

Prusky, D., Shalom, Y., Kobiler, I., Akerman, M. and Fuchs, Y. (2002) The level of quiescent infection of *Alternaria alternata* in mango fruits at harvest determines the postharvest treatment applied for the control of rots during storage. *Postharvest Biology and Technology* 25, 339–347.

Rodriguez, S., Casóliba, R.M., Questa, A.G. and Felker, P. (2005) Hot water treatment to reduce chilling injury and fungal development and improve visual quality of two *Opuntia ficus indica* fruit clones. *Journal of Arid Environments* 63, 366–378.

Saftner, R.A., Abbott, J.A., Conway, W.S. and Barden, C.L. (2003) Effects of 1-methyl-cyclopropene and heat treatments on ripening and postharvest decay in 'Golden Delicious' apples. *Journal of the American Society for Horticultural Science* 128, 120–127.

Schirra, M., Agabbio, M., D'Aquino, S. and McCollum, T.G. (1997a) Postharvest heat conditioning effects on early ripening 'Gialla' cactus pear fruit. *HortScience* 32, 702–704.

Schirra, M., Barbera, G., D'Hallewin, G., Inglese, P. and LaMantia, T. (1997b) Storage response of cactus pear fruit to CaCl$_2$ preharvest spray and postharvest heat treatment. *Journal of Horticultural Science* 72, 371–377.

Schirra, M., D'Hallewin, G., Inglese, P. and La Mantia, T. (1999) Epicuticular changes and storage potential of cactus pear (*Opuntia ficus-indica* Miller (L.)) fruit following gibberellic acid preharvest sprays and postharvest heat treatment. *Postharvest Biology and Technology* 17, 79–88.

Schirra, M., Cabras, P., Angioni, A., D'Hallewin, G. and Pala, M. (2002) Residue uptake and storage responses of tarocco blood oranges after preharvest thiabendazole spray and postharvest heat treatment. *Journal of Agricultural and Food Chemistry* 50, 2293–2296.

Schirra, M., Mulas, M., Fadda, A., Mignani, I. and Lurie, S. (2005) Chemical and quality traits of 'Olinda' and 'Campbell' oranges after heat treatment at 44 or 46°C for fruit fly disinfestations. *LWT-Food Science and Technology* 38, 519–527.

Sharp, J.L., Gaffney, J.J., Moss, J.I. and Gould, W.P. (1991) Hot-air treatment device for quarantine research. *Journal of Economical Entomology* 84, 520–527.

Smilanick, J.L., Sorenson, D., Mansour, M., Aieyabei, J. and Plaza, P. (2003) Impact of a brief postharvest hot water drench treatment on decay, fruit appearance, and microbe populations of California lemons and oranges. *HortTechnology* 13, 333–338.

Sopee, J. and Sangchote, S. (2005) Effect of heat treatment on the fungus Colletotrichum gloeosporioides and anthracnose of mango fruit. *Acta Horticulturae* 682, 2049–2053.

Suparlan-Itoh, K. (2003) Combined effects of hot water treatment (HWT) and modified atmosphere packaging (MAP) on quality of tomatoes. *Packaging Technology and Science* 16, 171–178.

Trierweiler, B., Schirmer, H. and Tauscher, B. (2003) Hot water treatment to control gloeosporium disease on apples during long-term storage. *Angewandte Botanik [Journal of Applied Botany]* 77, 156–159.

Tsang, M.M.C., Hara, A.H., Hata, T.Y., Hu, B.K.S., Kaneko, R.T. and Tenbrink, V. (1995) Hot-water immersion unit for disinfestation of tropical floral commodities. *Applied Engineering in Agriculture* 11, 397–402.

Vangdal, E., Nordbo, R. and Flatland, S. (2005) Postharvest calcium and heat treatment of sweet cherry (*Prunus avium* L.) *Acta Horticulturae* 682, 1133–1135.

Vicente, A.R., Martinez, G.A., Civello, P.M. and Chaves, A.R. (2002) Quality of heattreated strawberry fruit during refrigerated storage. *Postharvest Biology and Technology* 25, 59–71.

Vicente, A.R., Martinez, G.A., Chaves, A.R. and Civello, P.M. (2003) Influence of self-produced CO_2 on postharvest life of heat-treated strawberries. *Postharvest Biology and Technology* 27, 265–275.

Wang, C.Y. (2000) Effect of moist hot air treatment on some postharvest quality attributes of strawberries. *Journal of Food Quality* 23, 51–59.

Waskar, D.P. (2005) Hot water treatment for disease control and extension of shelf life of 'Kesar' mango (*Mangifera indica* L.) fruit. *Acta Horticulturae* 682, 1319–1323.

Whiting, E.C., Khan, A. and Gubler, W.D. (2001) Effect of temperature and water potential on survival and mycelial growth of *Phaeomoniella chlamydospora* and *Phaeoacremonium* spp. *Plant Disease* 85, 195–201.

Wijeratnam, R.S.W., Hewajulige, I.G.N and Abeyratne, N. (2005) Postharvest hot water treatment for the control of Thielaviopsis black rot of pineapple. *Postharvest Biology and Technology* 36, 323–327.

Wszelaki, A.L. and Mitcham, E.J. (2003) Effect of combinations of hot water dips, biological control and controlled atmospheres for control of gray mould on harvested strawberries. *Postharvest Biology and Technology* 27, 255–264.

8 Disinfestation of Stored Products and Associated Structures Using Heat

S.J. BECKETT,[1] P.G. FIELDS[2] AND BH. SUBRAMANYAM[3]

[1]CSIRO Entomology, Canberra, Australia; e-mail: stephen.beckett@csiro.au; [2]Agriculture and Agri-food Canada, Cereal Research Centre, Winnipeg, Canada; e-mail: pfields@agr.gc.ca; [3]Department of Grain Science and Industry, Kansas State University, Manhattan, Kansas, USA; e-mail: sbhadrir@ksu.edu

8.1 Introduction

Background

The term 'stored products' is broad. It refers generally to non-perishable crops and their dry durable products. The crops include grains such as wheat, barley, maize, oats, rye, rice, millet and sorghum; pulses such as peas, beans and lentils; and oil seeds such as those from canola and sunflower. Dried fruits such as raisins, sultanas and currants are often included, as well as herbs and spices, but will not be considered in this chapter. For the purposes of this discussion, the emphasis will be on grains, grain products and associated storage structures, handling equipment, processing facilities and machinery.

Insect pests have been associated with stored products since the dawn of settled agriculture some 10,000 years ago. Grains, skins and fibres stored in quantity made opportunistic habitats for insect species living in leaf litter, animal nests and under the bark of trees, or feeding on seeds, general plant material, carrion and fungi (Hill, 2002; Rees, 2004).

The invention of the McCormick reaper in the 1830s and threshing machines shortly after spurred modern storage methods (Drache, 1976). The first grain elevator built in the USA at Buffalo, New York, in 1842, combined with the reliability of the steam engine, ended former restrictions in the capacity of grain handling. For example, in Chicago, Illinois, annual shipment of grain and flour increased from around 150,000 t (6 million bushels) per year in 1855 to 3.75 million t (150 million bushels) thirty years later (Reed, 1992). However, it must be remembered that the changes in storage and handling over the last

150 years in the industrialized world have had little impact on the traditional ways practised in many underdeveloped countries.

With the establishment of stored-product research and development at the beginning of the 20th century (Munro, 1966) have come the many strategies for protection and disinfestation that we see today against the plethora of insect pests that confront the grain and food industry. Total world production of cereals in 2002 was about 2030 million t, of which 14% was traded internationally, and about half of that was accounted for by wheat and wheat flour (FAO, 2003, 2004). In response to market influence in developed and exporting countries, the current status of stored-product insect management is in a period of transition from chemical methods to non-chemical methods, and hence there is great potential for using heat in stored products and associated structures.

Stored-product Insect Pests

The majority of stored-product insects come from only two of the roughly 26 orders of the Class Insecta (see Fig. 8.1): Coleoptera and Lepidoptera. On the other hand, predators and parasitoids of stored-product insects come from the orders Hemiptera, Hymenoptera and Diptera (Rees, 2004). Stored-product insects are often classified by whether they develop and feed inside or outside grain kernels, and thus are referred to as either internal developers or external developers.

Internal developers

As the most destructive of the stored-product insects, six species from two families of Coleoptera and one family of Lepidoptera are recognized as internal developers, sometimes referred to as primary invaders.

These include the grain borers (Family: Bostrichidae), such as the lesser grain borer, *Rhyzopertha dominica* (Fabricius), which is one of the most damaging insects, and the larger grain borer, *Prostephanus truncatus* (Horn), a major tropical pest of maize; the grain weevils (Family: Curculionidae), such as the rice weevil, *Sitophilus oryzae* (L.), maize weevil, *S. zeamais* Motschulsky and the granary weevil, *S. granarius* (L.); and, finally, the angoumois grain moth (Family: Gelechiidae), *Sitotroga cerealella* (Olivier), which today is less destructive than the other internal developers because infestations commonly start in the field and can be minimized by modern methods of harvesting (e.g. combines) and storage.

External developers

External developers are sometimes referred to as secondary invaders because they tend to infest broken materials and grain dust created by primary

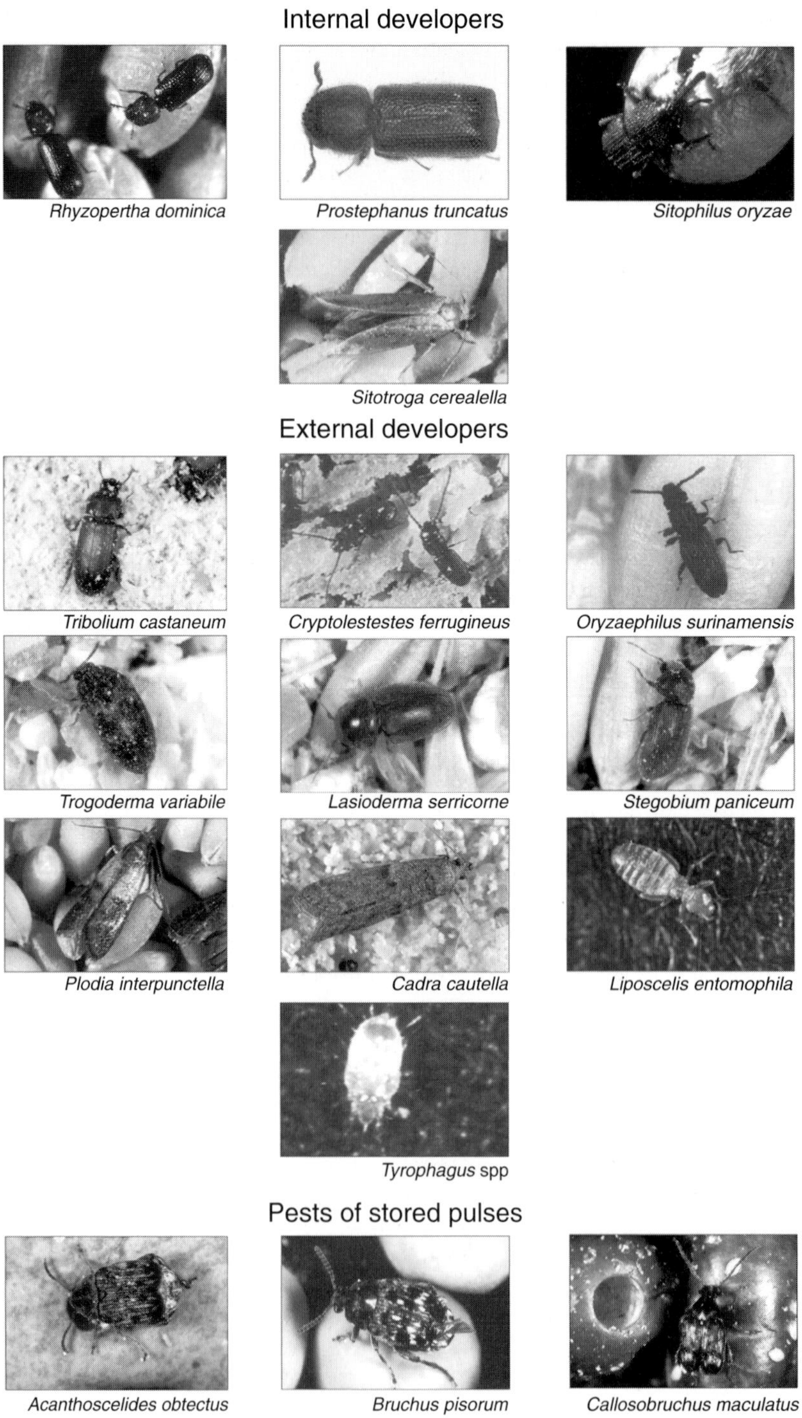

Internal developers

Rhyzopertha dominica

Prostephanus truncatus

Sitophilus oryzae

Sitotroga cerealella

External developers

Tribolium castaneum

Cryptolestestes ferrugineus

Oryzaephilus surinamensis

Trogoderma variabile

Lasioderma serricorne

Stegobium paniceum

Plodia interpunctella

Cadra cautella

Liposcelis entomophila

Tyrophagus spp

Pests of stored pulses

Acanthoscelides obtectus

Bruchus pisorum

Callosobruchus maculatus

Fig. 8.1. Major insect pests of stored products, pulses and storage and processing structures (in order of reference in text; photography by J. Green, D. McClenaghan and N. Starick, CSIRO, Canberra, Australia).

invaders. They are found in grain but are more common in flour, meals and other processed foods. They have also been reported in storages, handling equipment and processing facilities. External developers represent six families of Coleoptera and one family of Lepidoptera.

The families include flour beetles (Family: Tenebrionidae), such as the red flour beetle, *Tribolium castaneum* (Herbst) and the confused flour beetle *T. confusum* Jacquelin du Val; and the grain beetles (Families: Laemophloeidae and Silvanidae), which are generally smaller and flatter than other externally developing stored-product Coleoptera. The most commonly found species of the family Laemophloeidae are the flat grain beetle, *Cryptolestestes pusillus* (Schönherr), and the rusty grain beetle, *C. ferrugineus* (Stephens). The most common silvanid is the sawtoothed grain beetle, *Oryzaephilus surinamensis* (L.).

Another important group are the grain-infesting dermestids (Family: Dermestidae), which generally refers to several species of *Trogoderma*. The warehouse beetle, *Trogoderma variabile* (Ballion) and the khapra beetle, *T. granarium* (Everts), are the best known. The larvae of these species can diapause for many months, making them particularly difficult to control. Of the Coleoptera, the anobiid beetles (Family: Anobiidae) are the last main group worth mentioning. These include the cigarette beetle, *Lasioderma serricorne* (F.) and the drugstore beetle, *Stegobium paniceum* (L.).

Several moth species (Family: Pyralidae) infest cereal grains and grain products. Probably the most important of these are the Indianmeal moth, *Plodia interpunctella* (Hübner), almond moth, *Cadra cautella* (Walker), tobacco moth, *Cadra cautella* (Hübner) and Mediterranean flour moth, *E. kuehniella* (Zeller). The larvae of these species leave a trail of silk behind them that can completely soil the surface of food products as they feed.

Finally, there are several pests that flourish under very humid conditions. Probably the most important in this group are several species of psocids (Order: Pscoptera e.g. *Liposcelis entomophila*) and mites (Class: Arachnida e.g. *Tyrophagus* spp.), which struggle to survive at relative humidities (RH) < 60 and 75%, respectively.

Pests of stored pulses

Bruchids (Family: Bruchidae) specifically infest pulses as primary pests and do not attack cereal grains. Most commonly known members of this group are the bean weevil, *Acanthoscelides obtectus* (Say), various pea weevils (*Bruchus* spp.) such as *B. pisorum* (L.) and the cowpea weevil, *Callosobruchus maculatus* (F.). Members of this family lay their eggs on seed pods or exposed seeds. However, while other pests can continue to breed on dried seed in storage, *Bruchus* species are more important as a field pest and cannot reinfest once they emerge in storage.

Optimum conditions for population growth of most of the major pest species are between 28 and 33°C at 60–75% RH (see Table 8.1). However, several species grow well between 20 and 37°C and 25–80% RH. Adults at low

Table 8.1. Responses of stored-product insect pests to various temperature ranges (adapted from Fields, 1992).

Zone	Temperature range (°C)	Effect(s)
Lethal	> 62	Death in < 1 min
	50–62	Death in < 1 h
	45–50	Death in < 1 day
	35–42	Populations die out, mobile insects seek cooler environment
Suboptimal	35	Maximum temperature for reproduction
	32–35	Slow population increase
Optimal	25–32	Maximum rate of population increase
Suboptimal	13–25	Slow population increase
Lethal	5–13	Slowly lethal
	1–5	Movement ceases
	−10 to −5	Death in weeks, or months if acclimated
	−25 to −15	Death in < 1 h

suboptimum temperatures can live for many months; the minimum theoretical threshold for population growth being around 18°C.

An increase in the rate of egg-laying and a decrease in development time occur as temperatures increase, but populations often struggle to grow at 35°C unless humidity is high, and death in most species is practically inevitable at about 40°C. At optimum conditions, most species lay 100–400 eggs per female, and development time is between 20 and 40 days, which means that many species can more than double their population in about 1 month.

Stored-product insects are considered pests for a number of reasons. At high densities they can consume a considerable amount of food, which is an important problem in many developing countries where annual losses due to insects are estimated at 10–50% (USDA, 1965; Wolpert, 1967; Hall, 1970). At high densities, stored-product insects can cause heating of grain and encourage mould growth by initiating moisture migration and condensation (Longstaff, 1981). All of these processes can cause considerable losses and reduce the quality of finished products. Given the right conditions, insects can increase rapidly and cause extensive damage before they are detected.

Therefore, the threshold for stored-product insect management is very low or zero. For example, Canada and Australia stipulate that wheat at the point of export must be free of insects; Nigeria has similar restrictions with cocoa beans, as does South Africa with groundnuts (Snelson, 1987). According to Federal Grain Inspection Standards in the USA, stored wheat is considered infested if it has two or more live insects/kg.

Most Western countries have zero tolerance for insects in finished products such as flours or baked goods. Insect parts in processed food are also restricted. In the USA, wheat flour with 75 insect fragments/50 g of flour (Defect Action Level) cannot be commercially sold for human consumption (FDA, 1988). In some cases, buyers of grain and processed foods have the right to reject entire shipments based on the presence of insects or insect fragments. Buyers may also impose major penalties for insect infestation because of the cost of remediation and delay at port of entry (Snelson, 1987).

Management of Stored-product Insects

The development of relatively cheap and very effective chemical methods, which matured in the second half of the 20th century, led to chemicals being the method of choice for insect management. Two basic types of chemical insecticides are used on stored products, namely fumigants (Bond, 1984) and contact insecticides (Snelson, 1987).

Fumigants

Fumigants generally enter the insect through the respiratory system, and are toxic to all life stages. They are gaseous chemicals at ambient temperature and pressure, and can produce gas from a solid or liquid. They diffuse through air, permeate products and have little or no residual insecticidal effect (Bond, 1984; Harein and Davis, 1992).

Currently several fumigants are used against stored-product insects. For fumigation with phosphine (PH_3), solid metal phosphides are used in grain (aluminium phosphide) and warehouses (magnesium phosphide). Phosphine generators mix the metal phosphides with water to rapidly produce gaseous phosphine, which can be delivered to the site in cylinders and released directly into the commodity or structure. As phosphine is corrosive to copper and several other metals, it is not used extensively for the fumigation of structures. Methyl bromide (CH_3Br) is used extensively in cereal food-processing facilities for structural fumigation (flour mills, bakeries, pasta plants and breakfast cereal plants) and against quarantine pests, but is considered an ozone-depleting substance, so its production and use is now being phased out (Fields and White, 2002).

The resistance of insects to phosphine, and to a lesser extent to methyl bromide, is now an acute problem worldwide. For phosphine, this is in large part due to poor fumigation practices. Significant resistance was demonstrated even thirty years ago by Champ and Dyte (1976), who recorded phosphine resistance in 10% of 848 samples in over five stored-product insect species from 41% of 82 countries surveyed, and increased tolerance to methyl bromide in thirteen insect species in 5% of samples from 26% of countries.

Other fumigants include hydrogen cyanide, carbon disulphide, chloropicrin and ethyl formate. Sulphuryl fluoride has been used since the 1950s to control termites in the USA and Europe, and recently to manage insects in food-processing facilities (Drinkall *et al.*, 2002; Ducom *et al.*, 2002). Like sulphuryl fluoride, interest in hydrogen cyanide, carbon disulphide, and ethyl formate has been rekindled in recent years as alternatives to methyl bromide; carbonyl sulphide was patented in 1993 for this purpose (Banks *et al.*, 1993; Bengston and Strange, 1998).

Modified atmospheres, high carbon dioxide and high nitrogen (or low oxygen) have been used to a limited extent to disinfest grain or structures as alternatives to fumigation. In leaky structures, carbon dioxide is better than nitrogen, as concentrations of 35% are lethal to all life stages of stored-product insects. In contrast, nitrogen must reduce the level of oxygen to below 2%, requiring airtight storage to be cost-effective. Thus, the technology is hindered by the necessity for sealed storages and access to large amounts of cheap carbon dioxide or nitrogen. The use of modified atmospheres is discussed in detail by Banks and Annis (1990).

Contact insecticides

Contact insecticides, or protectants, generally enter the insect orally or across the cuticle. They are applied either to grain, floor–wall junctions, general surfaces or crevices in warehouses and food-processing facilities. Protectants are defined as insecticides that prevent infestations from becoming established in a commodity, but are less effective at managing a well-established infestation, and infested commodities and structures are often better treated by fumigation.

The most commonly used contact insecticides are: (i) organophosphates (malathion, dichlorvos, fenitrothion, pirimiphos-methyl, chlorpyrifos-methyl, chlorpyrifos-methyl with deltamethrin); (ii) pyrethroids (bioresmethrin, permethrin and deltamethrin); and (iii) synergized pyrethrins (Champ and Dyte, 1976; Snelson, 1987).

Because protectants vary in efficacy from species to species, they must be used judiciously. Moreover, as with fumigants, insect resistance is a major issue, which further adds to the complexity of application. Other less commonly used contact insecticides include diatomaceous earth (Ebeling, 1971; Desmarchelier and Dines, 1987; Paula *et al.*, 2002) and insect growth regulators such as methoprene and hydroprene (Strong and Diekman, 1973; Edwards *et al.*, 1987). Spinosad, an insecticide based on bacterial fermentation products (Spinosyns A and D), has been shown to be effective against stored-product insects in laboratory and field evaluations (Fang *et al.*, 2002a, b; Flinn *et al.*, 2004) and is registered in the USA as a grain protectant.

Physical disinfestation methods

Chemical methods of insect disinfestation face significant challenges. Methyl bromide is becoming more and more costly and is being phased out under the 1987 Montreal Protocol in both developed and developing countries. The use of phosphine is being increasingly regulated because of safety issues and insect resistance. Resistance to insecticides is widespread and chemical residues – owing to increased dosages – are becoming increasingly unacceptable to grain buyers and consumers. Not only do contracts with buyers now stipulate acceptable insect contamination and damage, but also which insecticides are acceptable and what constitutes the upper residue limit.

On the other hand, a physical method of grain protection and disinfestation would not carry these problems. One such method might be the use of mechanical methods such as impact (Bailey, 1962, 1969) or combinations of sieving, aspirating and blowing (Banks, 1987), which are used in food-processing facilities. One of the most widely researched physical methods, however, is the use of extreme temperatures for insect disinfestation in bulk-stored grain, associated structures and food-processing facilities.

Low temperatures are commonly used to manage stored-product insects. Few species can achieve a population increase < 18°C. *Sitophilus granarius* is one of the exceptions, as it can reproduce at temperatures down to 15°C. Between 1 and 5°C, depending upon acclimation and the species, stored-product insects are unable to move and reproduce. Temperatures < 0°C will kill insects; the lower the temperature, the faster the insects will succumb to cold injury (see Table 8.1).

There are a number of methods to reach these temperatures. Aeration of grain bulks with ambient air is used extensively immediately after harvest to cool the grain. The air, which is passed through the bulk at relatively low volume, has a minimal drying effect, but preserves grain quality by slowing population growth, minimizing moisture migration and preventing the build-up of hot spots (Darby, 1998). In conjunction with insecticides, aeration can delay the onset of resistance and reduce the amount of protectant used (Longstaff, 1984, 1986, 1988).

Aeration of grain with chilled or refrigerated air is also used (Fields, 1992; Burks *et al.*, 2000), and can be as cost-effective as phosphine fumigation. Low temperatures are sometimes used to manage insects within flour mills in temperate climates during cold winters, but this practice has fallen out of use because of the unpredictability of the low temperatures (−10°C) required, and the need to drain all water from the facility.

By contrast, the use of elevated temperatures has the major advantage of giving complete disinfestation while being comparatively rapid. Like the use of low temperatures, it is chemical-free, and insects are not as likely to develop resistance to it. Methods using heat have been developed that disinfest grain both at the on-farm and commercial storage levels, as well as storages, processing facilities and equipment. However, there are scientific, technical and economic issues still to overcome.

In this review of heat treatments to control stored-product insect pests, we cover the current thermal kinetic data for insect mortality, empirical methods used to obtain this information and common mathematical and statistical models designed to make treatment predictions as reliable as possible. We also examine the current status of heat treatment research and development, and include our outlook for thermal treatments of stored products and structures in terms of a solution to some of the problems confronting safe storage and disinfestation.

8.2 The Use of Heat for Insect Management

Heat Disinfestation of Stored Products

The concept of using thermal energy for the disinfestation of stored-product insects is not new. Traditional methods have made use of heat, particularly solar energy, for thousands of years. In China, heating grain in a thin layer on the ground to > 50°C and < 12% moisture content (MC) dates back 1500 years. The grain is then piled up to maintain that temperature for several hours before being stored in an insulated bin where the temperature drops to ambient over the following couple of months (Liu *et al.*, 1983). The first recorded occurrence of using heat to control stored-product insects was by Duhamel du Monceau and Tillet (1762) in western France. During a severe outbreak of the angoumois grain moth, *S. cerealella*, a temperature of 69°C was used to destroy caterpillars in grain spread out in ovens for 3 days.

The modern use of thermal energy has a history as long as other contemporary technologies for stored-product insect control. The first industrial-scale heat disinfesters were developed in Australia between 1915 and 1919 (Winterbottom, 1922). These early machines heated grain by conduction up to 60°C as the grain fell by gravity through a bank of steam-heated pipes. The units stood 6 m high and could treat 25 t (1,000 bushels)/h (see Fig. 8.2). Of the 12 such units built over this period, six processed over '10,000,000 bags of wheat' (816,000 t) (Winterbottom, 1922). However, as cheap chemical alternatives became more available, there was little interest in further pursuit of the use of thermal energy for grain disinfestation for the following 50 years or so.

With the impending phase-out of methyl bromide fumigation, there is now increasing urgency for an alternative rapid disinfestation system, particularly at point of export. While methyl bromide fumigation can be completed in < 2 days, current alternative fumigants require a minimum of 3 days, which is unacceptable to industry. By contrast, continuous flow heat treatment systems with a throughput up to 400 t/h are possible (Sutherland *et al.*, 1987), and there are also opportunities for farm-scale treatments.

Several factors must be addressed before thermal disinfestation of stored products becomes a significant part of current insect control practice. Most important is competitive cost, which is a combination of the lowest possible running and capital costs to challenge current chemical options, and includes issues such as throughput rate, thermodynamic design, energy costs, safety and versatility for a range of commodities. External forces, such as further deterioration in the efficacy of phosphine, may improve the cost balance in favour of heat disinfestation, but strategies for lowering cost must also be actively sought.

There is already considerable information on a range of laboratory-based methods of treatment, as well as some industrial-scale equipment. Our present understanding of the effects of thermal energy on insect mortality and grain quality will help improve the operation of current systems and the development of new ones.

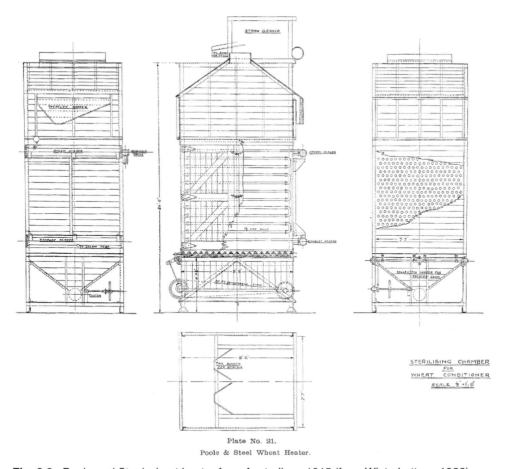

Plate No. 21.
Poole & Steel Wheat Heater.

Fig. 8.2. Poole and Steel wheat heater from Australia, *c.*1915 (from Winterbottom, 1922).

Heat Disinfestation of Structures

Heat disinfestation of structures is a simple concept involving raising the temperature of the whole or a portion of a facility to 50–60°C and maintaining these elevated temperatures for 24–36 h (Imholte and Imholte-Tauscher, 1999; Dowdy and Fields, 2002; Wright *et al.*, 2002; Dosland *et al.*, 2006). The minimum temperature for successful disinfestation is 50°C (Wright *et al.*, 2002; Mahroof *et al.*, 2003a, b; Roesli *et al.*, 2003); in portions of the facility where the temperature is < 50°C insect survival can be expected.

Structural heat treatments can be performed using gas, electric or steam heaters. Depending on the size and nature of the structure, long periods of heating may be needed for penetration of wall voids and equipment to kill insects harbouring in them (Fig. 8.3).

As hot air stratifies vertically and horizontally within structures, the use of air movers or fans is essential to ensure uniform heating of all portions of a building. Structural heat treatments are labour-intensive because the facility

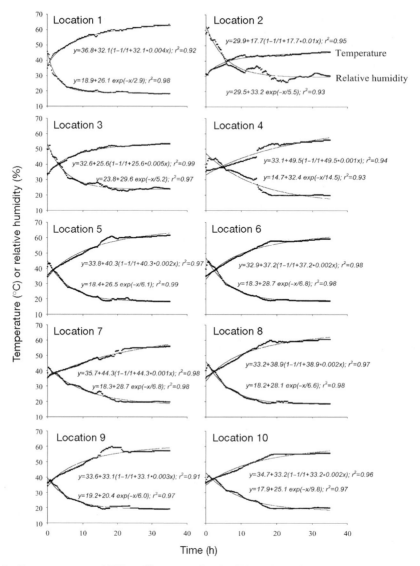

Fig. 8.3. Temperature and RH profiles at ten feed mill locations during heat treatment. Temperature and RH profiles were described by fitting two separate three-parameter nonlinear regression models to data. Each fitted line (solid line) was based on $n = 141$ temperature or RH observations. The adjusted r^2 values are presented in each graph (from Mahroof *et al.*, 2003a).

and equipment are slow absorbers of heat and need to be thoroughly cleaned. In addition, food, heat-susceptible materials and non-food materials (e.g. paper bags) should be removed from the facility because insects may seek refuge in these materials and escape treatment. Food products, if infested, should be removed and fumigated with phosphine before being returned to the structure after a heat treatment is finished. Good exclusion and sanitation practices are therefore essential for ensuring an effective heat treatment and prevention of reinfestation of structures.

In the USA and Canada, many flour mills used heat treatments as a pest management tool in the early 1900s (Dean, 1913), and in later years worker safety was often considered to be better with heat than with fumigants (Cotton, 1963). There exist anecdotal reports of adverse effects observed by millers, probably due to inadequate control of high temperatures, including warping of wood, stretching of line belts and degreasing of bearings (Imholte and Imholte-Tauscher, 1999).

Advances in building and equipment designs have greatly reduced many of the adverse effects of high temperatures on structures (Heaps, 1994; Imholte and Imholte-Tauscher, 1999; Wright *et al.*, 2002; Mahroof *et al.*, 2003a, b; Roesli *et al.*, 2003; Dosland *et al.*, 2006). The introduction of methyl bromide in the 1940s as a structural fumigant made the use of heat treatments almost obsolete. There is now a renewed interest in exploring heat treatments as alternatives to methyl bromide because of the impending phase-out of this fumigant (Makhijani and Gurney, 1995), although some critical uses of it (e.g. quarantine) may be retained for several years.

8.3 Effects of High Temperatures on Stored-product Insects

As temperature increases above the optimum of about 32°C for population growth, insect response goes through three stages (Fields, 1992; Dosland *et al.*, 2006). In the first stage (40–45°C), egg laying declines and ultimately halts, hatching and eclosion become difficult to complete, and with declining fecundity and shorter adult lifespan, the population starts to die out. In the second stage (45–55°C), individuals survive for several hours, experiencing severe water stress. High humidity (> 50% RH) can greatly extend survival in both of these stages. In the last stage, > 55°C, there is rapid mortality, where the entire population is dead in minutes to seconds. Disinfestation of structures and equipment generally makes use of temperatures associated with the second and third stages of mortality, while rapid disinfestation of bulk grain employs temperatures > 60°C (see Table 8.1).

The mortality of an insect stage at high temperatures is a function of temperature and exposure time. At any given temperature, mortality increases with an increase in exposure time and, at any given exposure time, mortality increases with an increase in temperature (Wright *et al.*, 2002; Mahroof *et al.*, 2003b; Boina and Subramanyam, 2004). High temperature causes a number of adverse biochemical changes in insects: lower ion concentrations (e.g. pH), inactivation of major glycolysis enzymes, disruption of plasma membranes and denatured proteins, nucleic acids, lipids and carbohydrates (Hochachka and Somero, 1984; Denlinger and Yocum, 1999; Neven, 2000).

High temperature can also have sublethal effects, such as reduced movement, fecundity and progeny survival. For example, the exposure of *T.*

castaneum pupae to high temperature prevented development to adults (Saxena *et al.*, 1992), or resulted in adults with separated elytra (A. Menon and Bh. Subramanyam, unpublished data, 2000), possibly due to chromosomal aberrations (Denlinger and Yocum, 1999). The growth and development of the mealworm, *Tenebrio molitor* (L.), and *T. confusum* and *E. kuehniella*, are also adversely affected by high temperature (Adler and Rassmann, 2000). *T. castaneum* exposed as pupae or adults had reduced fecundity, egg-to-adult survival and hence progeny production, even if only one of the mating pair had been exposed to heat (Mahroof *et al.*, 2005a).

8.4 Heat Tolerance in Stored-product Insects

Insect heat tolerance, or the ability to withstand high temperatures, is affected by, among other things, species, insect age, developmental stage and thermal acclimation (Dermott and Evans, 1978; Evans, 1981; Fields, 1992; Hallman and Denlinger, 1999). Insects exhibit physiological and biochemical adaptations to overcome injury caused by thermal stress (Evans, 1981), and are able to withstand some high temperatures because of induction or expression of heat shock proteins (HSPs), which protect cells – and consequently the organism – from heat stress by preventing aggregation or improper folding of proteins and resolubilizing and stabilizing proteins by targeting denatured proteins for degradation and removal (Currie and Tufts, 1997).

HSP 70s are constitutively present in all life stages of *T. castaneum*; however, the increased thermotolerance of young larvae (first-instars) is due to the relatively high levels of HSP 70s present when compared with the other stages. Time- and temperature-dependent expression of HSP 70 showed that the increased heat tolerance in young larvae lasted as long as 8 h at 40°C or 30 min at 46°C (Mahroof *et al.*, 2005b). Therefore, to kill young larvae of *T. castaneum*, heat treatments should target temperature and time combinations beyond those thresholds for heat tolerance.

Although insects are able to withstand high temperatures for brief periods of time, the temperatures used during structural heat treatments (50–60°C for 24–36 h), for example, kill even the most thermotolerant stages. Nevertheless, identifying species and stage-specific heat tolerance is important for determining the most heat-tolerant species of stored-product insects associated with structures and for identifying the most heat-tolerant stage of a species (e.g. flour beetles).

Using the minimum temperature–time combinations that ensure complete kill of the most heat-tolerant species or stage of a species will also ensure complete kill of the less heat-tolerant species or stages. To verify the effectiveness of structural heat treatments, cages or test arenas strategically placed within the heated structure and sampled at regular intervals of time should be used for the most heat-tolerant species and stages.

8.5 Survey of Current Thermal Kinetic Data: Empirical Methods and Common Models

Mortality Response Data

Fields (1992) conducted an extensive survey of the mortality data for stored-product insects exposed to moderate to high temperatures. The survey recorded data on 11 insect species, taking into consideration the following: (i) developmental stage; (ii) population strain; (iii) whether or not acclimation had occurred before treatment; (iv) heating method; and (v) RH. These factors were clearly shown to influence the magnitude of mortality responses to temperature over time.

Large data sets can often help in evaluating the influence that these factors have on mortality response. Some of these sets have been modelled to improve their predictive ability, such as:

- Developmental stages of *S. granarius* from 45 to 60°C at 13% grain moisture (Bruce *et al.*, 2004).
- Developmental stages of *T. castaneum* from 42 to 60°C at 22% RH (Mahroof *et al.*, 2003b).
- Developmental stages of *T. confusum* from 46 to 60°C at 22% RH (Boina and Subramanyam, 2004).
- Eggs of three species of psocids from 43 to 51°C at 70% RH (Beckett and Morton, 2003a).
- Developmental stages of *R. dominica* from 50 to 60°C at 12% grain moisture for four rates of heating (Beckett and Morton, 2003b).
- Large larvae of *T. variabile* from 50 to 56°C at 0% RH (Wright *et al.*, 2002).
- Developmental stages of *R. dominica* from 45 to 53°C and *S. oryzae* from 42 to 48°C at 9, 12 and 14% grain moisture for both species (Beckett *et al.*, 1998; Table 8.2).

Other data sets include:

- Adult *C. ferrugineus* from 45 to 50°C at 75% RH (Jian *et al.*, 2002).
- Developmental stages of *L. serricone* and *R. dominica* from 45 to 55°C at 14% grain moisture (Adler, 2002).
- Diapausing larvae of *E. elutella* from 40 to 45°C at 65% RH (Bell, 1983).

The great advantages that the continued acquisition of additional data ultimately bring are better targeted treatments based on good predictive models of mortality and the possibility that a better understanding of the processes at play brings new approaches to engineering solutions and treatment practices that are more cost-effective.

Table 8.2. LT 99s for a range of stored-product insect species predicted from several large data sets (Beckett et al., 1998; Wright et al., 2002; Beckett and Morton, 2003a, b; Mahroof et al., 2003a, b; Bruce et al., 2004; Boina and Subramanyam, 2004).

Species	Develop-ment stage	Moisture	Heating method	42	43	44	45	46	47	48	49	50	51	52	53	54	55	56	57	58	59	60
L. bostrichophila	e	70% rh	conductive		38.88	25.3	15.84	9.48	5.39	2.89	1.45	0.67	0.28									
L. decolor	e	70% rh	conductive		40.41	33.22	26.34	19.92	14.14	9.19	5.27	2.50	0.87									
L. paeta	e	70% rh	conductive				96.16	16.26	4.49	1.69	0.79	0.42	0.26									
R. dominica	5	12% mc	convective									6.84	3.45	1.75	0.90	0.46	0.24	0.12	0.07	0.03	0.02	0.01
	4	12% mc	convective									8.70	3.88	1.80	0.87	0.43	0.22	0.11	0.07	0.04	0.02	0.01
	3	12% mc	convective									14.29	5.96	2.61	1.20	0.58	0.29	0.15	0.08	0.04	0.03	0.01
	2	12% mc	convective									13.93	5.73	2.49	1.13	0.54	0.27	0.14	0.07	0.04	0.02	0.01
	1	12% mc	convective									15.71	6.18	2.58	1.14	0.53	0.26	0.13	0.07	0.04	0.02	0.01
	5	9% mc	conductive				49.09	31.03	19.37	11.95	7.27	4.36	2.58	1.50	0.86							
	4	9% mc	conductive				60.44	38.99	24.73	15.41	9.43	5.65	3.32	1.91	1.07							
	3	9% mc	conductive				55.05	37.99	25.65	16.91	10.85	6.77	4.09	2.39	1.34							
	2	9% mc	conductive				49.11	33.04	21.96	14.41	9.33	5.96	3.75	2.32	1.42							
	1	9% mc	conductive				41.59	29.62	20.74	14.26	9.61	6.34	4.09	2.57	1.57							
	5	12% mc	conductive				52.24	33.60	21.37	13.43	8.34	5.11	3.09	1.85	1.09							
	4	12% mc	conductive				60.89	39.53	25.28	15.91	9.85	5.99	3.58	2.09	1.20							
	3	12% mc	conductive				70.8	48.91	33.10	21.92	14.17	8.92	5.46	3.24	1.85							
	2	12% mc	conductive				62.9	42.86	28.83	19.14	12.52	8.07	5.12	3.19	1.96							
	1	12% mc	conductive				46.93	34.05	24.3	17.04	11.72	7.89	5.20	3.34	2.09							
	5	14% mc	conductive				54.32	35.32	22.72	14.45	9.09	5.65	3.46	2.10	1.25							
	4	14% mc	conductive				61.15	39.84	25.60	16.20	10.10	6.19	3.73	2.21	1.28							
	3	14% mc	conductive				82.21	56.8	38.49	25.54	16.56	10.48	6.46	3.86	2.24							
	2	14% mc	conductive				76.09	52.38	35.58	23.83	15.73	10.22	6.53	4.10	2.52							
	1	14% mc	conductive				51.04	37.52	27.14	19.30	13.46	9.20	6.15	4.01	2.55							
S. granarius	ol/p	~13% mc	convective				78.53			2.33		0.51					0.06		0.01			<0.01
			conductive				3.25			1.58		0.52					<0.01					

LT$_{99}$ (in hours), Temperature (°C)

Species	Stage			V1	V2	V3	V4	V5	V6	V7	V8	V9	V10	V11	V12	V13
S. oryzae	a	12% mc	conductive	13.42	8.26	5.06	3.78	2.39	1.32	0.64						
	5	12% mc	conductive	20.50	12.55	7.61	4.58	2.73	1.61	0.94						
	4	12% mc	conductive	25.98	16.95	10.70	6.50	3.79	2.10	1.10						
	3	12% mc	conductive	26.31	18.95	13.09	8.59	5.30	3.03	1.57						
	2	12% mc	conductive	25.63	18.05	12.52	8.53	5.71	3.74	2.40						
	1	12% mc	conductive	14.23	10.69	7.72	5.31	3.45	2.08	1.14						
T. castaneum	a	22% rh	convective	70.58	8.29	0.93	0.84	0.75	0.68	0.61	0.55	0.49	0.44	0.40	0.36	0.32
	p	22% rh	convective	83.70	13.01	1.45	1.21	1.01	0.84	0.69	0.58	0.48	0.40	0.33	0.28	0.23
	ol	22% rh	convective	76.32	9.19	1.26	1.16	1.07	0.98	0.90	0.83	0.76	0.70	0.64	0.59	0.54
	yl	22% rh	convective	47.50	7.18	7.21	5.22	3.78	2.74	1.98	1.43	1.04	0.75	0.54	0.39	0.29
	e	22% rh	convective	21.22	10.60	1.76	1.42	1.14	0.92	0.74	0.59	0.48	0.38	0.31	0.25	0.20
T. confusum	a	22% rh	convective	3.56	2.56	1.20			0.96			0.57				0.25
	p	22% rh	convective	4.19	2.54	0.81			0.43			0.20				0.17
	ol	22% rh	convective	4.99	2.94	1.50			0.93			0.64				0.40
	yl	22% rh	convective	3.14	2.25	0.74			0.26			0.20				0.10
	e	22% rh	convective	3.48	1.95	0.69			0.27			0.20				0.11
T. variabile	ll	0% rh	convective	3.65	0.41	0.19	0.13	0.09	0.07	0.06	0.05	0.04	0.05	0.04	0.04	

a: adult, p: pupal, ol: old larval, yl: young larval, e: egg 1-5: arbitrary immature development stages from egg to pupa

Factors affecting insect mortality

Species

Several species of insects may be found concurrently in grain and associated structures. Knowing which of them is least susceptible to heat will help in targeting treatments against the species most difficult to kill: Closely related species, or species with similar life histories, vary greatly in heat tolerance. In Fig. 8.4, the treatment required for 99.9% mortality of the most heat-tolerant development stage of *R. dominica* is compared with that of *S. oryzae* and three species of Psocoptera (Beckett and Morton, 2003a).

The two beetle species, though of different genera, have similar behaviours because both are internal developers. Nevertheless, eight times more treatment is required for *R. dominica* to reach the same level of mortality as *S. oryzae* across the same range of temperatures. The psocid species, *Liposcelis bostrichophila*, *L. paeta* and *L. decolor*, are of the same genus; while they are

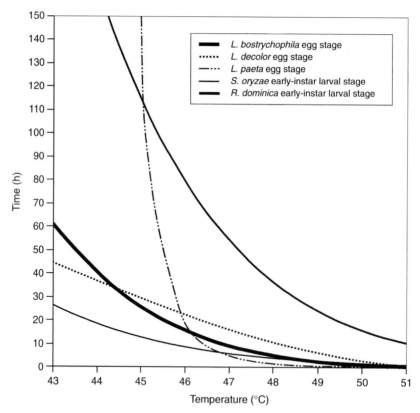

Fig. 8.4. A comparison of the times required to achieve $LT_{99.9}$ at a range of temperatures from 43 to 51°C for the most heat-tolerant stages of three species of Psocoptera, *Liposcelis bostrichophila*, *L. paeta* and *L. decolor*, and two species of Coleoptera, *Rhyzopertha dominica* and *Sitophilus oryzae*. RH was at 70% except for *S. oryzae*, where it was at 55% (from Beckett and Morton, 2003a).

far more susceptible than *R. dominica*, there is also considerable variation in heat-tolerance among them.

Liposcelis paeta, which survives at temperatures greater than most stored-product insects (Rees, 1998), is even more tolerant than *R. dominica* between 43 and 45°C. At 45°C, where times to mortality are still quite long, *R. dominica* takes roughly five times longer to kill than the most tolerant psocid. However, at 51°C, where times to mortality are short, it takes between nine and 36 times longer to kill than the psocids. *R. dominica* is thought to be the most heat-tolerant stored-product species (Dermott and Evans, 1978), and therefore is used as the reference point for maximum treatment conditions.

Roesli *et al.* (2003) gauged structural heat treatment effectiveness against insect species by examining insect numbers before and after heat treatment of a pilot feed mill, using commercial food-baited traps for several species of beetles and pheromone traps for *P. interpunctella*. The lack of or low numbers of captures of *L. serricorne* and *P. interpunctella* in commercial traps indicated that these species were extremely susceptible to elevated temperatures, while the effectiveness against *T. castaneum* did not last more than 4 weeks. Very limited data are currently available on possible reasons for reinfestation of insect species following an IPM intervention, including heat treatment. Understanding these reasons will help extend the duration of insect suppression obtained following a heat-treatment intervention.

Insect developmental stage

Several studies have shown that the heat tolerance of insects varies with developmental stage (Evans, 1981; Vardell and Tilton, 1981a; Fields, 1992). Beckett *et al.* (1998) quantified and compared variation between five arbitrary developmental stages of the internal developers *R. dominica* and *S. oryzae* over a range of temperatures. For example, in wheat at 14% MC, the time required to achieve 99.9% mortality of young larvae (the most heat-tolerant stage) of *R. dominica* required about 1.5 times more exposure than did old larvae and pupae of the same species at 45°C, and 2.5 times more exposure at 53°C. Because exposure time generally increases exponentially with a decrease in temperature, this finding indicates that the tolerant stage required about 45 h more treatment at 45°C and 2.5 h more at 53°C than did the susceptible stages.

This phenomenon is further illustrated by comparing the mortality responses of psocid eggs and adults (Beckett and Morton, 2003a). For example, at 47°C and 70% RH, the time required for 99.9% mortality of eggs of *L. bostrychophila* and *L. decolor* is 7.4 and 19.4 h, respectively, while that of adults is 1.7 and 3.3 h, respectively.

The relative heat tolerance among developmental stages can vary between closely related species (Mahroof *et al.*, 2003b; Boina and Subramanyam, 2004). Mahroof *et al.* (2003b) showed that young larvae (first-instars) of *T. castaneum* were generally more heat-tolerant at 50–58°C when compared with eggs, old larvae, pupae and adults, whereas Boina and Subramanyam (2004) found that the old larvae of *T. confusum* were more heat-tolerant than eggs, young larvae, pupae, and adults at 46–60°C.

Moisture

As mentioned above, environmental moisture at moderately high temperatures has a substantial impact on the chances of survival of stored-product insects. Environmental moisture refers to atmospheric water or, at the product level, the interstitial atmospheric moisture, which is chemically bound to the product and the moisture adsorbed by the product. Beckett *et al.* (1998) measured the impact of grain MC on the mortality response of *R. dominica* over a range of temperatures and found mortality to be less in grain at high MC than in grain at low MC at all temperatures from 45 to 53°C. For example, the increase in treatment time was 1.5 times greater for 99.9% mortality on 14%-moisture wheat when compared with 9%-moisture wheat. In other words, at 45°C the treatment takes 37 h longer for 99.9% mortality and, at 53°C, 1.7 h longer.

During heat treatment of structures, RH drops as temperature increases (Mahroof *et al.*, 2003a; Roesli *et al.*, 2003) and, at temperatures of 50°C and above, the RH during most of the heat treatment is around 20–25%. In unreplicated trials conducted during heat treatment of a flour mill (R. Roesli and Bh. Subramanyam, unpublished data, 2002), survival of *T. castaneum* adults occurred at higher humidities (> 60%), especially at temperatures < 50°C. However, at temperatures > 50°C, all adults were killed after 60 min of exposure irrespective of the humidity level. Temperatures for effective structural heat treatments should be ≥ 50°C for at least 24 h or more. Hence, the importance of insect survival due to high humidity levels should not be an issue during structural heat treatments.

Rate of heating

According to Evans (1987) and Sutherland *et al.* (1989), the rate of heating to a target temperature influences the rate of insect mortality. Beckett and Morton (2003b) found that, under particular conditions, this effect can be sizeable. Using hot air at different temperatures to produce different heating rates, the time required for 99.9% mortality of *R. dominica* at grain temperatures of 50–60°C nearly halved as the air temperature was increased from 80 to 100°C. However, as the rate of heating was increased further, the rate of reduction in treatment time decreased rapidly (see Fig. 8.5). In this case, the rate of convective heat penetration and the degree of homogeneous mixing may have been limiting factors.

Mahroof *et al.* (2003a, b) and Roesli *et al.* (2003) found that the rate of heating different areas of structures varied during structural heat treatment, from as low as 0.3°C/h up to 14°C/h for different floors of a food-processing facility. The differences in heating rates are related to horizontal and vertical stratification of temperatures. More work is needed to determine whether heating rate is an important factor that influences insect mortality in heat treatments due to limited data available.

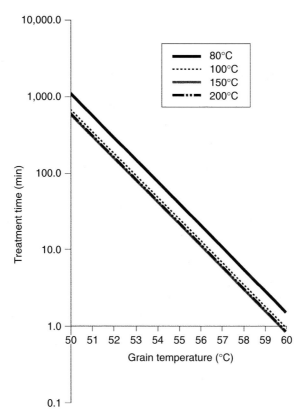

Fig. 8.5. The times required to achieve $LT_{99.9}$ at a range of rates of heating (80–200°C) based on air inlet temperature for a heat-sensitive immature stage of *Rhyzopertha dominica* (from Beckett and Morton, 2003b).

Empirical Methods

Obtaining temperature/mortality data

It is essential, when determining the heat tolerance of a species, that its developmental stages are considered individually, either specifically – which is possible with external developers – or arbitrarily, which is more practical with internal developers. Radiographic techniques can now be used to determine more accurately the age of stages developing within grain kernels (Sharifi and Mills, 1971). Heat should be generated and controlled with precision, and the temperatures measured in a way that is fast, reliable and accurate. The difference of one degree can have a substantial impact on the level of mortality. For example, for *R. dominica* at 14% moisture, the time difference for 99.9% mortality is 2.4 h between 52 and 53°C, 8.4 h between 49 and 50°C and 35.2 h between 45 and 46°C (Beckett *et al.*, 1998).

The method of heat generation is determined by the purpose of the study and the research objectives, be it commodity, storage facility, insect species, rate of heating, treatment temperature, microwave attenuation or infrared treatment. Various systems have been used to investigate rapid disinfestation of bulk grain: (i) fluid bed (Evans, 1981; Evans and Dermott, 1981; Vardell and Tilton, 1981b); (ii) spouted bed (Claflin *et al.*, 1986; Beckett and Morton, 2003b; Qaisrani and Beckett, 2003a); and (iii) pneumatic conveyor (Dzhorogyan, 1957; Fleurat-Lessard, 1980; Sutherland *et al.*, 1989) (see Figs 8.6a–e).

When considering heat dosage rates at moderate temperatures where treatment periods are protracted, controlled-environment cabinets or incubators are sufficient (Vardell and Tilton, 1981a). Beckett *et al.* (1998) combined the use of incubators with sealable containers to maintain constant MC However, the containers included preheated heat sinks to rapidly heat insect-infested grain to the target temperature. Another approach is to place insects confined in vials or bags into temperature-controlled oil baths (Thoroski *et al.*, 1985) or water baths (Bruce *et al.*, 2004).

Other systems rely on methods of heating such as electromagnetic energy in the form of micro- and radiowaves (Nelson and Whitney, 1960; Whitney *et al.*, 1961; Watters, 1976; Plarre *et al.*, 1997; Halverson *et al.*, 1998; Fig. 8.6f) and infrared radiation (Tilton and Schroeder, 1963). Yet another alternative is the high-temperature/short-time technique described by Mourier and Poulsen (2000), which uses equipment called a miroline toaster®, where hot air (150–750°C) is passed through a rotating drum. The grain, which is further heated by infrared radiation from the drum wall, is mixed as it passes through the drum (see Fig. 8.6g).

Another system uses a counter-flow heat exchange process, where grain is augered in the opposite direction to hot water in a surrounding jacket (Lapp *et al.*, 1986; Fig. 8.6h). Some of the technologies mentioned above are described in more detail later in this chapter. For heat treatment of structures, gas, electric or steam heaters can be used, which come with various heating capacities based on need. Equipment for structural treatments will also be described in more detail later.

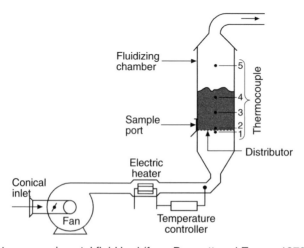

Fig. 8.6a. Batch-type experimental fluid bed (from Dermott and Evans, 1978).

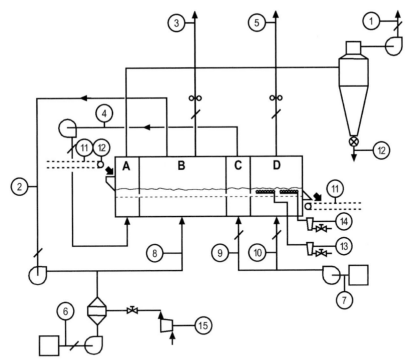

Fig. 8.6b. Continuous-flow fluid-bed process flow diagram (from Thorpe *et al.*, 1984). A, preheating and de-dusting; B, heating; C, cooling; D, evaporative cooling; 1, dust collection; 2, recirculated hot air; 3 and 5, exhaust gases; 4, heat recovery; 6, air intake for heating; 7, air intake for cooling; 8, hot air entry; 9 and 10, cool air entry; 11, grain conveyance; 12, grain inlet; 13 and 14, evaporative cooling system; and 15, LPG supply for burner.

Maintaining product quality and structural integrity

The maintenance of grain and stored-product quality, or the integrity of structures and equipment that experience a heat treatment, are of paramount importance. A heat treatment is not successful if these factors are compromised.

Ensuring product quality

Product quality is an issue when heat is used to disinfest grain, but not during the heat treatment of structures. In the latter case, all food products within structures should be removed to prevent insects from seeking refuge in these products and surviving the treatment. It is good practice to fumigate the removed products so that, if infested, they do not contribute to reinfestation of structures after the heat treatment.

Commodities that are heat-treated may respond in different ways. Wheat is generally considered fairly resilient, but this may not be true for other grains. Grain MC has also been shown to affect quality response to heat treatment

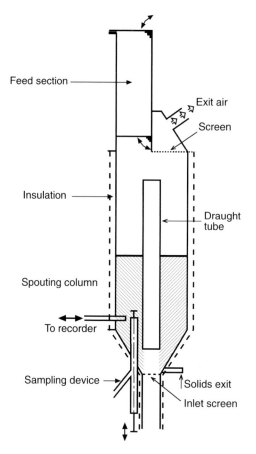

Fig. 8.6c. Batch-type experimental spouted bed (from Claflin *et al.*, 1986).

(Ghaly and Sutherland, 1984; Nellist and Bruce, 1987). An evaluation of germinative capacity provides a relatively quick initial measure of quality deterioration, but to identify any subtle modifications in grain composition, trials, such as dough and baking tests for wheat, micro-malting for malting barley and oil quality analysis for canola, are needed (Ghaly *et al.*, 1973; Dermott and Evans, 1978; Ghaly and Sutherland, 1984; Armitt and Way, 1990).

Dough and baking tests measure parameters such as flour yield, water absorption, dough development time, resistance to extension, extensibility, viscosity and bread volume. Micro-malting analysis measures parameters such as total extractable nutrient, wort colour, total nitrogen, soluble nitrogen, Kolbach index, wort viscosity, apparent attenuation limit, beta-glucan content, diastatic power and percentage malt yield (European Brewery Convention, 1998).

While germination or malt yield may not appear to be affected by heat treatment, other parameters may exist that could have a negative impact on

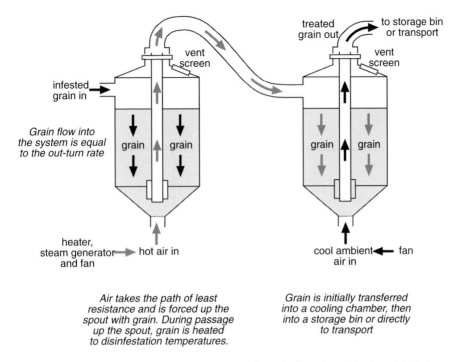

Fig. 8.6d. Continuous-flow spouted bed (adapted from Qaisrani and Beckett, 2003a).

brewing quality. Oil content analysis measures percentage oil, percentage free fatty acids, peroxide value, uric acid, glucosinolates, iodine value and percentage protein (Firestone, 1990). There may also be implications for quality maintenance under long-term storage.

Ensuring structural and equipment integrity

Data on the effects of heat treatments on structures are anecdotal; tests have not yet been developed or proposed to determine the effects on equipment and food-grade materials (gaskets, sifters, paints, lubricants, plastics, electronics, etc.) used in food-processing facilities. This is an area that requires further study.

During heat treatments of pilot flour and feed mills, no adverse effects were observed on the structures and no malfunction of the mill equipment was observed after multiple treatments (Mahroof *et al.*, 2003a; Roesli *et al.*, 2003). However, there have been sporadic reports of adverse effects such as overheating of certain areas due to improper air movement and temperature monitoring, and warping of metal or plastic following a heat treatment. Structural and equipment integrity can be ensured if proper planning and precautions are taken before, during and after a heat treatment (Dosland *et al.*, 2006).

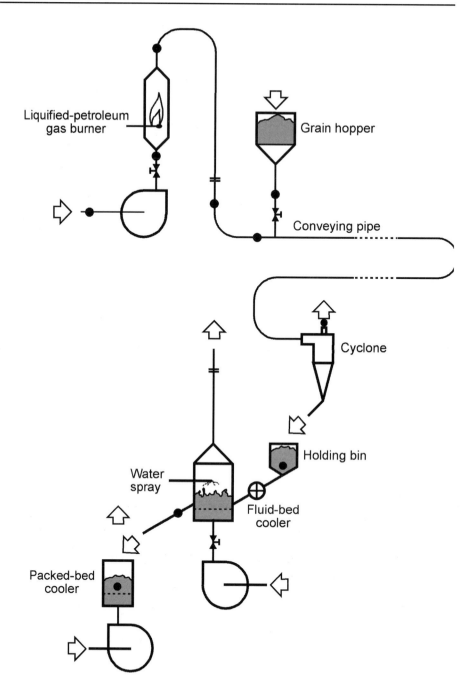

Fig. 8.6e. Pneumatic conveyor process flow diagram (from Sutherland *et al.*, 1989)

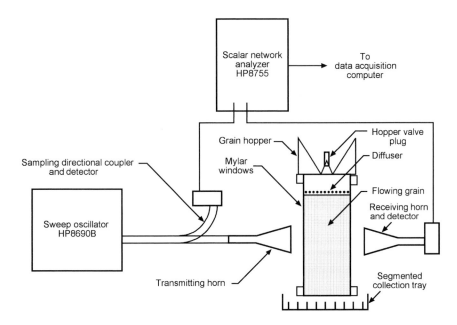

Fig. 8.6f. Microwave equipment for one-way path attenuation tests (from Halverson *et al.*, 1998).

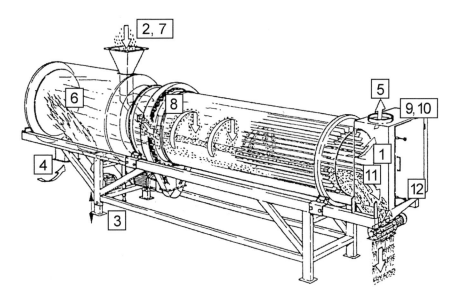

Fig. 8.6g. HTST Miroline toaster® (from Mourier and Poulsen, 2000). 1, air flow; 2, grain inlet; 3, drum inclination; 4, secondary air intake valve; 5, air outlet valve; 6, combustion air thermocouple; 7, material flow sensor; 8, drum rotation guard; 9, thermocouple for outlet air; 10, safety thermostat; 11, pressure guard; and 12, material overflow guard.

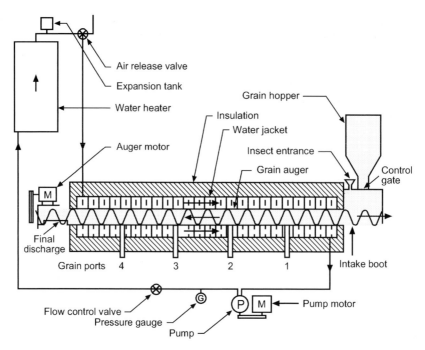

Fig. 8.6h. Counterflow heat exchanger (from Lapp *et al.*, 1986).

Simulation Models

Statistical transformations

Predictive models commonly rely on experimentation where mortality has been determined over a combination of temperatures (usually constant) and exposure times at a constant RH or grain MC. Obtaining sufficient data can be laborious, particularly when high levels of mortality are occurring, but such data are essential if meaningful predictions are to be made. Unfortunately, as mortality increases, the accuracy of predictions decreases and often a level of 99.0–99.9% mortality is considered a good compromise.

Predictions of 50% mortality are the most accurate and thus useful for comparative purposes. The estimation of 95% confidence limits is also useful for comparative purposes, but these expand as the level of mortality increases. They too should be considered with suspicion > 99% mortality, mainly because certainty about the scale of the statistical transformation being used decreases concomitantly.

Probit analysis is a satisfactory means of determining estimates and confidence limits of mortality, giving the linearized probit transformation of sigmoid distributions (Claflin *et al.*, 1986; Evans, 1987; Bruce *et al.*, 2004), which has been frequently used for evaluating lethal dosages of pesticides (Finney, 1971). A modified form of this analysis uses the inverse standard normal deviate (Wright *et al.*, 2002). Logit and complementary log–log transformations may at times fit the data better (Morgan, 1992), especially if mortality is being predicted as a function of temperature or time.

The difference between the three transformations is in the tails, or the first and last 10% of prediction. For heat treatments, interest rests in the upper tail because, the longer the tail, the longer the treatment time needed to kill the last of the population. A probit transformation gives the shortest upper tail, while those of logit and complementary log–log distributions are similar. However, this does not mean that the probit fit will automatically predict treatments for high mortality that are less than those from the other fits because, based on the distribution of the data, the whole curve can shift and the slope may change in an effort to obtain the overall best fit, including the tails.

Based on R^2 values, Beckett et al. (1998) found probits gave the best fits for R. dominica and S. oryzae over the conditions employed, while Beckett and Morton (2003a) found one or other of the other transformations gave better fits for three species of Psocoptera. Mahroof et al. (2003b) and Boina and Subramanyam (2004) preferred a complementary log–log transformation for estimating LT_{99} values for various life stages of T. castaneum and T. confusum exposed to several constant elevated temperatures.

Mortality models

Several models have been applied to temperature–mortality data. Tsuchiya and Kosaka (1943) found that, as temperature increased within the range 60–100°C, insect mortality was related to temperature via the hyperbola:

$$t_{x\%}(T - c) = K \tag{8.1}$$

where c and K are constants, $t_{x\%}$ is the treatment time to a given percent mortality, and T is the treatment temperature. Under circumstances of rapid convective heating, the inlet air temperature can be related to treatment temperature and time for a given mortality (Dermott and Evans, 1978; Fig. 8.7), with the linear equivalent being more convenient:

$$1 / t_{x\%} = a + bT \tag{8.2}$$

where $t_{x\%}$ is the treatment time to a given per cent mortality, T is the treatment temperature and a and b are the intercept and coefficient of a linear regression equation, respectively.

Alternatively, Fleurat-Lessard (1985) used the following equation:

$$T_{max} - T = ab^t x\% \tag{8.3}$$

where T_{max} is the inlet air temperature.

The Arrhenius equation, which describes the affect of temperature on the rate of chemical reactions, has been used to model mortality data from cold treatments (Brokerhof et al., 1992; Banks and Fields, 1995). This model is described as:

$$1/t_{x\%} = Ac^{-b/T} \tag{8.4}$$

where A is a constant.

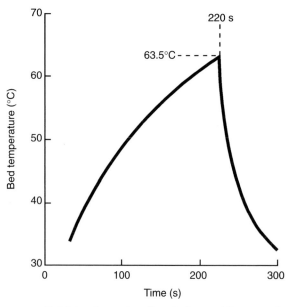

Batch heating and cooling of wheat to kill stages of *Rhyzopertha dominica*.

Air inlet temperature – 80°C

Air flow rate – 14.5 l/kg/s

Fig. 8.7. Heating and cooling curve for wheat exposed to 80°C for 220 s with an air flow of 14.5 l/kg/s at 1.6 m/s (from Dermott and Evans, 1978).

Mortality can also be modelled as a function of a temperature–time product that accrues above an arbitrary threshold temperature (Dermott and Evans, 1978; Banks and Fields, 1995; Wright *et al.*, 2002). This is similar to the concentration–time product concept used with fumigation and is described as:

$$t_{x\%}(T - T_0) = k \tag{8.5}$$

where $t_{x\%}$ is the treatment time for a given percentage mortality, T is the treatment temperature, T_0 is the threshold temperature and k is the accrued product constant. With this approach, changing temperature effects can be integrated and any statistical transformation can be used via:

$$y = a + bk \tag{8.6}$$

where y is the transformation corresponding to an observed mortality (Wright *et al.*, 2002). Wright *et al.* (2002) used this degree–minute approach for predicting the mortality of large larvae of *T. variabile* by obtaining time–mortality data at four constant temperatures between 50 and 56°C. The base temperature for accumulating degree–minutes, the intercept and slope of the linear regression of mortality (expressed as the inverse of the standard normal deviate) and the degree–minutes were different at each of the four temperatures. Despite these differences, Wright *et al.* (2002) pooled the data across 52, 54 and 56°C to describe the relationship between degree–minutes and mortality of *T. variabile* larvae. No statistical or biological basis was given for pooling the data across the three temperatures.

Subramanyam *et al.* (2003) developed a simple heat-accumulation model for predicting the mortality of first-instars of *T. castaneum* based on time–mortality data collected at six constant temperatures between 42 and 60°C (Mahroof *et al.*, 2003a). Independent data on first-instars, collected at the same constant temperatures, were used to validate the model. The base temperature for accumulating degree–minutes was 49.1°C and the model underestimated mortality by 25%, but explained about 70% of the variation in observed mortality of insects as a function of both temperature and time. These models need to be refined and validated under field conditions for predicting incremental mortality of insects at dynamically changing temperatures and times that occur during commodity or structural heat treatments.

Beckett *et al.* (1998) developed a model to cater for a range of grain temperatures and MC, and where control mortality is unknown, such as when immature stages are hidden within grain kernels or eggs are buried in substrate (Wadley's Problem). In this model, the expected survival of a given treatment is multiplied by the probability P, which is modelled by a linear predictor η on the scale of a given transformation (probit, logit, complementary log–log).

The confidence limits for the lethal time estimates (LTs) at a given percentage mortality (i.e. the time range in which one is confident that a treatment will give a certain level of mortality, say, 95% of the time) are obtained by a linear Taylor expansion of ln (LT). Thus, the model centres on the mid-point of the range of temperatures and grain *MC*, and a given LT is calculated as follows:

$$LT = \exp((\eta - \beta_1 - \beta_3(T - T_{mid}) - (\beta_5 + \beta_6(T - T_{mid}))$$
$$(MC - MC_{mid}))/\beta_2 + \beta_4(T - T_{mid}) + \beta_7(MC - MC_{mid}))) \tag{8.7}$$

where β_1 is the intercept, β_2 is the time coefficient, β_3 is the temperature coefficient and β_4 is the product term of time and temperature interactions. If there is only one moisture parameter, the last three regression coefficients are removed and the equation is simplified as:

$$LT = \exp((\eta - \beta_1 - \beta_3(T - T_{mid})) / \beta_2 + \beta_4(T - T_{mid})) \tag{8.8}$$

A novel model for predicting survival of heat-tolerant old or late-instar larvae of *T. confusum* during structural heat treatments was developed and validated at Kansas State University (Boina, 2004). The final model has the following form:

$$N_t = \frac{N_o}{10^{\left(\sum_0^t \frac{\Delta t}{D(T_t)}\right)}} \tag{8.9}$$

where N_t is the insect population after time t, N_0 is the initial insect population, Δt is the incremental exposure time (usually 1 min), $D(T_t)$ is the mean instantaneous value of D as a function of temperature and T_t is the time-dependent temperature profile. *D*-value, in microbiological applications, is the time required for one log reduction in population.

The model was developed for heat-tolerant *T. confusum* old larvae (Boina and Subramanyam, 2004) based on data at constant temperatures between 46 and 60°C, validated for the species in a structural heat treatment, and found to be satisfactory in describing insect survival at nine different heating rates. Figure 8.8 shows model predictions at two heating rates.

8.6 Current Status of Research and Development in Heat Disinfestation of Stored Products

Many approaches to heat disinfestation of stored products have been developed and advanced over the last 50 years, with technologies borrowed from a range of industries and governed by some combination of the physical principles of

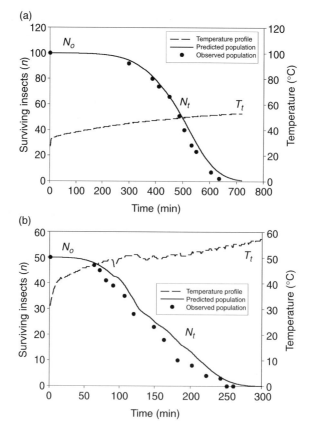

Fig. 8.8. Observed and predicted survival of *Tribolium confusum* old larvae during heat treatment at 2.44 °C/h (a) and 5.50 °C/h (b) at Kansas State University feed mill (from Boina, 2004).

mechanics, fluid dynamics, thermodynamics, heat and mass transfer and electromagnetic radiation. Several of these approaches are discussed below.

Convective Heating

Fluid bed

A fluid bed for the purpose of heat disinfestation is comprised of a column of grain supported by a perforated distributor plate. Air is passed through the distributor at a velocity sufficient to suspend, separate and mix the grain, which is called fluidization. Air at high temperatures can therefore be used, allowing rapid and uniform heat transfer and precise temperature treatment of the grain mass (see Fig. 8.6a, b). This technology serves both as an experimental tool to study heat dose mortality and a potential practical means of heat disinfestation (Dermott and Evans, 1978; Lim et al., 1978; Fleurat-Lessard, 1980; Evans, 1981; Evans and Dermott, 1981; Vardell and Tilton, 1981a, b; Evans et al., 1983).

The rapid disinfestation of large grain volumes requires a continuous flow process that depends on consideration of longitudinal dispersion of grain through a bed during treatment; hence, the residence time of the grain in the bed must be sufficient to achieve optimum disinfestation temperature (Evans et al., 1983). The same principle is used to cool the grain, either with ambient air or air in combination with water spray for evaporative cooling. Relative to the other convective systems, cooling is considered easy and efficient.

There are several commercial prototypes, from a capacity of 10 t/h to the largest at 150 t/h. The latter was built and trialled in Victoria, Australia in the 1980s (Thorpe et al., 1982, 1984; Evans et al., 1984; Fig. 8.9). At full capacity, the plant achieved complete disinfestation of R. dominica in wheat at a air flow rate of 2.1 kg/s/m^2, an air inlet temperature of 230°C, a grain residence time of 2.2 min and a grain temperature of 70°C. Wheat quality as determined by dough and baking tests was unaffected.

At an initial grain temperature of 25°C, the energy requirement was 1.13 kWh of electricity and 26.6 kWh of gas/t. Heating efficiency was increased by recirculating hot air from the top of the bed back into the heating system, which caused blockage of the distributor plate with entrained husks, largely overcome by using a recirculating fan as a centrifugal separator and inserting a cyclone in the recirculating duct.

Plans for a 500 t/h plant were considered, dependent on how hot the inlet air could be before affecting grain quality. A hybrid bed consisting of multiple hexagonal apertures to overcome the likelihood of blockage resulted (Gray and Darby, 1996). Capital costs and the cheaper cost of fumigation have halted further developments for the time being.

Fig. 8.9. Continuous-flow fluid bed heat disinfestation plant (150 t/h capacity) in Victoria, Australia.

Spouted bed

A spouted bed also uses hot air to lift, mix and heat grain. However, it differs from the fluid bed in that the air enters the column by passing at relatively high velocity through a nozzle at the bottom of a conical floor that causes the grain to 'spout' above it rather than 'fluidize' (see Fig. 8.6c, d). Thus, the grain moves in a more regular circular motion during heating, up the spout and down the annulus, rather than undergoing the bubbling motion of the fluid bed. The spouted bed also uses lower air velocity and operating pressures, and has better thermal economy (Mathur and Epstein, 1974).

Another advantage of the spouted bed is its ability to handle larger grains such as maize. It was initially developed in Canada for grain drying (Mathur and Gishler, 1955), but was subsequently investigated for its heat disinfestation potential (Claflin and Fane, 1981). Claflin *et al.* (1986) reported extensive trials with a 24 kg-capacity batch system, where the inclusion of a draught tube above the inlet nozzle gave more even heat treatment, which is essential for disinfestation.

The operating costs of the two systems are similar (Claflin *et al.*, 1986), and there is no inherent problem of air flow blockage by entrained debris. However, the difficulty with spouted beds is in scaling up to deal with large throughputs. This could partially be overcome by employing a multi-spouted system where a bed would have several cells, each up to a 1 m², with a throughput of 50–100 t/h operating in parallel and in series (Sutherland *et al.*, 1987). However, if total throughput requirements become too large (e.g. 400 t/h), the system becomes too complex for practical purposes.

In response to the commercial potential for farm-scale spouted bed use, a prototype continuous-flow system was constructed that can successfully operate at a rate of 8 t/h (see Fig. 8.10). The system consists of a heating and cooling chamber (0.75 × 0.75 × 1.5 m), each containing a draught tube to transport the grain out. Hot air at 220–250°C, heated via a LPG burner, enters the heating chamber at 0.25 m³/s. Grain rises in the spout and passes up the draught tube into the cooling chamber, while being heated to just above 60°C. The same process is then repeated with ambient air in the cooling chamber. Cost of treatment in 2002 was about US$0.90/t (Qaisrani and Beckett, 2003a).

Pneumatic conveying

This means of convective heating takes advantage of a common method of handling, except that the air used to transport the grain is heated. Disinfestation by pneumatic conveying has been considered for some time (Dzhorogyan, 1957; Fleurat-Lessard, 1980), including as an alternative to fluid- and spouted-bed processes (Sutherland *et al.*, 1986, 1989).

This involved a detailed assessment of a system composed of a 30 m long × 50 mm diameter conveying pipe with air supplied by a centrifugal fan and

Fig. 8.10. Prototype of continuous-flow spouted bed heat disinfester (8 t/h capacity).

heated by an 80 MJ/h capacity liquefied petroleum gas burner giving air temperatures up to 320°C and a grain throughput capacity of up to 1 t/h (see Fig. 8.6e). Grain was fed into the conveying pipe via a venturi feeder and removed from the air at the end via a high-efficiency cyclone. The grain was then cooled by either a column cooler or fluid bed (Sutherland *et al.*, 1986).

The advantages of pneumatic conveying are: (i) grain heating is very rapid and not constrained by air velocity as it is with fluidization; (ii) there is no dispersion of grain residence time that gives uneven treatment; and (iii) chaff and dust remain with the grain. A grain-holding stage can also be included before cooling, which allows for lower grain treatment temperatures. The disadvantages are: (i) conveying time may be too quick for grain temperature to equilibrate; (ii) grain temperatures are harder to measure and control; (iii) thermal efficiency is poor without air recirculation; and (iv) cooling as a separate process is inefficient (Sutherland *et al.*, 1987). Because of the short particle residence times and containment of chaff and dust, pneumatic conveying is also not practical for either grain drying or grain cleaning, and it remains unconfirmed for large grain throughputs.

Computer simulations of the three convective systems show that the energy costs of the fluid and spouted beds are proportional to grain flow rate, but slightly lower for the spouted bed. If inlet air temperature is increased from 100 to 300°C, there is a clear cost advantage, but a further increase to 500°C does not overcome the extra capital cost involved. Although the energy costs of the pneumatic conveyor system are larger than those of the other systems and are not proportional to flow rate, they decrease as throughput increases. None the less, the pneumatic conveyor is thermally less efficient.

Grain dryers

Much of the data on grain viability and quality come from work on grain drying and storage (Ellis and Roberts, 1980; Nellist, 1981; Dickie *et al.*, 1990). Probit, or the inverse standard normal deviate can be applied equally well to measuring decline in seed viability as insect mortality. Several studies have also assessed the feasibility of using grain dryers for heat disinfestation, as such equipment is often readily available and would negate capital cost constraints.

Bruce *et al.* (2004) conducted detailed trials with a popular type of mixed-flow dryer in the UK, for which good drying simulation models exist. Qaisrani and Beckett (2003b, c) assessed the potential of using a commonly used cross-flow and radial-flow dryer in Australia. Results generally point to uneven heating and less-than-perfect mixing of grain. However, high levels of insect mortality can be achieved, and easily identifiable modifications to existing equipment could improve results considerably.

Conductive Heating

Developed and used at the beginning of the century (Winterbottom, 1922), the Poole and Steel wheat heater (see Fig. 8.2) was noteworthy not only because it marked the beginning of interest in thermal energy as an industrial disinfestation process, but also because it relied on conductive heating, which has the potential to be thermodynamically more efficient than convective heating in terms of thermal load, driving force and heat contact. Moreover, energy is not wasted moving and mixing the grain because it moves by gravity and, by falling over a series of steam-heated metal rods in alternating alignment, good mixing is ensured. Issues of dust generation, condensation and combustion may also be less problematic. While cooling was not included in the Poole and Steel rig, a similar process to heating could be envisioned using coolant-filled rods refrigerated by a heat pump.

Electromagnetic Energy

Micro- and radiowaves

The use of microwaves (MW) (Watters, 1976; Halverson et al., 1997, 1998; Plarre et al., 1997) and radiowaves (RF) (Nelson and Whitney, 1960; Whitney et al., 1961; Nelson and Kantack, 1966) for heat disinfestation of grain has been extensively researched, mainly as a more efficient method of eradicating insect pests. Nelson (1973, 1987, 1996) evaluated the potential of this method of generating thermal energy as a treatment and reviewed the status of the technology.

Generally speaking, radiowaves range from very low frequency/long wavelength (10^4 Hertz/10^4 m) to extreme high frequency/short wavelength (10^{11} Hertz/10^{-3} m). However, the portion of the electromagnetic spectrum between 30 and 300 MHz with wavelengths from 10 to 1.0 m, respectively, defines the specific radio frequency (RF) range, while microwaves are around 10^{10} Hz/0.01 m.

Two kinds of physical processes describe how RF and MW energy is absorbed at the molecular level in the material: (i) heating of the material by ionic conductivity; and (ii) heating by dielectric polarization (where electric dipoles of molecules interact in an oscillating electric field). The main significant differences between RF and MW heating are power density and wavelength. Since the power density within the material is proportional to frequency, power densities for MW heating are considerably higher than those for RF. Depth of penetration is directly proportional to wavelength, and since RF wavelength is greater than MW, RF may be used to treat much thicker materials.

Any energy efficiency advantage from electromagnetic heating depends on the potential use of selective heating due to the significant dielectric difference between insects and grain, particularly between 10 and 40 MHz. At these conditions, the contaminating insects should experience lethal temperature before significant power is wasted heating the grain bulk. The dielectric properties of several stored-product species were presented by Nelson et al. (1998) and Nelson (2001), and wheat permittivity measurements by Kraszewski et al. (1996). The frequency dependence of the dielectric loss factor of bulk samples of adult S. oryzae and hard red winter wheat at 24°C and 10.6% MC is shown in Fig. 8.11 (Nelson, 1996).

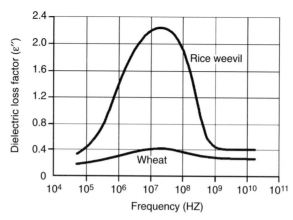

Fig. 8.11. Frequency dependence of the dielectric loss factor of bulk samples of the rice weevil, *Sitophilus oryzae*, and of hard red winter wheat at 24°C with 10.6% MC (from Nelson, 1996).

Nelson (1996) found no cost-saving advantage from the phenomenon because, once significant heat transfer to bulk grain begins in the RF heating process, selective heating cannot work to economic advantage and treatments ultimately generate bulk grain temperatures equivalent to other methods.

In comparison, Wang *et al.* (2001, 2002, 2003) had some success using RF to treat walnut pests. Not only can RF be used to rapidly heat walnut kernels to treatment temperatures that can then be maintained with hot air, but RF may also be an efficient method of drying after a common water bleaching process. Furthermore, clear differential heating was demonstrated using a surrogate gel material with similar dielectric properties to the codling moth (*Cydia pomonella* (L.)) pest. When placed inside the walnut and the kernel heated from 20 to 53°C at 27 MHz, the gel had a heating rate 1.4–1.7 times faster than the kernel, reaching temperatures 12.6–21.2°C higher than it. These results, along with no evidence of significant effects on quality, hold out promise that progress may also be made for grain disinfestation with RF.

Infrared radiation

Infrared radiation (non-ionizing radiation) is electromagnetic energy with wavelengths (0.075–1000 μm) longer than visible light (380–750 nm) and shorter than MW (0.1–100 cm; Penner, 1998). This energy is transferred to whatever material absorbs it, and the absorbed energy causes a measurable change in the material's temperature. This radiant 'heat energy transfer' depends on how readily the molecularly bonded atoms in the material convert the incident radiant energy into molecular vibrational energy that in turn raises the temperature of the absorbing material and its surroundings. Water readily absorbs mid-infrared radiant energy by the symmetric and asymmetric stretching of molecular bonds between oxygen and hydrogen atoms and by bending of the same bonds (Wehling, 1998). The wavelengths most associated with these absorption mechanisms fall between about 2.8 and 7.0 μm.

The unique nature of infrared radiation absorption by water has been used for rapid drying of cereal commodities, especially wheat (Bradbury et al., 1960) and rice (Schroeder, 1960; Schroeder and Rosberg, 1960; Faulkner and Wratten, 1969). Blazer, in 1942, was the first person to suggest direct application of infrared irradiation to control stored-grain insects (Frost et al., 1944). Because insects have a higher percentage of water (> 60) compared with grain, it is likely that they would absorb more infrared radiation than grain and heat up faster.

The selective rapid heating of insects as opposed to grain – presumably without loss of grain quality – makes the use of infrared radiation an appealing technology for stored-grain insect control. Frost et al. (1944) measured internal temperatures of four species of stored-grain insects exposed to various wavelengths, intensities and exposure time, and attributed their mortality to increased body temperature.

The use of intense infrared radiation was evaluated two to three decades ago in the USA for killing immature stages of stored-grain insects developing within kernels and adults (Tilton and Schroeder, 1963; Cogburn, 1967; Cogburn et al., 1971; Kirkpatrick and Tilton, 1972; Kirkpatrick et al., 1972; Kirkpatrick, 1973; Tilton et al., 1983). In all these tests, infrared radiation sources used natural gas or propane combusted over ceramic panels in the presence of oxygen. These gas-fired radiation sources were of high intensity (3360 g-cal/s or 48,000 BTU/h), producing temperatures close to 926°C. The infrared radiation wavelength produced was 2.5 μm; small amounts of carbon dioxide and water vapour were also produced.

Kirkpatrick and Tilton (1972) exposed 100 adults of 12 stored-product insects in 150 g of soft red winter wheat (13.5% moisture) placed below the gas-fired heater in a single kernel thickness layer, 65 cm from the radiation source. From an initial temperature of 26°C, the grain attained a final temperature of 49°C in 20 s, 57°C in 32 s and 65.5°C in 40 s. A 40-s exposure was necessary to kill 99.6% of 12 species of adult stored-grain insects.

Control of immature stages developing within kernels requires higher temperatures than that required for killing exposed adults. Schroeder and Tilton (1961) reported complete control of S. oryzae and R. dominica stages developing inside kernels of rough rice at final grain temperatures averaging 56 and 68°C, respectively. The final temperature of rough rice (41–63°C) was inversely related to its distance from the heater (15.2–50.8 cm) and increased with an increase in exposure time at any given distance from the heater (Tilton and Schroeder, 1963).

The mortality of immature stages of R. dominica, S. oryzae and S. cerealella, based on emergence of adults from irradiated samples, varied between the species, and the treatments did not completely prevent adult emergence. The order of species susceptibility (high to low) was S. oryzae, S. cerealella and R. dominica. The authors extrapolated the adult emergence curves as a function of temperature and recommended a final grain temperature of 65–70°C for killing immatures of all three species. A final temperature of 59°C was attained after a 29-s exposure of soft red winter wheat at 65 cm distance from the infrared source. At this temperature, adult *R. dominica* mortality was 61% and adult emergence from immatures developing within kernels was 55% (Kirkpatrick *et al.*, 1972). A comparison of infrared and MW radiation against *S. oryzae* in soft red winter wheat revealed that infrared radiation was superior to MW in killing both adults and immatures developing within the kernels (Kirkpatrick *et al.*, 1972).

Potential for Heat Disinfestation of Stored Products

Opportunities to reduce costs may come from increasing the rate of heating and decreasing the treatment temperature (Beckett and Morton, 2003b; Fig. 8.12). However, decreased treatment temperature means an increase in treatment time, which has logistical and quality implications. It should be remembered that, for any given treatment time, the temperature required for most external developing species and *Sitophilus* spp. is relatively low compared with *R. dominica* (see Table 8.2).

So if this species is not present, the operating costs may be competitive and the capital cost recoverable. New developments in the application of electromagnetic energy may also lead to a reduction in treatment costs, and methods that rely more on conductive heat transfer could theoretically be promising. Another possible approach to cost-effective heat disinfestation is lower application temperatures, but in conjunction with an additional treatment such as an insect growth regulator.

8.7 Heat Disinfestation of Structures

The goal of a heat treatment is to control insects within a structure without damaging any product left in the equipment or the structure itself. Heat treatments are a well-established method of insect control in the food-processing industry, and the current practice is to warm the building up at 5°C/h to a target temperature of 50–57°C, hold this temperature for 24–36 h and, finally, cool the building at 5–10°C/h.

This intensity and duration of heat may seem excessive given that insects die in only a few minutes at 50°C (Fields, 1992; Strang, 1992; Table 8.1), but because the heat is generated from a few point sources, different parts of facilities heat up at radically different rates. For example, the metal surface of a roll stand will heat up rapidly, but a pile of flour inside a cinderblock wall will not. Therefore, the 24–36 h hold time at 50°C is needed to allow for the heat to reach all parts of the building (Dowdy, 1999; Fig. 8.13).

The two basic approaches to heat treatment are either heating the entire facility or only a section (e.g. a room with roll stands or sifters). There are several advantages in heating the entire facility, primarily that insects cannot escape to other, untreated parts of the building. Depending upon the design of the facility, if one section is not in production, the other areas also cannot

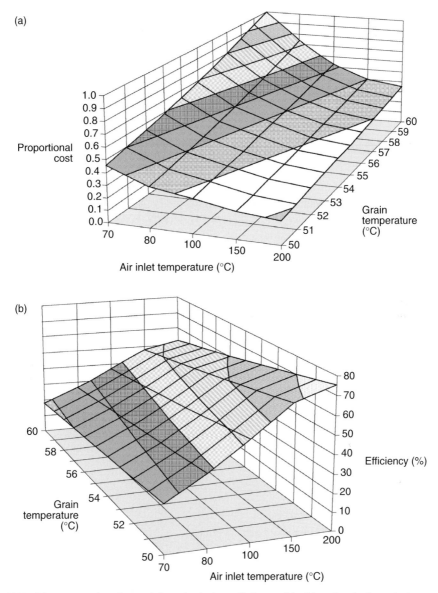

Fig. 8.12. The proportional cost (a) and relative efficiency (b) of heating bulk grain to a range of temperatures (50–60°C) using air inlet temperatures at 70–200°C.

function, so it is best to cover all parts of the facility at once to maximize production. It may also be more efficient for heating, as heat escaping from one area may move to another heated area. Finally, a whole-structure heat treatment is preferable because it is difficult to know all of the locations where insects are present.

Several food-processing companies choose to heat treat only part of their

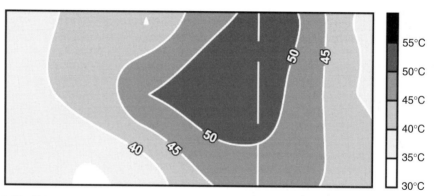

Fig. 8.13. Temperature gradients that develop during the heat treatment of a flour mill. The target temperature is 50–57°C for 24–36 h (from A.K. Dowdy, 1999, personal communication).

facilities at any one time (Heaps, 1996; Norstein, 1996), to reduce the number of heaters required and thus lower the capital cost for heat treatment. Some facilities are designed so that production can continue while the other sections of the facility are stopped for servicing. The size of the area of partial heat treatment may be as large as half of the facility, or part of a floor sectioned off with tarpaulins from floor to ceiling, or a single piece of equipment covered with a tarpaulin to contain the heat.

Unlike a fumigation, where the entire facility must be vacated, heat treatment of part of a facility does not force the rest of the facility to be vacated. Spot-heat treatment also allows for the rapid treatment of specific pieces of equipment, which can be heated and cooled more rapidly than entire structures, so that an individual machine need only be off-line for half a day (Norstein, 1996).

Equipment

Steam, gas or electrical heaters can be used for heat treatment. Steam is the least expensive form of heat. Some food-processing facilities (breakfast food, pasta, bakeries and pet food) use heat as part of their production. These facilities often use steam heat with boilers on-site. For facilities without boilers, portable boilers are available that provide sufficient heat for heat treatments. Steam heaters specifically built for heat treatments can be either fixed or portable (see Fig. 8.14a, b, c), but need pipes or hoses to carry the steam to the heater and return lines for the distillate (see Fig. 8.14a).

Gas- or propane-fired heaters are also popular (see Fig. 8.14b, c) and come in many sizes, usually placed on the outside of the building, forcing hot air inside (Norstein, 1996; Johnson and Danely, 2003) to maintain a positive pressure in the building, thereby carrying the hot air throughout the structure. This method requires more heat than heaters within the building. The most expensive form of heat is electric, so it is used only on a trial basis on a small-

Fig. 8.14a. Fixed steam heater used in food-processing facilities.

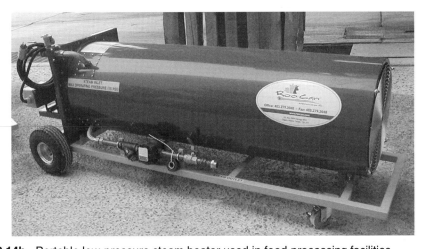

Fig. 8.14b. Portable low-pressure steam heater used in food-processing facilities.

scale (i.e. not to heat treat entire structures; Imholte and Imholte-Tauscher, 1999).

Air circulation is almost as important as heating, but is often neglected in heat treatments (Dosland *et al.*, 2006). To distribute heat throughout an entire structure, eliminate cool spots (see Fig. 8.13) and avoid the potential for damage from overheating; good air circulation is essential. Swivel fans provide

Fig. 8.14c. Propane-fired heater used in food-processing facilities.

better circulation than fixed fans, but both must be able to operate at > 50°C for 36 h. Extra fans are usually installed, as some can be expected to fail during the treatment.

Just as measuring the concentration of methyl bromide is essential for a successful fumigation, so is measuring temperature key for a successful heat treatment. Temperatures should be recorded at least every hour or at shorter time intervals during the entire heat-up and cool-down. Some companies that regularly conduct heat treatments measure air temperatures with a hand-held thermometer at chest height once an hour in each corner of each room. Once all readings in a room attain 50°C, they maintain at least that temperature for 24 h. Fans on heaters may be shut off to prevent overheating. At the same time temperatures are taken, the facility is inspected for problems such as activated sprinkler systems, fan and heater operation, oil spills, etc.

Infrared thermometers allow for quick measurements of hard-to-reach surfaces. Care must be taken, as objects give different temperature readings due to different infrared emissivity even though they are at the same temperature. One way to prevent this is to place pieces of tape throughout the facility and take most readings off the tape, but note that this will yield only surface temperatures, whereas insects may be deep within the fabric of the building. Alternatively, thermocouple wires allow temperatures to be monitored deep in equipment, and can be run to the outside of the building to minimize worker exposure to heat.

Several small portable data loggers are available (Fields, 2005). Certain systems use RF to upload data to a central computer. Data loggers allow for continuous monitoring, but only a few allow for real-time data acquisition, which is needed to adjust the heat during treatment.

Many electronic thermometers may go out of calibration when held at the high temperatures typical in a heat treatment, and while glass thermometers are inexpensive and accurate, they are not allowed in food-processing facilities because of the risk of glass entering into the food stream.

Determining Heat Energy Requirements for Structural Treatments

It is important to determine the amount of heat energy required to raise the temperature of a building to provide lethal temperatures for disinfestation. An approximate estimation of the amount of heat energy required in British Thermal Units (BTU)/h is generally calculated by engineers working for the food-processing facility or by the companies that perform the heat treatment. One BTU is approximately equivalent to the following: 252 calories, 778 foot-pounds, 1055 joules, 108 kg-m or 0.0003 kW-h.

Estimating the amount of heat lost due to infiltration, steel components within a structure and exposed surfaces such as floors, walls, doors, windows and ceilings is the first step in determining heat energy requirements. A simple 'rule-of thumb' to determine whether or not enough heat energy is being used for heating a structure is that if the target temperature of 50°C is not attained within 6–8 h, it is safe to assume that additional heaters are needed.

Imholte and Imholte-Tauscher (1999) provided basic calculations to accomplish this task, which requires the measurement of surfaces in each room of a facility, estimating the weight of steel components and determining infiltration losses. Losses due to infiltration are generally estimated to be one to two complete air exchanges per hour with recirculating air. This number may be greater than two if outside air is heated and purged into a facility – for example, when using gas-fired heaters.

Knowledge of the type of construction material used for various surfaces is essential for arriving at a reasonable estimate of the heat energy required. These calculations help in determining if the current heating capacity is adequate or whether additional heating sources are needed to achieve the desired heat treatment temperature within the required heat treatment time of 24–36 h.

The Heat Treatment Calculator (HTC)

The Heat Treatment Calculator (HTC) is a software program recently developed at Kansas State University (Subramanyam, 2003) that calculates the amount of energy needed to heat a facility to the required temperature and maintain it for a given amount of time. Using this energy requirement, the HTC also calculates the amount of fuel needed for the heat treatment, which in turn provides a cost estimate. To calculate the energy requirements, the HTC uses the coefficients of heat transfer of the building materials used in the facility, the target temperatures and the dimensions of the facility (e.g. height, length and width of walls, windows, floors and roofs, etc.) to calculate the amount of heat dissipated through conduction, convection and radiation. The equations used for the various calculations are as follows:

Surfaces: (equation applied to walls, windows, doors, ceilings, and floors)

$$q_S = \Delta T \times Area \times \left(\frac{1}{\dfrac{x}{k}} \right) \tag{8.10}$$

$$q_S = \Delta T \times Area \times U \tag{8.11}$$

where ΔT is the temperature difference of the exposed wall, U is the heat transfer coefficient of the material, k is the thermal conductivity and x is the thickness.

Infiltration:

$$q_I = \Delta T \times 0.018 \times Volume \times aircirculations \tag{8.12}$$

Steel:

$$q_{St} = \Delta T \times 0.12 \times Steelweight \tag{8.13}$$

Hence, the total heat energy requirement is:

$$q_{Total} = \sum q_S + q_I + q_{St} \tag{8.14}$$

Fuel consumption:

$$Fuel = \left(\frac{q_{Total}}{q_{unit}} \right) \times efficiency \tag{8.15}$$

where q_{unit} is the amount of energy produced per unit of the fuel.

Using the current temperature profile and facility specifications, the HTC calculates the energy requirements for the given period and converts it to corresponding fuel requirements. The benefits and uses of the HTC include:

- Studying the variations of energy, fuel and cost with respect to changes in the heating requirements and weather conditions (outside and inside temperatures).
- Arriving at a price estimate of heat treatment under any given climatic conditions.
- Determining optimal conditions for cost-effective heat treatment without actually performing the heat treatment.
- Calculating the amount of heat lost due to various surfaces (floors, walls, doors, windows) and equipment, or any bottlenecks in the heat treatment. Areas where heat loss is severe can be better insulated using heat blankets and other heat-insulating materials.

To our knowledge, there is no published information on the costs associated with structural heat treatments, even though this technology is now becoming popular as a methyl bromide alternative. Increased use of this technology in the future, coupled with the use of empirical insect survival models, HTC and fuel costs may shed additional light on the actual cost/benefit relationships of heat treatment. Anecdotal reports from current users of this technology suggest that heat treatment is cost-competitive with existing structural fumigation treatments.

Common Problems

Although heat treatments are used extensively today in modern food-processing facilities, there are still several problems that may arise (Heaps, 1996; Imholte and Imholte-Tausher, 1999). Fortunately, many of the problems can be addressed with careful planning, but some will only be resolved after a company has gained experience in heat treatments.

Sprinkler heads should be rated for not lower than 95°C. Electronic equipment should be heat-tolerant, removed or jacketed and supplied with cool air. High temperature damages some plastics found in water lines, air hoses, brushes and brooms. Other items that may be damaged by heat include carbon dioxide fire extinguishers, adhesives and aerosol cans. Rubber conveyer belts should also be slackened to avoid damage.

One common problem with a first-time heat treatment is that not enough heat is supplied to a building to quickly obtain a target temperature of 50°C. Heating engineers should be consulted as to the amount of heat required. Heating during the hottest months will make it easier to obtain 50°C. The processing equipment within a facility often generates heat that is exhausted outside of the building. These exhaust systems may be able to be turned off before the heat treatment begins, allowing the buildup of heat in anticipation of the heat treatment.

Some areas of a building are difficult to heat, such as sub-floors, basements, outside walls and internal walls attached to non-heated sections of the building; extra heaters may be needed in these areas. Another problem is overheating; this can be prevented by good air circulation, monitoring temperatures during the heat treatment, judicious placement of heaters, moving portable heaters and/or shutting off heaters periodically.

Although heat treatments are safer than fumigations, some basic precautions should be taken to avoid workers experiencing heat stress (National Institute for Occupational Safety and Health, 1992). It is important to minimize the time spent inside a building during heat treatment. There should be at least two people entering a heated area at a time. Workers should drink plenty of fluids. People, who have medical problems such as a heart condition, are elderly or overweight may be more affected by heat stress.

Checklist for Heat Treatments

Checklists are a useful tool to make sure a heat treatment is conducted efficiently. The following is an abbreviated list taken from Dosland *et al.* (2006),

Temp-Air, Division of Rupp Industries Inc. and Roo-Can International, Inc. Prior to heat treatment:

- Estimate the heat required in terms of heaters and fans.
- Set sprinkler heads to at least 95°C.
- Make sure there are enough electrical circuits and they can function at 60°C for 48 h.
- Remove heat-sensitive equipment, food products, additives, oils and glues.
- Close doors, seal intake and exhaust fans and place sand snakes at door jams.
- Thoroughly clean facility, open equipment and remove rubbish and product lines.
- Turn off power to electronic equipment, compressed, intake and exhaust air and loosen belts.
- Test temperature-monitoring equipment.
- Treat the non-heat-treated area adjacent to the heat-treated section with contact insecticide to prevent insect migration.
- Establish a worker safety plan, including a central area out of heat for rest, data sheets and emergency telephones.
- Place caged test insects in 'hard-to-heat' areas or in areas where insect problems are generally noticed.

During heat treatment:

- Set target temperatures at 50–57°C for 24–36 h with a heating rate of 5°C/h.
- Monitor temperature throughout the facility at least once per hour.
- Tour heat-treated areas once per hour to check for problems, adjust fans and heaters or lower temperatures if > 60°C.
- Note where there are accumulations of insects as a means of finding hidden infestations and target future control measures.

After heat treatment:

- Shut off all heaters, leave fans on, open doors and windows, cool down at a rate of 5–10°C/h.
- Return removed equipment and products.
- Collect test insects and remove data loggers and temperature wires.
- Write report summarizing problems, temperature data and caged insect survival.
- Discard first run of product, as it will contain high amounts of dead insects and insect parts.
- Perform required maintenance on heaters.

8.8 Conclusions

Many stored-product insects are associated with raw and processed cereal commodities and contribute to significant quantitative and qualitative losses.

Management of such species is a challenge, primarily because they are able to feed on a variety of food products and develop resistance to pesticides generally used to combat them. In addition, our limited ability to determine infestation sources and understand infestation patterns makes it even more difficult to understand the impact of our pest management methods. Most companies use integrated pest management to keep insects below economically damaging levels. There is a general shift in the grain and food industry from chemical to non-chemical methods.

One viable alternative to pesticides is the use of heat to disinfest grain and structures, which can be carried out with several methods. Heat disinfestation is relatively rapid and does not leave any insecticide residues but, like fumigants, grain or structures could become reinfested after treatment. Quick processing of disinfested grain or sanitation and exclusion practices in structures help minimize reinfestation.

The use of heat is gaining popularity in the grain and food industries. As chemical use is restricted or phased out due to government regulations and customer specifications, we predict there will be an increased adoption of heat for grain treatment and structural treatment. Heat is an environmentally friendly technology, and can be used in conjunction with other chemical (fumigants) and non-chemical (diatomaceous earth dusts) methods for insect management. However, this overview of technologies and methods used for grain and structural disinfestation includes several gaps in our knowledge and practice that need to be filled by additional research to make this technology more widely acceptable to the grain and food industries.

8.9 References

Adler, C. (2002) Efficacy of heat treatments against the tobacco beetle *Lasioderma serricone* F. (Coleoptera: Anobiidae) and the lesser grain borer *Rhyzopertha dominica* F. (Coleoptera: Bostrichidae). In: Credland, P.F., Armitage, D.M., Bell, C.H. Cogan, P.M. and Highley, E. (eds) *Advances in Stored Product Protection. Proceedings of the 8th International Working Conference on Stored Product Protection*, 22–26 July 2002. CAB International, Wallingford, UK, pp. 617–621.

Adler, C. and Rassmann, W. (2000) Utilization of extreme temperatures in stored product protection. *IOBC Bulletin* 23, 257–262.

Armitt, J.D.G. and Way, H.G. (1990) Breeding and assessment of new malting barley varieties in Australia. In: *Proceedings of the 21st Convention of the Institute of Brewing (Australian and New Zealand Section)*, Auckland, New Zealand, pp. 57–64.

Bailey, S.W. (1962) The effects of percussion on insect pests in grain. *Journal of Economic Entomology* 55, 301–304.

Bailey, S.W. (1969) The effects of physical stress in the grain weevil *Sitophilus granarius*. *Journal of Stored Products Research* 5, 311–324.

Banks, H.J. (1987) Impact, physical removal and exclusion for insect control in stored products. In: Donahaye, E. and Navarro, S. (eds) *Proceedings of the 4th International Working Conference on Stored-product Protection*, September, 1986, Tel Aviv, Israel. Maor-Wallach Press, Jerusalem, Isreal, pp. 165–184.

Banks, H.J. and Annis, P.C. (1990) Comparative advantages of high CO_2 and low O_2 types of controlled atmospheres for grain storage. In: Calderon, M. and Barkai Golan, R. (eds) *Food Preservation by Modified Atmospheres.* CRC Press, Roca Baton, Florida, pp. 93–122.

Banks, H.J., Desmarchelier, J.M. and Ren, Y.L. (1993) Carbonyl sulfide fumigant and method of fumigation. *International Application Published under the Patent Cooperation Treaty (PCT)*, WO93/13659.

Banks, J. and Fields, P. (1995) Physical methods for insect control in stored-grain ecosystems. In: Jayas, D.S., White, N.D.G. and Muir, W.E. (eds) *Stored-grain Ecosystems.* Marcel Dekker Inc., New York, pp. 353–409.

Beckett, S.J. and Morton, R. (2003a) The mortality of three species of Psocoptera, *Liposcelis bostrichophila* Badonnel, *Liposcelis decolor* Pearman and *Liposcelis paeta* Pearman, at moderately elevated temperatures. *Journal of Stored Products Research* 39, 103–115.

Beckett, S.J. and Morton, R. (2003b) Mortality of *Rhyzopertha dominica* (F.) (Coleoptera: Bostrychidae) at grain temperatures ranging from 50°C to 60°C obtained at different rates of heating in a spouted bed. *Journal of Stored Products Research* 39, 313–332.

Beckett, S.J., Morton, R. and Darby, J.A. (1998) The mortality of *Rhyzopertha dominica* (F.) (Coleoptera: Bostrychidae) and *Sitophilus oryzae* (L.) (Coleoptera: Curculionidae) at moderate temperatures. *Journal of Stored Products Research* 34, 363–376.

Bell, C.H. (1983) Effects of high temperatures on larvae of *Ephestia elutella* (Lepidoptera: Pyralidae) in diapause. *Journal of Stored Products Research* 19, 153–157.

Bengston, M. and Strange, A.C. (1998) Report of the national working party on grain protection. In: Banks, H.J., Wright, E.J. and Damcevski, K.A. (eds) *Grain in Australia. Proceedings of the Australian Postharvest Technical Conference*, Canberra, 26–29 May. CSIRO, Canberra, Australia, pp. 145–147.

Boina, D. (2004) Time–mortality relationships of *Tribolium confusum* (Jacquelin du Val)

life stages exposed to elevated temperatures and a dynamic model for predicting survival of old larvae during heat treatments. MS Thesis, Kansas State University, Manhattan, Kansas, 98 pp.

Boina, D. and Subramanyam, Bh. (2004) Relative susceptibility of *Tribolium confusum* (Jacquelin du Val) life stages to elevated temperatures. *Journal of Economic Entomology* 97, 2168–2173.

Bond, E.J. (1984) *Manual of Fumigation for Insect Control.* FAO Plant Production and Protection Paper 54, Food and Agricultural Organization of the United Nations, Rome.

Bradbury, D., Hubbard, J.E., MacMasters, M.M. and Senti, F.R. (1960) *Conditioning Wheat for Milling: a Survey of Literature.* USA Department of Agriculture, Agricultural Research Service, Miscellaneous Publication No. 824, US Government Printing Office, Washington, DC.

Brokerhof, A.W., Banks, H.J. and Morton, R. (1992) A model for time–temperature–mortality relationships for eggs of the webbing moth, *Tineola bisselliella* (Lepidoptera: Tineidae) exposed to cold. *Journal of Stored Products Research* 28, 269–277.

Bruce, D.M., Hamer, P.J.C., Wilkinson, D.J., White, R.P., Conyers, S. and Armitage, D.M. (2004) *Disinfestation of Grain using Hot-air Dryers: Killing Hidden Infestations of Grain Weevils without Damaging Germination.* Project Report No. 345, The Home-Grown Cereals Authority, UK.

Burks, C.S., Johnson, J.A., Maier, D.E. and Heaps, J.W. (2000) Temperature. In: Subramanyam, Bh. and Hagstrum, D.W. (eds) *Alternatives to Pesticides in Stored-Product IPM.* Kluwer Academic, Boston, Massachusetts, pp. 73–104.

Champ, B.R. and Dyte, C.E. (1976) *Report of the FAO Global Survey of Pesticide Susceptibility of Stored Grain Pests.* FAO Plant Production and Protection Series No. 5, Food and Agriculture Organization of the United Nations, Rome.

Claflin, J.K. and Fane, A.G. (1981) The use of spouted beds for the heat treatment of grains. In: *CHEMECA 81: 9th Australasian Conference on Chemical*

Engineering, Christchurch, New Zealand, 30 August–4 September. The Chemical Engineering Group, New Zealand Institution of Engineers, Christchurch, New Zealand, pp. 65–72.

Claflin, J.K., Evans, D.E., Fane, A.G. and Hill, R.J. (1986) The thermal disinfestation of wheat in a spouted bed. *Journal of Stored Products Research* 22, 153–161.

Cogburn, R. (1967) Infrared radiation effect on reproduction by three species of stored-product insects. *Journal of Economic Entomology* 60, 548–550.

Cogburn, R., Brower, J.H. and Tilton, E.W. (1971) Combination of gamma and infrared radiation for control of Angoumois grain moth in wheat. *Journal of Economic Entomology* 64, 923–925.

Cotton, R.T. (1963) *Pests of Stored Grain and Grain Products*. Burgess Publishing Company, Minneapolis, Minnesota.

Currie, S. and Tufts, B. (1997) Synthesis of stress protein 70 (Hsp 70) in rainbow trout (*Oncorhynchus mykiss*) red blood cells. *Journal of Experimental Biology* 200, 607–614.

Darby, J. (1998) Putting grain aeration in order with generalized aeration categories. In: Banks, H.J., Wright, E.J. and Damcevski, K.A. (eds) *Stored Grain in Australia. Proceedings of the Australian Postharvest Technical Conference*, Canberra, 26–29 May 1998. CSIRO, Canberra, Australia, pp. 203–208.

Dean, D.A. (1913) Further data on heat as a means of controlling mill insects. *Journal of Economic Entomology* 6, 40–53.

Denlinger, D.L. and Yocum, G.D. (1999) Physiology of heat sensitivity. In: Hallman, G. and Denlinger, D. (eds) *Temperature Sensitivity in Insects and Application in Integrated Pest Management*. Westview Press, Boulder, Colorado, pp. 7–53.

Dermott, T. and Evans, D.E. (1978) An evaluation of fluidised-bed heating as a means of disinfesting wheat. *Journal of Stored Products Research* 14, 1–12.

Desmarchelier, J.M. and Dines, J.C. (1987) Dryacide treatment of stored wheat: its efficacy against insects, and after processing. *Australian Journal of Experimental Agriculture* 27, 309–312.

Dickie, J.B., Ellis, R.H., Kraak, H.L., Ryder, K. and Tompsett, P.B. (1990) Temperature and seed storage longevity. *Annals of Botany* 65, 197–204.

Dosland, O., Subramanyam, Bh., Sheppard, K. and Mahroof, R. (2006) Temperature modification for insect control. In: Heaps, J. (ed.) *Insect Management for Food Storage and Processing*, 2nd edn. American Association for Clinical Chemistry, St Paul, Minnesota, pp. 89–103.

Dowdy, A.K. (1999) Heat sterilization as an alternative to methyl bromide fumigation in cereal processing plants. In: Jin, Z., Liang, Q., Liang, Y., Tan, X. and Guan, L. (eds) *Proceedings of the 7th International Working Conference on Stored Product Protection*, Beijing, China, pp. 1089–1095.

Dowdy, A.K. and Fields, P.G. (2002) Heat combined with diatomaceous earth to control the confused flour beetle (Coleoptera: Tenebrionidae) in a flour mill. *Journal of Stored Products Research* 38, 11–22.

Drache, H.M. (1976) *Beyond the Furrow*. Interstate, Danville, Illinois.

Drinkall, M.J., Zaffagnini, V., Süss, L. and Locatelli, D.P. (2002) Efficacy of sulfuryl fluoride on stored-product insects in a semolina mill trial in Italy. In: Credland, P.F., Armitage, D.M., Bell, C.H., Cogan, P.M. and Highley, E. (eds) *Advances in Stored Product Protection. Proceedings of the 8th International Working Conference on Stored Product Protection*, 22–26 July. CABI Publishing, Wallingford, UK, pp. 884–887.

Ducom, P., Dupuis, S., Stefanini, V. and Guichard, A.A. (2002) Sulfuryl fluoride as a new fumigant for the disinfestation of flour mills in France. In: Credland, P.F., Armitage, D.M., Bell, C.H., Cogan, P.M. and Highley, E. (eds) *Advances in Stored Product Protection. Proceedings of the 8th International Working Conference on Stored Product Protection*, 22–26 July. CABI Publishing, Wallingford, UK, pp. 900–903.

Duhamel du Monceau, H.L. and Tillet, M. (1762) Histoire d'un insecte qui dévore les grains de l'Angoumois, avec les moyens que l'on peut employer pour le détruire. Gurin et Delatour, avec privilege du Roi, Paris, 1–314.

Dzhorogyan, G.A. (1957) Disinfestation of grain in a flow of hot gases (in Russian). *Mukomol'no-Elevatornaya Promyshlennost* 8, 29–30.

Ebeling, W. (1971) Sorptive dusts for pest control. *Annual Review of Entomology* 16, 123–158.

Edwards, J.P., Short, J.E. and Abraham, L. (1987) New approaches to the protection of stored products using an insect juvenile hormone analogue. *British Crop Protection Council Monograph No. 37: Stored Products Pest Control*, pp. 197–205.

Ellis, R.H. and Roberts, E.H. (1980) The influence of temperature and moisture on seed viability period in barley (*Hordeum distichum* L.). *Annals of Botany* 45, 31–37.

European Brewery Convention (1998) *Analytica-EBC*, Nurnberg, Germany. Verlag Hans Carl Getranke-Fachverlag, Germany.

Evans, D.E. (1981) The influence of some biological and physical factors on the heat tolerance relationships for *Rhyzopertha dominica* (F.) and *Sitophilus oryzae* (L.) (Coleoptera: Bostrychidae and Curculionidae*). Journal of Stored Products Research* 17, 65–72.

Evans, D.E. (1987) The influence of rate of heating on the mortality of *Rhyzopertha dominica* (F.) (Coleoptera: Bostrychidae). *Journal of Stored Products Research* 23, 73–77.

Evans, D.E. and Dermott, T. (1981) Dosage–mortality relationships for *Rhyzopertha dominica* (F.) (Coleoptera: Bostrychidae) exposed to heat in a fluidized bed. *Journal of Stored Products Research* 17, 53–64.

Evans, D.E., Thorpe, G.R. and Dermott, T. (1983) The disinfestations of wheat in a continuous-flow fluidized bed. *Journal of Stored Products Research* 19, 125–137.

Evans, D.E., Thorpe, G.R. and Sutherland, J.W. (1984) Large scale evaluation of fluid-bed heating as a means of disinfest-

ing grain. In: *Proceedings of the 3rd International Working Conference of Stored-product Entomology*, Kansas State University, Manhattan, Kansas, USA, pp. 523–530.

Fang, L., Subramanyam, Bh. and Arthur, F.H. (2002a) Effectiveness of spinosad on four classes of wheat against five stored-product insects. *Journal of Economic Entomology* 95, 640–650.

Fang, L., Subramanyam, Bh. and Dolder, S. (2002b) Persistence and efficacy of spinosad residues in farm stored wheat. *Journal of Economic Entomology* 95, 1102–1109.

FAO (2003) *Production Yearbook Vol. 56 – 2002.* FAO Statistics Series No. 176, Food and Agriculture Organization of the United Nations, Rome.

FAO (2004) *Trade Yearbook Vol. 56 – 2002.* FAO Statistics Series No. 178, Food and Agriculture Organization of the United Nations, Rome.

Faulkner, M.D. and Wratten, F.T. (1969) Lousiana State University infrared preheat rice dryer. In: *61st Annual Progress Report Rice Experiment Station.* Louisiana State University Agricultural Experiment Station, Crowley, Louisiana, pp. 101–122.

FDA (1988) Wheat flour adulterated with insect fragments and rodent hairs. *Compliance Policy Guides, Processed Grain Guide 7104.06. January 20.* Chapter 4. Food and Drug Administration (FDA), Rockville, Maryland.

Fields, P.G. (1992) The control of stored-product insects and mites with extreme temperatures. *Journal of Stored Products Research* 28, 89–118.

Fields, P.G. (2005) Temperature measurement. *Bulletin of the Entomological Society of Canada* 37, 20–22.

Fields, P.G. and White, N.D.G. (2002) Alternatives to methyl bromide treatments for stored product and quarantine insects. *Annual Review of Entomology* 47, 331–359.

Finney, D.J. (1971) *Probit Analysis.* Cambridge University Press, London.

Firestone, D. (1990) *Official Methods and Recommended Practices of the American*

Oil Chemists Society, 4th edn. AOCS, Champaign, Illinois.

Fleurat-Lessard, F. (1980) Lutte physique par l'air chaud ou les hautes fréquencies contre les insectes des grains et des produits céréaliers. *Bulletin Technique d'Information de Desinsectisation pour les Cereales Stockistes* 349, 345–352.

Fleurat-Lessard, F. (1985) Les traitements thermiques de désinfestation des céréales et des produits céréaliers: possibilité d'utilisation pratique et domaine d'application. *Bulletin OEPP/EPPO* 15, 109–118.

Flinn, P.W., Subramanyam, Bh. and Arthur, F.H. (2004) Comparison of aeration and spinosad for suppressing insects in stored wheat. *Journal of Economic Entomology* 97, 1465–1473.

Frost, S.W., Dills, L.E. and Nicholas, J.E. (1944) The effects of infrared radiation on certain insects. *Journal of Economic Entomology* 37, 287–290.

Ghaly, T.F. and Sutherland, J.W. (1984) Heat damage to grain and seeds. *Journal of Agricultural Engineering Research* 30, 337–345.

Ghaly, T.F., Edwards, R.A. and Ratcliffe, J.S. (1973) Heat-induced damage in wheat as a consequence of spouted bed drying. *Journal of Agricultural Engineering Research* 18, 95–106.

Gray, D.S. and Darby, J.A. (1996) *A Model of the Heating of Wheat in a Hexagonal 'V-section' Distributor Fluid Bed.* Consultative Report No. 30. CSIRO Canberra, Australia.

Hall, D.W. (1970) *Handling and Storage of Food Grains in Tropical and Sub-tropical Areas.* FAO Agricultural Development Paper (90), Rome, Italy.

Hallman, G.J. and Denlinger, D.L. (1999) Introduction: temperature sensitivity and integrated pest management. In: Hallman, G.J. and Denlinger, D.L. (eds) *Temperature Sensitivity in Insects and Application in Integrated Pest Management.* Westview Press, Boulder, Colorado, pp. 1–5.

Halverson, S.L., Plarre, R., Burkholder, W.E., Booske, J.H., Bigelow, T.S. and Misenheimer, M.E. (1997) Recent advances in the control of insects in stored products with microwaves. In: *ASAE Annual International Meeting, Paper No. 976098,* 10–14 August, Minneapolis, Minnesota, USA.

Halverson, S.L., Plarre, R., Bigelow, T.S. and Lieber, K. (1998) Recent advances in the use of EHF energy for the control of insects in stored products. In: *ASAE Annual International Meeting, Paper No. 986052,* 12–15 July, Orlando, Florida, USA.

Harein, P.K. and Davis, R. (1992) Control of stored-grain insects. In: Sauer, D.B. (ed.) *Storage of Cereal Grains and Their Products.* American Association of Cereal Chemists, St Paul, Minnesota, pp. 491–534.

Heaps, J.W. (1994) Temperature control for insect elimination. *Association of Operative Millers Bulletin,* 6467–6470.

Heaps, J.W. (1996) Heat for stored product insects. *IPM Practitioner* 18, 18–19.

Hill, D.S. (2002) *Pests of stored foodstuffs and their control.* Kluwer Academic Publishers, Boston, Massachusetts, p. 496.

Hochachka, P.W. and Somero, G.N. (1984) *Biochemical Adaptation.* Princeton University Press, Princeton, New Jersey.

Imholte, T.J. and Imholte-Tauscher, T.K. (1999) *Engineering for Food Safety and Sanitation,* 2nd edn. Technical Institute of Food Safety, Woodinville, Washington, 382 pp.

Jian, F., Jayas, D.S., White, N.D.G. and Muir, W.E. (2002) Temperature and geotaxis preference by *Cryptolestes ferrugineus* (Coleoptera: Laemophloeidae) adults in response to 5°C/m temperature gradients at optimum and hot temperatures in stored wheat and their mortality at high temperature. *Population Ecology* 31, 816–826.

Johnson, R.D. and Danely, T.T. (2003) *Pest Control System.* USA patent 6,588,140.

Kirkpatrick, L.R. (1973) Gamma, infra-red and microwave radiation combinations for control of *Rhyzopertha dominica* in wheat. *Journal of Stored Products Research* 9, 19–23.

Kirkpatrick, L.R. and Tilton, W.E. (1972) Infrared radiation to control adult stored product Coleoptera. *Journal of the Georgia Entomological Society* 7, 73–75.

Kirkpatrick, L.R., Brower, H.J. and Tilton, W.E. (1972) A comparison of microwave and infrared radiation to control rice weevils in wheat. *Journal of the Kansas Entomological Society* 45, 434–438.

Kraszewski, A.W., Trabelsi, S. and Nelson, S.O. (1996) Wheat permittivity measurements in free space. *Journal of Microwave Power and Electromagnetic Energy* 31, 135–141.

Lapp, H.M., Madrid, F.J. and Smith, L.B. (1986) *A Continuous Thermal Treatment to Eradicate Insects from Stored Wheat.* American Society of Agricultural Engineers, St Joseph, Michigan, Paper 86-3008, 14 pp.

Lim, G.S., Tee, S.P., Ong, I.M. and Lee, B.T. (1978) Problems and control of insects in rice packing. *Malaysian Agricultural Research and Development Institute Research Bulletin* 6, 119–128.

Liu, R., Xiong, Y. and Wang, C. (1983) Studies on heat treatment of wheat grains in China. In: Sheritt, L.W. (ed.) *Chemistry and World Food Supplies: the New Frontier.* Pergamon Press, Oxford, UK, pp. 443–452.

Longstaff, B.C. (1981) The effects of aeration upon insect population growth. In: Champ, B.R. and Highley, E. (eds) *Proceedings of the Australian Development Assistance Course on the Preservation of Stored Cereals.* CSIRO, Canberra, pp. 459–465.

Longstaff, B.C. (1984) An analysis of the demographic consequences of temperature manipulation upon the efficacy of certain pesticides used in the control of stored-product pests. In: Bailey, P. and Swincer, D. (eds) *Proceedings of the Fourth Australian Applied Entomological Research Conference,* September 1984. South Australian Department of Agriculture, Adelaide, Australia.

Longstaff, B.C. (1986) Environmental manipulation as a physiological control measure. In: Donahaye, E. and Navarro, S. (eds) *Proceedings of the 4th International Working Conference on Stored-product Protection,* September 1986, Tel Aviv,

Israel. Maor-Wallach Press, Jerusalem, Isreal, pp. 47–61.

Longstaff, B.C. (1988) Temperature manipulation and the management of insecticide resistance in stored grain pests: a simulation study for the rice weevil, *Sitophilus oryzae. Ecological Modelling* 43, 303–313.

Mahroof, R., Subramanyam, Bh. and Eustace, D. (2003a) Temperature and relative humidity profiles during heat treatment of mills, and its efficacy against *Tribolium castaneum* (Herbst) life stages. *Journal of Stored Products Research* 39, 555–569.

Mahroof, R., Subramanyam, Bh., Throne, J.E. and Menon, A. (2003b) Time–mortality relationships for *Tribolium castaneum* (Coleoptera: Tenebrionidae) life stages exposed to elevated temperatures. *Journal of Economic Entomology* 96, 1345–1351.

Mahroof, R., Subramanyam, Bh. and Flinn, P. (2005a) Reproductive performance of *Tribolium castaneum* (Herbst) (Coleoptera: Tenebrionidae) exposed to the minimum heat treatment temperature as pupae and adults. *Journal of Economic Entomology* 98, 626–633.

Mahroof, R.M., Zhu, K.Y. and Subramanyam, Bh. (2005b) Changes in expression of heat shock proteins in *Tribolium castaneum* (Herbst) in relation to developmental stage, exposure time and temperature. *Annals of the Entomological Society of America* 98 (1), 100–107.

Makhijani, A. and Gurney, K.R. (1995) *Mending the Ozone Hole: Science, Technology and Policy.* MIT Press, Massachusetts.

Mathur, K. and Epstein, N. (1974) *Spouted Beds.* Academic Press, New York.

Mathur, K. and Gishler, P. (1955) A study of the application of the spouted bed technique to wheat drying. *Journal of Applied Chemistry* 5, 624.

Morgan, B.J.T. (1992) *Analysis of Quantal Response Data.* Chapman and Hall, London.

Mourier, H. and Poulsen, K.P. (2000) Control of insects and mites in grain using a high temperature/short time (HTST) technique. *Journal of Stored Products Research* 36, 309–318.

Munro, J.W. (1966) *Pests of Stored Products.* Hutchinson, London.

National Institute for Occupational Safety and Health (1992) *Working in Hot Environments.* National Institute for Occupational Safety and Health, Atlanta, Georgia, 6 pp.

Nellist, M.E. (1981) Predicting the viability of seeds dried with heated air. *Seed Science and Technology* 9, 439–455.

Nellist, M.E. and Bruce, D.M. (1987) Drying and cereal quality. *Aspects of Applied Biology* 15, 439–456.

Nelson, S.O. (1973) Insect-control studies with microwaves and other radiofrequency energy. *Bulletin of the Entomological Society of America* 19, 157–163.

Nelson, S.O. (1987) Potential agricultural applications for RF and microwave energy. *Transactions of the American Society of Agricultural Engineers* 30, 818–822, 831.

Nelson, S.O. (1996) Review and assessment of radio-frequency and microwave energy for stored-grain insect control. *Transactions of the American Society of Agricultural Engineers* 39, 1475–1484.

Nelson, S.O. (2001) Radio frequency and microwave dielectric properties of insects. *Journal of Microwave Power and Electromagnetic Energy* 36, 47–56.

Nelson, S.O. and Kantack, B.H. (1966) Stored-grain insect control studies with radio-frequency energy. *Journal of Economic Entomology* 59, 588–594.

Nelson, S.O. and Whitney, W.K. (1960) Radio-frequency electric fields for stored grain insect control. *Transactions of the American Society of Agricultural Engineers* 3, 133–137, 144.

Nelson, S.O., Bartley, Jr., P.G. and Lawrence, K.C. (1998) RF and microwave dielectric properties of stored-grain insects and their implications for potential insect control. *Transactions of the American Society of Agricultural Engineers* 41, 685–692.

Neven, L.G. (2000) Physiological responses of insects to heat. *Postharvest Biology and Technology* 21, 103–111.

Norstein, S. (1996) *Heat Treatment in the Scandinavian Milling Industry: Heat Treatment as an Alternative to Methyl Bromide.* SFT Miljøteknologi 96, Norwegian Pollution Control Authority, Olso.

Paula, M.C.Z., Lazzari, F.A. and Lazzari, S.M.N. (2002) Use of diatomaceous earth for insect control in paddy rice stored in silos. In: Credland, P.F., Armitage, D.M., Bell, C.H., Cogan, P.M. and Highley, E. (eds) *Advances in Stored Product Protection. Proceedings of the 8th International Working Conference on Stored Product Protection*, 22–26 July, CAB International, Wallingford, Oxon, UK, pp. 896–899.

Penner, M.H. (1998) Basic principles of spectroscopy. In: Nielsen, S.S. (ed.) *Food Analysis*, 2nd edn. Aspen Publications, Gaithersburg, Maryland, pp. 387–396.

Plarre, R., Halverson, S.L., Burkholder, W.E., Bigelow, T.S., Misenheimer, M.E., Booske, J.H. and Nordheim, E.V. (1997) Effects of high-power microwave radiation on *Sitophilus zeamais* Motsch. (Coleoptera: Curculionidae) at different frequencies, Volume 3. In: *ANPP – 4th International Conference on Pests in Agriculture*, Montpellier, France, 6–8 January. Association Nationale de Protection des Plantes, Paris, pp. 819–828.

Qaisrani, R. and Beckett, S. (2003a) Heat disinfestation or wheat in a continuous-flow spouted bed. In: Credland, P.F., Armitage, D.M., Bell, C.H., Cogan, P.M. and Highley, E. (eds) *Advances in Stored Product Protection. Proceedings of the 8th International Working Conference on Stored Product Protection*, York, UK. CAB International, Wallingford, Oxon, UK, pp. 622–625.

Qaisrani, R. and Beckett, S. (2003b) Possible use of cross-flow dryers for heat disinfestation of grain. In: Wright, E.J., Webb, M.C. and Highley, E. (eds) *Stored Grain in Australia 2003. Proceedings of the Australian Postharvest Technical Conference*, Canberra, 25–27 June. CSIRO, Canberra, Australia, pp. 30–35.

Qaisrani, R. and Beckett, S. (2003c) Possible use of radial-flow dryers for heat disinfestation of grain. In: Wright, E.J., Webb,

M.C. and Highley, E. (eds) *Stored Grain in Australia 2003. Proceedings of the Australian Postharvest Technical Conference*, Canberra, 25–27 June. CSIRO, Canberra, Australia, pp. 36–42.

Reed, C. (1992) Development of storage techniques: a historical perspective. In: Sauer, D.B. (ed.) *Storage of Cereal Grains and their Products*. American Association of Cereal Chemists, St Paul, Minnesota, pp. 143–156.

Rees, D.P. (1998) Psocids as pests of Australian grain storages. In: Banks, H.J., Wright, E.J. and Damcevski, K.A. (eds) *Stored Grain in Australia. Proceedings of the Australian Postharvest Technical Conference*, Canberra, 26–29 May. CSIRO, Canberra, Australia, pp. 46–51.

Rees, D.P. (2004) *Insects of Stored Products.* CSIRO Publishing, Collingwood, Victoria, Australia.

Roesli, R., Subramanyam, Bh., Fairchild, F. and Behnke, K. (2003) Insect numbers before and after heat treatment in a pilot feed mill. *Journal of Stored Products Research* 39, 521–540.

Saxena, B.P., Sharma, P.R., Thappa, R.K. and Tikku, K. (1992) Temperature-induced sterilization for control of three stored grain beetles. *Journal of Stored Products Research* 28, 67–70.

Schroeder, H.W. (1960) Infra-red drying of rough rice. II. Short grain type: Calrose and Caloro. *Rice Journal* 63, 25–27.

Schroeder, H.W. and Rosberg, D.W. (1960) Infra-red drying of rough rice. I. Long grain type: Rexoro and Bluebonnet 50. *Rice Journal* 63, 3–5.

Schroeder, H.W. and Tilton, E.W. (1961) *Infrared Radiation for the Control of Immature Insects in Kernels of Rough Rice.* US Department of Agriculture, AMS-445, 10 pp.

Sharifi, S. and Mills, R.B. (1971) Developmental activities and behaviour of the rice weevil inside wheat kernels. *Journal of Economic Entomology* 64, 1114–1118.

Snelson, J.T. (1987) *Grain Protectants*. Monograph No. 3, Australian Centre for International Agricultural Research, Canberra, Australia.

Strang, T.J.K. (1992) A review of published temperatures for the control of pest insects in museums. *Collection Forum* 84, 1–67.

Strong, R.G. and Diekman, J. (1973) Comparative effectiveness of fifteen insect growth regulators against several pests of stored products. *Journal of Economic Entomology* 66, 1167–1173.

Subramanyam, Bh. (2003). Heat treatment calculator. *Milling Journal*, Third Quarter, pp. 46–48.

Subramanyam, Bh., Flinn, P.W. and Mahroof, R. (2003) Development and validation of a simple heat accumulation model for predicting mortality of *Tribolium castaneum* (Herbst) first instars exposed to elevated temperatures. In: Credland, P.F., Armitage, D.M., Bell, C.H., Cogan, P.M. and Highley, E. (eds) *Proceedings of the 8th International Working Conference on Stored Product Protection*, 22–26 July 2002, York, UK. CAB International, Wallingford, UK, pp. 369–374.

Sutherland, J.W., Thorpe, G.R. and Fricke, P.W. (1986) Grain disinfestation by heating in a pneumatic conveyor. In: *Proceedings of the Conference on Agricultural Engineering*, 24–28 August, 1986, Adelaide. The Institution of Engineers, Australia. Canberra, Australia, pp. 419–425.

Sutherland, J.W., Evans, D.E., Fane, A.G. and Thorpe, G.R. (1987) Disinfestation of grain with heated air. In: Donahaye, E. and Navarro, S. (eds) *Proceedings of the 4th International Working Conference on Stored-product Protection*, September 1986, Tel Aviv, Israel. Maor-Wallach Press, Jerusalem, Israel, pp. 261–274.

Sutherland, J.W., Fricke, P.W. and Hill, R.J. (1989) The entomological and thermodynamic performance of a pneumatic conveyor wheat disinfestor using heated air. *Journal of Agricultural Engineering Research* 44, 113–124.

Thoroski, J.H., Lapp, H.M. and Madrid, F.J. (1985) *Stored wheat insect responses to changes in their physical environment.* Paper No. NCR 85-503 presented to the Annual Meeting of American Society of Agricultural Engineers, Bismarck, North Dakota.

Thorpe, G.R., Fricke, P.W. and Goldsworthy, I.D. (1982) An instrumentation system for a 50 tph fluidized bed high temperature wheat disinfestation plant. In: *Proceedings of the Scientific Instruments in Primary Production Conference, Australian Scientific Industry Association*, Brisbane, Australia, pp. 151–161.

Thorpe, G.R., Evans, D.E. and Sutherland, J.W. (1984) The development of a continuous flow fluidized bed high temperature disinfestation process. In: Ripp, B.E. (ed.) *Proceedings of the International Symposium, Practical Aspects of Controlled Atmosphere and Fumigation in Grain Storage*, Perth, Australia, 1983. Elsevier, Amsterdam.

Tilton, W.E. and Schroeder, W.H. (1963) Some effects of infrared irradiation on the mortality of immature insects in kernels of rough rice. *Journal of Economic Entomology* 56, 720–730.

Tilton, E.W., Vardell, H.H. and Jones, R.D. (1983) Infrared heating with vacuum for the control of the lesser grain borer (*Rhyzopertha dominica* F.) and rice weevil (*Sitophilus oryzae* L.) infesting wheat. *Journal of the Georgia Entomological Society* 18, 61–64.

Tsuchiya, T. and Kosaka, K. (1943) Experimental studies on the resistance of the rice weevil, *Calandra oryzae* L., to heat. II. Lethal effects of heat upon the larvae and pupae. *Berichte d Ohara Institut Landwirtscharftlich Forschung* 9, 191–207.

USDA (1965) *Losses in Agriculture*. Agriculture Handbook No. 291, USDA, Washington, DC, 120 pp.

Vardell, H.H. and Tilton, E.W. (1981a) Heat sensitivity of the lesser grain borer, *Rhyzopertha dominica* (F.). *Journal of the Georgia Entomological Society* 16, 116–121.

Vardell, H.H. and Tilton, E.W. (1981b) Control of the lesser grain borer, *Rhyzopertha dominica* (F.) and the rice weevil, *Sitophilus oryzae* (L.), in wheat with a heated fluidized bed. *Journal of the Kansas Entomological Society* 54, 481–485.

Wang, S., Ikediala, J.N., Tang, J., Hansen, J.D., Mitcham, E., Mao, R. and Swanson, B. (2001) Radio frequency treatments to control codling moth in in-shell walnuts. *Postharvest Biology and Technology* 22, 29–38.

Wang, S., Tang, J., Johnson, J.A., Mitcham, E., Hansen, J.D., Cavalieri, R.P., Bower, J. and Biasi, B. (2002) Process protocols based on radio frequency energy to control field and storage pests in in-shell walnuts. *Postharvest Biology and Technology* 26, 265–273.

Wang, S., Tang, J., Cavalieri, R.P. and Davis, D.C. (2003) Differential heating of insects in dried nuts and fruits associated with radio frequency and microwave treatments. *Transactions of the American Society of Agricultural Engineers* 46, 1175–1182.

Watters, F.L. (1976) Microwave radiation for the control of *Tribolium confusum* in wheat and flour. *Journal of Stored Products Research* 12, 19–25.

Wehling, R.L. (1998) Infrared spectroscopy. In: Nielsen, S.S. (ed.) *Food Analysis*, 2nd edn. Aspen Publications, Gaithersburg, Maryland, pp. 413–424.

Whitney, W.K., Nelson S.O. and Walkden, H.H. (1961) *Effects of High-frequency Electric Fields on Certain Species of Stored-grain Insects*. Marketing research report No. 455, USDA, Washington, DC.

Winterbottom, D.C. (1922) *Weevil in Wheat and Storage of Grain in Bags: a Record of Australian Experience During the War Period (1915 to 1919)*. R.E.E. Rogers, Adelaide, Australia.

Wolpert, V. (1967) Needless losses. *Far Eastern Economic Review* 55, 411–412.

Wright, E.J., Sinclair, E.A. and Annis, P.C. (2002) Laboratory determination of the requirements for control of *Trogoderma variabile* (Coleoptera: Dermestidae) by heat. *Journal of Stored Products Research* 38, 147–155.

9 Considerations for Phytosanitary Heat Treatment Research

G.J. HALLMAN

USDA-ARS Subtropical Agricultural Research Center, Weslaco, Texas, USA; e-mail: ghallman@weslaco.ars.usda.gov

9.1 Introduction

The efficacy of heat treatments is more susceptible to variations in biological and physical factors than are all other major phytosanitary treatments. The other major treatments – cold storage, methyl bromide fumigation and ionizing irradiation – can be applied in a generic sense; i.e. the treatment does not differ between different species of pests and commodities.

For example, a cold treatment at 1.1°C against Mediterranean fruit fly (*Ceratitis capitata* Wiedemann) requires 14 days for fruits ranging in size from grape to grapefruit (APHIS, 2005). That is because, though fruit size does affect the time required to bring fruits down to the treatment temperature, cold treatments are usually initiated when the warmest areas within the load reach the treatment temperature. Likewise, the same methyl bromide fumigation parameters are used against *Anastrepha* spp. fruit flies on citrus fruit from Mexico, whether the host is a small tangerine or a large grapefruit (APHIS, 2005). With irradiation, the same dose per pest is used regardless of the commodity (Hallman, 2002).

Minor differences in temperature ranges of heat that kill insects but are tolerated by fresh agricultural commodities (~40–50°C) can result in significant changes in mortality of quarantined pests. For example, Jang (1986) found that decreases of one degree between 43 and 48°C could more than double lethality times for three species of tephritid fruit flies (see Fig. 9.1).

Changes in the atmospheric composition of the treatment chamber can also affect heat treatment efficacy. The time for achieving 99% mortality in lightbrown apple moth (*Epiphyas postvittana* Walker) subjected to 40°C ambient air was reduced by half when the treatment was done under an atmosphere of 2% oxygen and 5% carbon dioxide (Whiting *et al.*, 1999).

Failures of commercial heat treatments have shown that small changes in temperature and atmospheric composition indicate problems with research

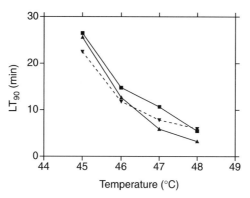

Fig. 9.1. LT$_{90}$ (time to achieve 90% mortality) for first-instar Mediterranean fruit fly (Medfly, ■), oriental fruit fly (Orifly, ▼) and melon fly (▲) at different temperatures (from Jang, 1986).

methodology and regulatory implementation (*Zee et al.*, 1989; Shellie and Mangan, 2002a). The objective of this chapter is to highlight and discuss issues related to heat treatment research that might impact disinfestation efficacy so that researchers may avoid the repetition of mistakes, and regulators and industry may make the most efficacious application of developed treatments. The discovery during commercial application that a treatment does not provide the required level of efficacy can result in great interruption of exports with much economic loss, not to mention an infestation of new pests.

9.2 Source of Research Organisms

Most postharvest disinfestation research is performed with laboratory colonies that have been under captivity for many generations and are thus adapted to environmental conditions that are acceptable to humans; these also provide abundant numbers for rapid testing of the large numbers of uniform organisms required in quarantine treatment research.

However, laboratory-reared organisms may differ from feral ones. If these differences are in the susceptibility of the species to the phytosanitary treatment, a potential problem exists. If the laboratory organisms are more tolerant of the treatment than are feral organisms, the devised treatment will be more severe than it needs to be. If, however, the test organisms are more susceptible than feral ones, the stage is set for an ineffective treatment.

Fortunately, finding a quarantine pest in a shipped commodity is not the norm, so the treatment may not be challenged often. But given the large quantities and frequencies of some commodities shipped, the cumulative effect of a sub-efficacious treatment could eventually result in a pest infestation on those occasions when pests are actually present. Mangan *et al.* (1997) calculated that, under certain conditions, Mexican fruit fly infestation levels in mangoes from Mexico were sufficiently high to challenge the security of approved quarantine treatments for importation into the USA. This prediction was fulfilled when live larvae were found in hot water-treated mangoes from Mexico in 2000 (Scruton, 2000).

Few data illuminate the potential of this problem. Mangan and Hallman (1998) analysed two papers on tephritid fruit flies and concluded that differences in response to heat treatments between the feral and colony flies were not of concern in these cases. Wang *et al.* (2005) found no significant differences in mortality between two colonies of fifth-instar navel orangeworm (*Amyelois transitella* Walker) that had been in culture for five and 100 generations when they were subjected to temperatures of 48–52°C.

However, insects in captivity may evolve rapidly to the laboratory-rearing environment, which is almost always different to the natural setting. Five generations might be enough to see significant change compared with the founding generation. For example, Economopoulos and Loukas (1982) measured genetic change after only four generations of rearing olive fruit fly (*Dacus oleae* Gmelin) on a semi-artificial diet.

Reproduction within laboratory colonies under unchanging, mild temperature conditions allows for the loss of tolerance to temperature extremes if that tolerance exerts a developmental cost and is selected against. *Drosophila melanogaster*, reared at 28°C for 15 years, received more heat shock protection than those reared at 18 or 25°C even when reared at the same temperature for one generation before testing, demonstrating that the population reared at consistently higher temperatures evolved to be more resistant to higher temperatures (Cavicchi *et al.*, 1995).

Therefore, it would be advisable to test a research colony against available feral organisms for differences in reaction to heat in the early stages of the development of a phytosanitary treatment, especially if the pest exists in the wild at temperatures consistently higher than those used in the rearing facility. This is seldom performed, in part because it is often difficult to obtain and manipulate sufficient feral organisms for adequate testing.

9.3 Rearing Conditions

Not only the genotype, but also the physical rearing conditions of a colony may alter its susceptibility to heat (see Fig. 9.2). Survival of Caribbean fruit fly (*Anastrepha suspensa* Loew) third-instars reared inside carambolas at 20, 25 and 30°C and then immersed within the ambient temperature (24°C)-acclimated fruit in 46°C water for 25 min was 0.00, 0.083 and 0.59%, respectively, showing that rearing temperature had a direct effect on heat treatment survival (Hallman, 1994).

In the previously cited study on *Drosophila melanogaster* (Cavicchi *et al.*, 1995), the rearing temperature of individuals from the population kept for many generations at 18°C had a great effect on increasing thermotolerance. Adults reared at 18°C had a mean survival rate of 10% when subjected to 41°C for 30 min, and a survival rate of 70% when reared at 28°C and then subjected to 41°C for 30 min. The survival rate of the population kept for many

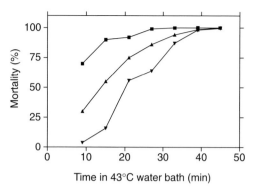

Fig. 9.2. Mortality of third-instar Caribbean fruit fly reared at three different temperatures (20°C, ■; 25°C, ▲; 30°C, ▼) and immersed in water at 43°C (from Hallman, 1994).

generations at 28°C was no different when individuals were reared at 18 and 28°C, but the population kept for many generations at 25°C showed variable survival rates of 34, 55 and 14.7%, respectively, when individuals were reared at 18, 25 and 28°C, indicating complex responses to temperature conditioning. Chapter 11 of this volume analyses the role of heat shock proteins in the thermotolerance of pests and their fruit hosts.

The significant differences in response to heat indicate that the temperature regime a pest is exposed to during rearing – even if it does not consist of heat spikes that may elicit the formation of heat shock proteins – can affect thermotolerance and should be taken into consideration when developing heat quarantine treatments. Rearing in facilities exposed to ambient temperatures of the areas from where commodities are to be exported is preferable to controlled laboratory temperatures. Whenever that cannot be done because of quarantine restrictions, temperatures in the laboratory facilities could be made more like ambient conditions if experience shows that significant induced thermotolerance occurs.

Denlinger and Yocum (1998) discussed numerous physiological responses to heat and their potential contributions to the heat-induced death of an organism. These differences in the amounts and qualities of these physiological components could affect the susceptibility of a pest to heat quarantine treatments. For example, host could theoretically affect thermotolerance; this possibility has not been studied.

9.4 Methods of Infesting Commodities for Disinfestation Research

An effective phytosanitary treatment must achieve the required level of lethality to feral organisms naturally infesting the host commodity. Because of the low level of infestation naturally found in most quarantine pests and lack of uniformity of stages, phytosanitary research is usually conducted on hosts infested in a controlled environment, which may affect efficacy.

A more natural method of infestation is to place the host commodity in a cage with adult pests and let them oviposit on the commodity. The immatures are reared until they reach the desired stage of development, usually the one most tolerant of the treatment. If the most tolerant stage is a later one, such as fifth-instar codling moth (*Cydia pomonella* L.) (Yokoyama *et al.*, 1991), the commodity may have to remain infested at a moderate temperature for many days until that stage is reached. An advanced instar may not be the most thermotolerant stage but, because it penetrates deepest into the commodity, it is most likely to survive a conductive heat treatment (Sharp, 1988).

By the time a pest feeding on a fresh commodity reaches a later instar, the commodity may be degraded and not heat at the same rate as sound produce. Shellie and Mangan (2002b) found that mangoes infested with Mexican fruit fly eggs (*Anastrepha ludens* Loew) lost 30% of their weight by the time the insects reached the late third-instar and heated faster than sound mangoes, requiring 74% of the time to reach the same seed-surface temperature. On the contrary, grapefruits infested with Mexican fruit fly at the egg stage and then subjected to radio frequency (RF) heating (27 MHz) when the insects reached the late third-instar heated more slowly than sound grapefruits artificially infested with third-instars immediately before heating (Hallman, unpublished data).

Because of the poor quality of fresh commodities after heavy infestations of pests through late instars, other techniques have been developed to treat sound fruit infested with late instars (see Fig. 9.3). Most of these techniques involve boring a hole with a cork borer, drill or similar device to place late-instar pests in the centre of fruits (Mangan and Ingle, 1994; Neven *et al.*, 1996). A fruit plug covers the insects and is secured with a glue or tape-like substance. This results in a worst-case scenario for convective heating (conventional hot water and heated air treatments) of all test subjects being placed in the centre of the fruit, where the heat load is the mildest. The theory is that, if all of these insects can be killed, then insects anywhere in the fruit will be killed.

None the less, this technique also assumes that larvae reared on a diet and placed in fruit near the end of the larval stage are not more susceptible to heat than are larvae that had developed from the egg stage inside of the fruit. Shellie and Mangan (2002b) noted that cage oviposition and artificial infestations with third-instars of mangoes yielded similar numbers of West Indian (*Anastrepha obliqua* Macquart) fruit fly puparia after immersion in 46.1°C water for 70 min.

However, cage infestation resulted in only a fraction of the number of larvae present in the fruit compared with artificial infestation; based on percentage survival, cage and artificial infestations resulted in 42 and 3.7% survival, respectively. Additionally, the cage-infested mangoes heated much faster, reaching a maximum mean seed surface temperature of 42.6 and 38.5°C for cage- and artificially-infested fruit, respectively. Furthermore, in the cage technique the larvae were spread among the fruit pulp; those larvae closer to the surface received a greater heat load.

In the artificial technique, the larvae were all placed at the seed surface where the heat load was the lowest. Therefore, one should conclude that larvae from the cage infestation technique were much more thermotolerant than those

Fig. 9.3. An example of artificial infestation of fruit for phytosanitary treatment research. Fruit fly third-instars reared on meridic diet (in plastic cups) are inserted into the fruit centre via a cork borer (left-hand fruit). Fruit plugs are reinserted and sealed with hot-melt glue or another similar substance (right-hand fruit shows glue cover).

from artificial infestation. The cage infestation technique resulted in larvae closer to what would be expected in a feral situation. Unfortunately, artificial infestation was then used to develop mango hot water immersion treatments, thus leaving the treatment open to further failure in the future if, indeed, fruit-reared larvae are more thermotolerant than diet-reared larvae subsequently placed in fruit.

A modification of artificial infestation is to allow diet-reared, artificially infested insects to feed within the commodity for a day or so after infestation in order to achieve some semblance of natural infestation before the treatment is performed (Neven et al., 1996).

Artificial infestation with tephritid fruit flies typically involves rearing larvae on a meridic diet,[1] washing them from the diet via a sieve and placing them in cavities made in fruit. For large-scale testing performed to confirm phytosanitary treatments, larvae may remain in water or another medium until they are placed in the fruit, and it may be hours between removal from the diet and initiation of the treatment. Tephritid larvae go through physiological changes involving irreversible cessation of feeding, emergence from the fruit, increased mobility and entering the soil before pupation in the hours after removal from a food source (Fraenkel and Bhaskaran, 1973). These physiological changes may be associated with changes in thermotolerance.

[1] A diet consisting of both chemically defined compounds, such as sucrose, and undefined compounds, such as corn meal.

The effect of infestation method on the thermotolerance of quarantine pests is one area of heat quarantine treatments that requires further research to prevent treatment failure. It is known that low oxygen tension synergizes the effect of heat on insects (see Chapter 10, this volume). Perhaps insects placed in the centre of fruit suffer from a greater oxygen deficit than those naturally infesting fruit and are thus easier to kill. Possibly the competition of 25–50 larvae placed together in a fruit cavity causes stress that lessens their thermotolerance. As larvae develop in a fruit they may modify their immediate environment, making it more favourable to their survival, which may mean an increase in thermotolerance compared with larvae reared elsewhere and artificially placed in a fruit immediately before treatment. Conversely, some of these changes resulting from artificial infestation could increase thermo-tolerance.

Another failure of commercial heat quarantine treatment illustrates potential pitfalls in research methodology. A two-stage hot water treatment was developed to kill only tephritid fruit fly eggs and early-instars near the fruit surface of papaya (Couey and Hayes, 1986). It was concluded that no stage beyond first-instar would be found in papayas harvested no riper than one-quarter skin colour development, so treatment need not kill later instars. Within a couple of years of the treatment's implementation, live late-instar oriental fruit fly (*Bactrocera dorsalis* Hendel) were found in treated papayas, and the treatment was withdrawn. It was subsequently discovered that some papaya fruit had an opening at the blossom end (see Fig. 9.4) that allowed fruit flies to oviposit in the papaya at an earlier stage than quarter-ripe; the resulting larvae could enter the centre of the fruit, where lethal temperatures might not be reached during hot water treatment (Zee et al., 1989).

Why was this opening not detected during the research? A survey in 1987, shortly after and in the same district where the research was conducted, revealed that fruit from 5–31% of the local papaya trees had that opening (Zee

Fig. 9.4. Pole-to-pole cross-section of developing papaya showing abnormal small opening from blossom end to centre (left). Fruit flies may lay eggs at the blossom end and larvae enter the centre of fruit where they are harder to kill via convective heat than if they were more superficially located in the pulp.

et al., 1989). Commercial confirmation of this treatment used fruit from one orchard without an estimate of the level of infestation. It is possible that there was little or no infestation, hence not a robust test. More troubling, during the laboratory-testing phase with cage-infested fruit, some survivors were found but dismissed as 'accidental reinfestations' (Couey and Hayes, 1986).

The lessons from this research are that fruits used in establishing a quarantine treatment should broadly represent the population of fruit to be treated commercially and that one should not so easily dismiss unexpected results. In fact, one could argue that unexpected results are what cause science to advance. Another lesson is that no entomologist was included among the developers of this treatment, which supports the contention that multi-disciplinary science is important and should be adequately represented in any serious research endeavour.

9.5 Determination of Disinfestation Efficacy

Researchers may determine disinfestation efficacy in ways that are not feasible for inspectors, resulting in possible treatment failures. For example, in 2000, live Mexican fruit fly larvae were found upon cutting open hot water-treated mangoes imported into the USA from Mexico (Scruton, 2000). In the research used to develop the treatment, treated mangoes were placed in trays and the larvae therein were not counted as survivors unless they emerged from the fruit by their own power or were washed out weeks later and resulted in 'normally formed' puparia (Sharp, 1988). Larvae may have been alive for days after the treatment but did not emerge from the fruit. Fully formed third-instar tephritids often remain in a quiescent stage inside fruit, apparently waiting for certain conditions before exiting the fruit and pupariating in the soil.

Although the way that this research was performed may (or may not) have been sufficient to prevent fruit fly infestations – as live larvae that cannot emerge from fruit would not complete development unless the fruit was opened – it did not satisfy the requirements of phytosanitary inspectors. Regulatory officials cannot wait for live larvae to form normal-looking puparia before declaring treatment failure: they do not have the facilities, mandate or time. Fresh produce must be on its way as soon as possible to maintain quality; it cannot be quarantined for several days to see if live larvae found will develop further. Phytosanitary inspectors should not keep quarantine pests alive at ports of entry and may be prohibited by regulations from doing so.

Furthermore, in the case of mango hot water immersion, only 'normal' puparia were counted as survivors (Sharp, 1988). Thomas and Mangan (1995) found that normal adults emerged from abnormal Mexican and West Indian fruit fly puparia formed by heated third-instars. This is understandable when one considers that the puparium is the hardened, dead, cast skin of the third-instar and simply the case in which the fly pupates. It need not be entirely normal-looking for the fly to successfully pupate and develop to the adult inside.

A quarantine treatment must produce the desired effect by the time inspectors assay the treated produce. For heat treatments, as well as most other quarantine treatments (except ionizing irradiation), pests should be dead upon examination by regulatory officials. A good rule of thumb is that pests should not be detectably alive (not moving when prodded) when left at ambient temperatures 24 h after the termination of treatment. Pests may seem dead immediately after a heat treatment, but recover hours later. Alternatively, if it is proved that live insects observed until a defined time after treatment will not result in an infestation, live insects need not jeopardize treatment efficacy, which is how irradiation phytosanitary treatments are evaluated (Hallman, 2002).

9.6 Commodity Conditioning

Subjecting fresh commodities to pre-treatment cold or heat regimes that are not lethal to quarantine organisms but condition the commodities to better tolerate the subsequent lethal heat treatment may condition the pests also, thereby increasing their tolerance to the treatment.

Analysis of the influence of heat shock proteins on pests and commodities is found in Chapter 11, this volume. No preconditioning temperature treatment is currently used commercially to lessen heat damage to a commodity, but quarantine treatments that do so should also test the effect of this preconditioning on the tolerance of quarantine pests to the treatment and make adjustments as required to ensure that it continues to provide the necessary level of control.

Hydrocooling of mangoes immediately after hot water treatment was used for several years, ostensibly to reduce damage to fruit despite evidence that this could increase the ability of tephritid fruit flies to survive treatment (see Fig. 9.5; Hallman and Sharp, 1990; Mangan and Hallman, 1998; Shellie and Mangan, 2002a). This practice may have had a role in the finding of live Mexican fruit fly larvae in mangoes in 2000. In response, the protocol was changed to allow immediate hydrocooling at $\geq 21°C$ if the treatment temperature was extended by 10 min. An alternative is to wait 30 min before hydrocooling.

9.7 Commercial Possibilities

Industry must be able to implement treatments economically for them to be feasible. Heat treatments are especially susceptible to problems resulting from scaling up a process that works on an experimental scale to industrial quantities, because the treatments are relatively short and heating is modified by several factors such as: (i) initial commodity temperature; (ii) volume and velocity of heat applied; and (iii) rate of heat loss during treatment. The level of difficulty is higher for heated air treatments compared with hot water because of the compressibility of air and variability of humidity within it. Williamson and

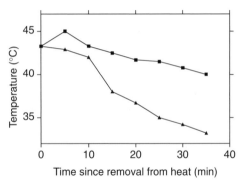

Fig. 9.5. Temperature at seed surface of heated mangoes allowed to cool in still air at 21°C (■) or placed in circulating water at 27°C (▲) (from Hallman and Sharp, 1990). Note that the air-cooled mango seed surface temperature continued to increase for up to 5 min following removal from heat due to latent heat in the pulp.

Winkleman (1994) offer suggestions for the development of commercial heated air treatment facilities.

Commercial implementation of novel heat treatments such as RF and ohmic might be expected to be more difficult still because of additional complicating factors such as effects of: (i) product size, shape and internal morphology; (ii) orientation in relation to the energy fields involved; and (iii) surface moisture on heating uniformity. Chapter 12 of this volume discusses the challenges in designing commercial RF heat treatments.

Commercial-scale heated air treatment facilities may not be able to achieve the same heating rate as in research. Neven (1998) noted that the heating rate was positively related to mortality in codling moth, meaning that a commercial facility using a lower heating rate might not achieve the same degree of efficacy as research done at a higher heating rate (see Fig. 9.6).

Prolonging the time to reach 44°C from 15 to 120 min resulted in greater thermotolerance in Mexican fruit fly (Thomas and Shellie, 2000). A slower heating rate yielded lower mortality of navel orangeworm heated to 48°C, but at no other temperatures between 46 and 52°C (Wang *et al.*, 2005). Conversely, a slower heating rate may result in a greater level of mortality simply because the total treatment time is longer. For example, a slower heating rate resulted in greater fruit fly kill (Armstrong *et al.*, 1995). Likewise, a slower heating rate resulted in higher mortality for codling moth subjected to 46–52°C (Wang *et al.*, 2002). These examples show that the effect of heating rate on mortality is not uniform. As mentioned before, Chapter 6, this volume, covers this topic in greater depth.

9.8 Conclusions and Recommendations

Heat disinfestation treatments remain the most studied of all phytosanitary treatments (Mangan and Hallman, 1998) and the type presenting the most

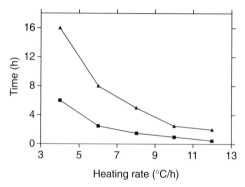

Fig. 9.6. Heating rate *versus* time required at final treatment temperature (44, ▲ and 46°C, ■) to achieve 95% mortality of fifth-instar codling moth in meridic diet in glass vials immersed in heated water (from Neven, 1998).

difficulties in research and commercial application. Although much research has been conducted, fundamental questions remain.

Research methodology can affect the efficacy of heat quarantine treatments in ways that may lead to failures of the treatment when applied commercially, resulting in – at best – an interruption of trade in the affected commodities and – at worst – the introduction and establishment of a new pest. Heat treatments seem to be more susceptible to problems than other major treatments such as cold storage, fumigation or irradiation. Sources of problems may be: (i) the population of organism used; (ii) rearing conditions; (iii) infestation methods; (iv) methods of determining efficacy; (v) commodity conditioning; and (vi) differences between research and commercial applications.

Researchers should be mindful of the differences between laboratory environments and agroecosystems and how these differences may affect the results, test for possible effects and compensate for them when designing commercial treatments. Studies of the physiological effects of heat on insects can illuminate factors that might lead to the formulation of more effective treatments (Neven, 2000).

9.9 References

APHIS (Animal and Plant Health Inspection Service) (2005) *Treatment Manual* (http://www.aphis.usda.gov/ppq/manuals/port/Treatment_Chapters.htm).

Armstrong, J.W., Hu, B.K.S. and Brown, S.A. (1995) Single-temperature forced hot-air quarantine treatment to control fruit flies (Diptera: Tephritidae) in papaya. *Journal of Economic Entomology* 88, 678–682.

Cavicchi, S., Guerra, D., La Torre, V. and

Huey, R.B. (1995) Chromosomal analysis of heat-shock tolerance in *Drosophila melanogaster* evolving at different temperatures in the laboratory. *Evolution* 49, 676–684.

Couey, H.M. and Hayes, C.F. (1986) Quarantine procedure for Hawaiian papaya using fruit selection and a two-stage hot-water immersion. *Journal of Economic Entomology* 79, 1307–1314.

Denlinger, D.L. and Yocum, G.D. (1998) Physiology of heat sensitivity. In: Hallman, G.J. and Denlinger, D.L. (eds) *Temperature Sensitivity in Insects and Application in Integrated Pest Management.* Westview Press, Boulder, Colorado, pp. 7–53.

Economopoulos, A.P. and Loukas, M.G. (1982) Selection at the alcohol dehydrogenase locus of the olive fruit fly under laboratory rearing: effect of larval diet and colony temperature. In: Cavalloro, R. (ed.) *Fruit Flies of Economic Importance; Proceedings of the CEC/IOBC International Symposium,* Athens, Greece, 16–19 November 1982, pp. 178–181.

Fraenkel, G. and Bhaskaran, G. (1973) Pupariation and pupation in cyclorrhaphous flies (Diptera): terminology and interpretation. *Annals of the Entomological Society of America* 66, 418–422.

Hallman, G.J. (1994) Mortality of third-instar Caribbean fruit fly (Diptera: Tephritidae) reared at three temperatures and exposed to hot water immersion or cold storage. *Journal of Economic Entomology* 87, 405–408.

Hallman, G.J. (2002) Irradiation as a quarantine treatment. In: Molins, R. (ed.) *Food Irradiation: Principles and Applications.* Wiley Interscience, New York, pp. 113–130.

Hallman, G.J. and Sharp, J.L. (1990) Mortality of Caribbean fruit fly (Diptera: Tephritidae) larvae infesting mangoes subjected to hot-water treatment, then immersion cooling. *Journal of Economic Entomology* 83, 2320–2323.

Jang, E.B. (1986) Kinetics of thermal death in eggs and first instars of three species of fruit flies (Diptera: Tephritidae). *Journal of Economic Entomology* 79, 700–705.

Mangan, R.L. and Hallman, G.J. (1998) Temperature treatments for quarantine security: new approaches for fresh commodities. In: Hallman, G.J. and Denlinger, D.L. (eds) *Temperature Sensitivity in Insects and Application in Integrated Pest Management.* Westview Press, Boulder, Colorado, pp. 201–234.

Mangan, R.L. and Ingle, S.J. (1994) Forced hot-air quarantine treatment for grapefruit infested with Mexican fruit fly (Diptera: Tephritidae). *Journal of Economic Entomology* 87, 1574–1579.

Mangan, R.L., Frampton, E.R., Thomas, D.L. and Moreno, D.S. (1997) Application of the maximum pest limit concept to quarantine security standards for the Mexican fruit fly (Diptera: Tephritidae). *Journal of Economic Entomology* 90, 1433–1440.

Neven, L.G. (1998) Effects of heating rate on the mortality of fifth-instar codling moth (Lepidoptera: Tortricidae). *Journal of Economic Entomology* 91, 297–301.

Neven, L.G. (2000) Physiological responses of insects to heat. *Postharvest Biology and Technology* 21, 103–111.

Neven, L.G., Rehfield, L.M. and Shellie, K.C. (1996) Moist and vapour forced air treatments of apples and pears: effects on the mortality of fifth instar codling moth (Lepidoptera: Tephritidae). *Journal of Economic Entomology* 89, 700–704.

Scruton, T. (2000) Inspectors find larvae in Mexican mangoes. *The Packer* 20 March, A7.

Sharp, J.L. (1988) Status of hot water immersion quarantine treatment for Tephritidae immatures in mangoes. *Proceedings of the Florida State Horticultural Society* 101, 195–197.

Shellie, K.C. and Mangan, R.L. (2002a) Cooling method and fruit weight: efficacy of hot water quarantine treatment for control of Mexican fruit fly in mango. *HortScience* 37, 910–913.

Shellie, K.C. and Mangan, R.L. (2002b) Hot water immersion as a quarantine treatment for large mangoes: artificial *versus* cage infestation. *Journal of the American Society of Horticultural Science* 127, 430–434.

Thomas, D.B. and Mangan, R.L. (1995) Morbidity of the pupal stage of the Mexican and West Indian fruit flies (Diptera: Tephritidae) induced by hot-water immersion in the larval stage. *Florida Entomologist* 78, 235–246.

Thomas, D.B. and Shellie, K.C. (2000) Heating rate and induced thermotolerance in Mexican fruit fly (Diptera: Tephritidae) larvae, a quarantine pest of citrus and mangoes. *Journal of Economic Entomology* 93, 1373–1379.

Wang, S., Ikediala, J.N., Tang, J. and Hansen, J.D. (2002) Thermal death kinetics and heating rate effects for fifth-instar *Cydia pomonella* (L.) (Lepidoptera: Tortricidae). *Journal of Stored Products Research* 38, 441–453.

Wang, S., Johnson, J.A., Tang, J. and Yin, X. (2005) Heating condition effects on thermal resistance of fifth-instar *Amyelois transitella* (Walker) (Lepidoptera: Pyralidae). *Journal of Stored Products Research* 41, 469–478.

Whiting, D.C., Jamieson, L.E., Spooner, K.J. and Ly-Yee, M. (1999) Combination high-temperature controlled atmosphere and cold storage as a quarantine treatment against *Ctenopseustis obliquana* and *Epiphyas postvittana* on 'Royal Gala'

apples. *Postharvest Biology and Technology* 16, 119–126.

Williamson, M.R. and Winkleman, P.M. (1994) Heat treatment facilities. In: Paull, R.E. and Armstrong, J.W. (eds) *Insect Pests and Fresh Horticultural Products: Treatments and Responses*. CAB International, Wallingford, UK, pp. 249–271.

Yokoyama, V.Y., Miller, G.T. and Dowell, R.V. (1991) Response of codling moth (Lepidoptera: Tortricidae) to high temperature, a potential quarantine treatment for exported commodities. *Journal of Economic Entomology* 84, 528–531.

Zee, F.T., Nishina, M.S., Chan, H.T., Jr. and Nishijima, K.A. (1989) Blossom end defects and fruit fly infestation in papayas following hot water quarantine treatment. *HortScience* 24, 323–325.

10 Heat with Controlled Atmospheres

E.J. MITCHAM

Department of Plant Sciences, University of California, Davis, California, USA; e-mail: ejmitcham@ucdavis.edu

10.1 Introduction

Arthropod pest mortality is accelerated by combining heat treatments with controlled atmospheres (CA) – generally elevated CO_2 and/or reduced O_2. Controlled atmospheres treatments can also be insecticidal by themselves. Low O_2 and elevated CO_2 atmospheres have been used for many years to control stored-product pests in grains (De Lima, 1990) and dried nuts and to retain the quality of fresh fruits and vegetables in extended storage or transportation. Although CA are not yet commercially used for insect control in fresh fruit and vegetables, there has been considerable research in this area (Mitcham *et al.*, 2001).

Controlled atmospheres that are insecticidal generally contain $\geq 20\%$ CO_2 and/or $\leq 1\%$ O_2, depending on the temperature, with the remainder of the atmosphere composed of N_2 gas. These atmospheres are outside the optimum range used for storage or transport of nearly all fresh fruits and vegetables, and generally induce stress in the commodity. Overcoming this hurdle is generally the limiting factor in developing an effective insecticidal CA treatment.

Insecticidal CA treatments at either ambient temperatures or typical storage temperatures generally require days to completely control arthropod pests, while control occurs within hours at high temperatures (Mitcham *et al.*, 2001). The mechanisms by which CA control arthropod pests are only partly understood. This chapter covers the effects of CA on arthropods and commodities and the synergistic effects of heat in combination with CA on arthropod mortality and commodity tolerance.

10.2 Mode of Action of Controlled Atmospheres on Insects

Arthropods cope with reduced O_2 and elevated CO_2 atmospheres by a reduction in metabolic rate, also called metabolic arrest (Hochachka, 1986). This reduction in metabolism lessens the pressure on the organism to initiate anaerobic metabolism, which could lead to accumulation of toxic byproducts, but also to a reduction in ATP production. Despite the similarities in response, high CO_2 generally yields greater arthropod mortality than low O_2. There appears to be a greater decrease in ATP and energy charge in arthropods exposed to high CO_2 as compared with low O_2, possibly due to greater membrane permeability under CO_2 atmospheres that causes inefficient production of ATP (Mbata *et al.*, 2000; Zhou *et al.*, 2001; Mitcham *et al.*, 2006). Reduced O_2 and elevated CO_2 atmospheres can have an additive effect in some cases, depending on the concentrations used. The effect of these atmospheres on arthropods also depends on temperature, species and life stage (Mitcham *et al.*, 2006).

Schroeder and Eidman (1986) found that CA containing high CO_2 were more effective against bark beetles than those containing mainly N_2 – and therefore low O_2. Similar results were obtained by Soderstrom *et al.* (1996) when treating diapausing *Cydia pomonella* and by Zhou *et al.* (2001) with *Platynota stultana* (see Fig. 10.1), where O_2 concentrations as low as 1% resulted in lower mortality than in a 20% CO_2 atmosphere at 20°C.

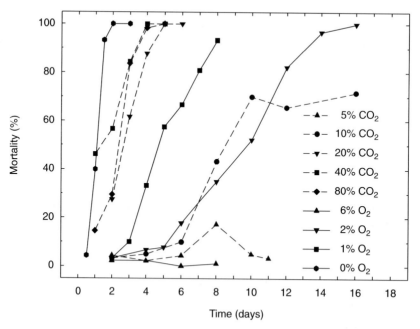

Fig. 10.1. The metabolic heat rate of four leafroller (*Platynota stultana*) pupae under extended exposure at 20°C to air; 5, 10, 20, 40 and 80% CO_2 (in 21% O_2); and 6, 2, 1 and 0% O_2. All data points represent the mean of three replications. The metabolic heat rates were weight-corrected by the mean weight of all samples (from Zhou *et al.*, 2001).

However, the opposite trend applied to treatments of *C. pomonella* adults and eggs (Soderstrom *et al.*, 1991). Mortality may be affected by the immediate or latent effects of CO_2, such as anaesthesia (Nicolas and Sillans, 1989). In some insect species, the anaesthetic effect of CO_2 decreased as temperature increased (Dentener *et al.*, 1999).

The reasons for this variability in response to low O_2 and elevated CO_2 are unknown. Zhou *et al.* (2001) suggested that failure of membrane function occurs more quickly under hypercarbia than under hypoxia, contributing to the higher susceptibility of omnivorous leafroller pupae to hypercarbia. Membrane function fails under low O_2 mainly because not enough energy can be supplied to maintain membrane gradients (Hochachka, 1986). However, elevated CO_2 could directly increase the permeability of membranes (Friedlander, 1983; Hochachka, 1986).

Carbon dioxide has also been shown to increase intercellular Ca^{2+} concentrations by decreasing pH (Lea and Ashley, 1978). According to Hochachka (1986), a high concentration of Ca^{2+} in the cytosol can cause cell and mitochondrial membranes to become more permeable. Therefore, the failure of membranes under high CO_2 could result from both energy insufficiency and increased membrane permeability. This double effect is more likely to kill arthropod pest cells.

Species and Life Stage Differences

The species of arthropod has a tremendous influence on its susceptibility to CA treatment. This large variability in response has necessitated empirical testing of CA for a wide range of insect and mite pests. Mites are among the most resistant species, requiring a longer exposure to similar CA for 99% mortality (*see* Table 10.1). Life stage also has a large influence on susceptibility to CA treatment.

The relative order of decreasing LT_{99} for life stages of codling moth (*Cydia pomonella*), light brown apple moth (*Epiphyas postvittana*), greenheaded leafroller (*Planotortix octo*) and brownheaded leafroller (*Ctenopseustis obliquana*) was: fifth-instars ≥ third-instars ≥ 3-day-old eggs ≥ first-instars, respectively, for exposure to 0.4% O_2 and 5% CO_2 at 20°C (Whiting *et al.*, 1992b).

However, the relative order of susceptibility between life stages is not consistent for all species. For example, for Pacific spider mites exposed to 45%

Table 10.1. Susceptibility to controlled atmospheres (CA) treatment of selected arthropods.

Arthropod	Treatment (% CO_2)	LT_{99} (days)
Two-spotted spider mite	50	11.9
Pacific spider mite	45	8.1
Western flower thrips	45	5.8
Omnivorous leafroller	45	4.1

CO_2 at 0°C, the relative order of decreasing LT_{99} mortality was: adult $\geqslant$ larva (instar) $\geqslant$ egg (Mitcham *et al.*, 1997). Life stage responses are also different for various types of control treatments such as fumigation, heat and CA.

10.3 Effects of Controlled Atmospheres on Commodities

Although the slowed ripening and reduced decay benefits of CA allow long-distance shipping of commodities with typically short postharvest life, when the CA are outside the range tolerated by the commodity, damage can occur and the benefit is then lost. The likelihood of commodity damage depends on the commodity, concentrations of CO_2 and O_2, temperature and duration of exposure (Kader, 1995).

Benefits

When optimal levels of O_2 and CO_2 are used for fruit storage, reductions in respiration rate and ethylene production and a slowing of compositional changes associated with colour, firmness, flavour and nutritional quality occur (Kader, 1980). Elevated CO_2 atmospheres can reduce decay and the damaging effects of chilling temperatures (Wang, 1990). Atmospheres with less than 5% O_2 reduce ethylene production (Kader, 1995). Ethylene promotes natural ripening and senescence processes in fruits, vegetables, ornamental flowers and plants. Elevated CO_2 atmospheres inhibit the activity of ACC synthase, a key regulatory enzyme of ethylene biosynthesis (Chavez-Franco and Kader, 1993). Ethylene action is also inhibited by elevated CO_2 atmospheres (Sisler and Wood, 1988). Controlled atmospheres decrease the activity of cell-wall degrading enzymes, thereby slowing softening in fruits and vegetables (Wang, 1990). Flavour quality is maintained by slowing the loss of acidity and biosynthesis of flavour volatiles (Kader, 1995).

Drawbacks

Insecticidal CA are often limited in their success as quarantine treatments because of negative effects on the commodity. Fresh fruits and vegetables vary in their tolerance to low O_2 and elevated CO_2 (Kader, 1986; Weichmann, 1986). Threshold concentrations are the levels of O_2 below which or of CO_2 above which physiological damage is expected (Kader, 1995). Exposure of fresh fruits and vegetables to atmospheres outwith these limits can result in various physiological disorders, including: (i) impaired ripening of climacteric fruits such as tomatoes, melons and plums; (ii) internal browning of lettuce, celery, cabbage, pears, apples and peaches; (iii) external browning of tomato skin, pepper calyx and lettuce; and (iv) pitting of cucumbers, mushrooms, apples, and pears (Kader, 1986).

Accumulations of acetaldehyde, ethanol, ethyl acetate and/or lactate, which can lead to off-flavours or toxic effects, are enhanced at O_2 concentrations < 0.5% and/or CO_2 concentrations > 20% (Kader, 1995). Preclimacteric fruit are more tolerant to extreme atmospheres and have more potential to repair damage than postclimacteric fruit (Kader, 1995). Stress from elevated CO_2 is often additive or synergistic to exposure to other environmental factors outside the optimum range, such as O_2 concentration, temperature, relative humidity (RH) and ethylene, leading to a greater negative impact on the commodity.

These limits of tolerance vary at temperatures above or below the optimum for each commodity. Also, a given commodity may tolerate a short exposure to high CO_2 or low O_2 concentrations and show damage only after longer periods. The tolerance to low O_2 decreases as storage temperature or length of storage increases, since O_2 requirements for aerobic respiration increase incrementally as the temperature increases (Boersig et al., 1988).

10.4 Effects of Heat and Controlled Atmospheres on Arthropod Pests

When the added stresses of heat and CA are combined, effective treatments for arthropod control can be accomplished in hours instead of days (Mitcham et al., 2001). The efficacy of CA treatments for control of insect and mite pests varies with temperature. In general, the higher the temperature, the faster mortality is achieved under a given atmosphere (Paul and Armstrong, 1994; Yahia, 1998).

Reducing the availability of O_2 during heat stress hinders the insect's ability to support elevated metabolic demands. Carpenter (1997) found that increases in time and temperature had a greater effect on mortality of thrips and aphids than increases in CO_2 or reductions in O_2. At very low temperatures, mortality can also increase. For example, LT_{99} for the pupae of the omnivorous leafroller (P. stultana) exposed to 45% CO_2 decreased from 3.8 to 2.4 days as temperatures decreased from 5 to 0°C, respectively (Mitcham et al., 1997), while LT_{99} with exposure to 40% CO_2 decreased from 7.6 to 5.9 to 1.5 days with increasing temperatures from 10 to 20 to 30°C, respectively (Zhou et al., 2000; Fig. 10.2).

Insect Response to High-temperature Controlled Atmospheres

The temperature during exposure to CA has a great effect on arthropod mortality. In general, susceptibility to CA is greater at higher temperatures due to enhanced respiratory demand; however, temperatures outside the optimum range for the arthropod can be an added stress (Mbata and Phillips, 2001).

The normal metabolic rate in air of P. stultana pupae tripled from 10 to 20°C and doubled again from 20 to 30°C, reflecting the huge impact of temperature on arthropod metabolism (Zhou et al., 2000). Temperature also

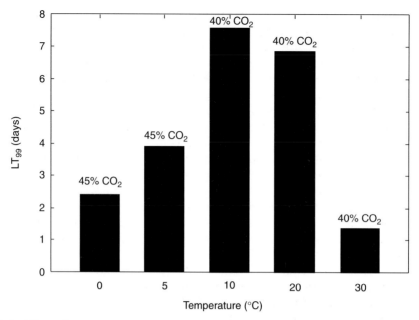

Fig. 10.2. Effect of temperature on LT_{99} estimates for days to mortality of omnivorous leafroller (*Platynota stultana*) pupae exposed to 40 and 45% CO_2.

had a slight but significant effect on the metabolic response of *P. stultana* to both reduced O_2 and elevated CO_2, but the percentage decrease in metabolism by a given low O_2 concentration was higher at higher temperatures, whereas the percentage decrease in metabolism through a certain elevated CO_2 concentration was lower at higher temperatures. However, the response patterns with varying O_2 or CO_2 concentrations at different temperatures were similar.

Huhu (*Prionoplus reticularis*) larvae and eggs treated with CA alone (100% N_2, 100% CO_2 or 50% N_2/50% CO_2) at 20°C had < 36% mortality after 11 days (Dentener *et al.*, 1999). Increasing the treatment temperature to 40°C resulted in LT_{99} figures of 8.3, 6.9 and 7.6 h, respectively. The CA treatments were more effective than the air treatments at 40°C. Similar reductions in LT values by increasing temperatures during CA treatments have been observed for various coleopteran species in stored products (Donahaye *et al.*, 1994), *Cydia pomonella* L. in walnuts (Soderstrom *et al.*, 1996) and *Epiphyas postvittana* (Walker) (Whiting *et al.*, 1991) and *Tetranychus urticae* Koch (Whiting and van den Heuvel, 1995) on fresh produce.

Shellie *et al.* (1997) found that low O_2 was more effective than 20% CO_2 in controlling Mexican fruit fly (*Anastrepha ludens* Loew) larvae during forced hot air heating of grapefruit at 46°C. Reducing the O_2 concentration from 21 to 1% during a heat treatment resulted in 30% shorter exposure time (3.5 *versus* 5 h) to achieve 100% larval mortality. The concentrations of O_2 and CO_2 inside the grapefruit during exposure to 46°C air were significantly different when the fruit were exposed to low O_2 and/or high CO_2 atmospheres during heating. In fact, the internal atmosphere with heat plus CA treatment was similar to that found inside the fruit during hot water exposure (see Fig. 10.3).

However, other researchers have found that CO_2 is more effective than low O_2 in killing internal feeding insects at ambient or refrigerated temperatures. For example, Soderstrom et al. (1990) found 60% CO_2 more effective at 25°C than 0.5% O_2 at killing codling moth larvae. When Benschoter (1987) evaluated the effectiveness of 2, 10 and 20% O_2 and 20, 50 and 80% CO_2 at 10 and 15°C, he found the mortality of Caribbean fruit fly larvae to vary with CO_2 concentration and time of exposure. Neven (2005) and Whiting et al.'s (1992a) research with Lepidopteran species showed that CO_2 concentrations of up to 20% were relatively inconsequential to treatment efficacy at 40°C.

The general pattern of increased treatment efficacy with decreases in O_2 concentrations and increases in treatment temperatures up to 30°C is well established for a range of fresh fruit and stored product-infesting arthropod pests (Hallman, 1994). Whiting et al. (1995) found the importance of atmosphere composition was reduced as the high-temperature tolerance limit of the insect species was approached. For obscure mealy bug (*Pseudococcus affinis* Maskell) exposed to various O_2 concentrations (0.4–20.9%) at high temperatures (35–45°C), the time required for 99% mortality decreased with increasing temperatures and decreasing O_2 concentrations at 35 and 40°C. But at 45°C, the effect of reducing O_2 concentrations was diminished (Whiting and Hoy, 1997).

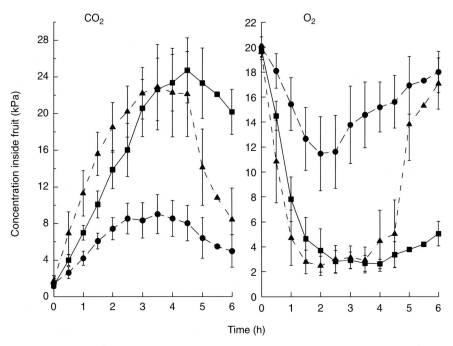

Fig. 10.3. Concentrations of CO_2 (a) and O_2 (b) inside grapefruit during heating in air, ●; controlled atmosphere (1 kPa O_2 + 20 kPa CO_2), ▲; and water at 44°C, ■ (from Shellie et al., 1997).

Zhou *et al.* (2000) found a greater difference in the rate of increase in mortality of *Platynota stultana* pupae with time of exposure to 20, 40 or 79% CO_2 at 10°C than at 20 or 30°C (*see* Fig. 10.4). However, the increase in mortality still occurred faster at higher concentrations of CO_2 within each temperature tested. There can also be species differences in response to low O_2 and elevated CO_2 (Shellie *et al.*, 1997); however, these differences are often minimized when CA are applied with a heat treatment. For example, Yahia *et al.* (1999) observed no difference in the response of *Anastrepha ludens* and *A. oblique* to high-temperature CA treatments.

Life Stage Response

Obscure mealy bug *Pseudococcus affinis* Maskell LT_{99} levels were reduced with increasing temperatures (35–40°C) and decreasing O_2 concentrations from 20.9 to 0.4% (Whiting and Hoy, 1997). Life stage differences at these temperatures, such as the greater tolerance of female adult instars compared with male instars, were distinguished only in air (*see* Fig. 10.5). However, *E. postvittana* exposed to 2% O_2 with 5% CO_2 at 40°C exhibited a significant trend of increasing LT estimates with advancing developmental stages: 3-day-old eggs < first-instar < third-instar < fifth-instar (Hoy and Whiting, 1998).

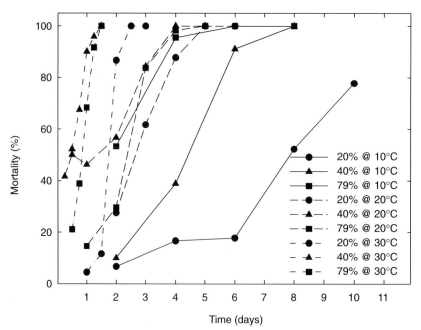

Fig. 10.4. Percentage mortality of 1–2-day-old omnivorous leafroller (*Platynota stultana*) female pupae under 20, 40 and 79% CO_2 at 10, 20 and 30°C. Means represent three to four replications (from Zhou *et al.*, 2000).

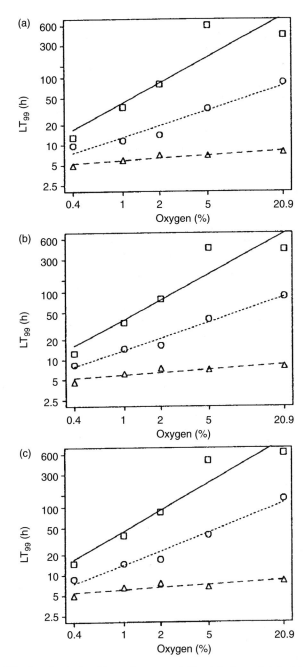

Fig. 10.5. Mean LT$_{99}$ levels derived from time–mortality data from (a) first-instar, (b) second/third-instar, and (c) female adult obscure mealy bug (*Pseudococcus affinis*) exposed to atmospheres containing different O$_2$ concentrations at three high temperatures (□, 35°C; ○, 40°C; △, 45°C. Each error bar represents twice the appropriate overall standard error (from Whiting and Hoy, 1997).

Other species, such as *Cnephasia jactatana* (Walker), had similar mortality responses to CA at each life stage. Bear in mind, however, that the CA used in these studies were very mild compared with most insecticidal atmospheres. Neven (2005) found that codling moth egg stages were less tolerant to high-temperature CA treatments (45°C, 1% O_2 + 15% CO_2) than all larval stages, and third- and fourth-instar larvae were the most tolerant.

Diapausing insects can be particularly difficult to kill. At 20°C, an O_2 concentration of 0.4% combined with a CO_2 concentration of 20% resulted in an LT_{99} of 113 h (4.7 days) for adult diapausing two-spotted spider mites (*T. urticae*) (Whiting and van den Heuvel, 1995). An increase in treatment temperature under this atmosphere significantly decreased the lethal time estimates. At 40°C (the highest temperature tested), the mean LT_{99} was 15.5 h. The substantial reduction in time to kill *T. urticae* with a 0.4% O_2 + 20% CO_2 atmosphere and increase in treatment temperature (> 20°C) may be caused by a partial reactivation of the diapausing adults (Veerman, 1977). If adults were induced to terminate diapause, they would be more susceptible to CA treatment, as non-diapausing *T. urticae* are more susceptible to high temperature disinfestations than diapausing adults (Waddell and Birtles, 1992).

10.5 Commodity Response to High-temperature Controlled Atmospheres

Exposure of fruits and vegetables to high temperatures or insecticidal CA results in stress on the commodity. Combining high temperature with low O_2 or high CO_2 creates a dual stress. However, the treatment time required to kill pests is significantly shorter when heat and CA are used together, and time is a critical factor in determining the tolerance of fruits and vegetables to a particular treatment.

For sweet cherries, it took 50% less time to control fifth-instar codling moth larvae at 45 and 47°C in 1% O_2 and 15% CO_2 than in air (Neven and Mitcham, 1996). Sweet cherry tolerance to heat with CA treatment was as good as with heat treatment alone, and therefore the shorter treatment was advantageous for fruit quality. This reduction in treatment time is particularly important for temperate fruit that are less adapted to high temperatures than subtropical and tropical fruit. However, there was an unacceptable increase in surface pitting of the sweet cherries subjected to these treatments. Later studies by Shellie *et al.* (2001) indicated that sweet cherries can tolerate high-temperature CA treatments designed to control codling moth larvae without significant fruit damage. The difference in tolerance may be due to the initial quality and hardiness of the cherries at harvest.

In the case of Bartlett pears, fruit were affected more by the combination of heat and CA than heat alone; however, tolerance was good to short (hours) exposure to atmospheres of 15% CO_2 compared with 1.5–3.0% CO_2 thresholds during long-term (months) storage at refrigeration temperatures

(Mitcham *et al.*, 1996). Bartlett pears were treated with 1% O_2 and 15% CO_2 during forced-air heating at 46°C for 1, 2 and 3 h for control of codling moth (*Cydia pomonella*) larvae (Mitcham *et al.*, 1999; Neven, 2004). The presence of CA greatly increased larval mortality compared with heat alone (see Table 10.2).

Fruit were treated in two seasons with slow and fast heating rates. Heat treatment – and especially heat with CA – slowed fruit ripening even after 3 weeks of storage; however, this could be an advantage in some export markets where temperature management is less than ideal. The longer the treatment, the greater the ripening inhibition. The mortality of codling moths was 100% after 2.5 h with the fast heating rate and 3 h with the slower heating rate when CA were included. The pears tolerated the slower heating rate better, showing less skin browning and less inhibition of ripening.

The tolerance of Hayward kiwi fruit to heat treatments at 40°C was greatly decreased by adding CA with 0.4% O_2 + 20% CO_2 (Lay-Yee and Whiting, 1996), but resulted in 100% mortality for non-diapausing and diapausing two-spotted spider mites after 5.4 and 8.1 h, respectively. The kiwi fruit tolerated a 7 h exposure to heat and CA if hydrocooled after treatment, but not the 10 h exposure. The damage observed included vascular browning, increased decay and cavitation. After storage for 8 weeks at 0°C, flesh firmness in fruit treated in air at 40°C for 10 h or CA at 40°C for 7 or 10 h was lower than for control fruit.

The tolerance of Packham's Triumph pears to temperatures from 30 to 40°C combined with < 1% O_2 for 16–48 h followed by 1 month of storage in air at 0°C was explored (Chervin *et al.*, 1997). At 30°C, pears withstood up to 30 h of hypoxia. The fruits ripened slightly faster than controls after storage, but were undamaged; at 35°C, there was slight skin browning; at 40°C, there was substantial skin blackening.

A combined treatment of 30 h at 30°C under 1% O_2 plus 1 month of cold storage in air at 0°C killed 100% light brown apple moth pupae, fifth-instar larvae of codling moth (*C. pomonella*) and oriental fruit moth (*Grapholita molesta* (Busck)) and was also tolerated by the fruit.

Table 10.2. Mortality of fifth-instar codling moth (*Cydia pomonella*) in pears following heat treatments of 100°C/h to a final temperature of 46°C in air and in controlled atmospheres (CA) of 1% O_2 + 15% CO_2 (adapted from Neven, 2004).

Treatment time (h)	Mortality (%)	
	Air	CA
1.0	19.5	42
1.5	30	90
2.0	80	98
2.5	78	100

Exposure to one stress can provide plant protection against a subsequent different stress (Vierling, 1991), but this may not be the case with insects (Neven, 2004). Chervin et al. (1996) suggested that plants could be adapted to extreme CA by exposure to moderate O_2 or CO_2 concentrations, thereby increasing their tolerance of these stress atmospheres. Plant cells clearly sense and respond to O_2 levels that are well in excess of those that limit terminal oxidase. However, for insect disinfestation, it seems better to use immediate stress conditions, as arthropods are also able to adapt to CA and therefore become more resistant to disinfestation treatments (Chervin et al., 1996).

Adaptation to low oxygen concentrations may involve chaperonins that stabilize enzymes under stress conditions. These chaperonins assist the folding of newly synthesized proteins and assembly into larger structures (Chervin et al., 1996), are involved in crustacean cell responses to anoxia, heat shock and chilling temperatures (Neven et al., 1992; Clegg et al., 1994) and protect plant proteins against heat shock (Ellis, 1991). Whether they protect against anoxia remains to be determined (Chervin et al., 1996), and it may be an even greater challenge to protect against the double stress of heat and CA.

10.6 Synergistic Effects of Heat and Controlled Atmospheres

In the case of high-temperature CA treatments, both the sequence of and time to establish target temperature and CA conditions can be altered. The rate of temperature increase during heat treatments can affect pest response due to the establishment of thermotolerance (Neven and Drake, 1996). Lay-Yee et al. (1997) found that shortening the duration of the temperature rise made fifth-instar E. postvittana more susceptible to treatment, possibly by reducing the opportunity to adapt to the high-temperature conditions. Acquired thermo-tolerance was correlated with de novo synthesis of certain proteins, suggesting physiological adaptation (Lester and Greenwood, 1997).

However, Whiting and Hoy (1998) found no increase in thermotolerance from varying the heating rate during a high-temperature CA treatment. Their results suggest that the additional exposure time to kill E. postvittana larvae with a slow heating rate compared with a fast heating rate was less than or equal to the duration of the temperature delay, not the development of thermotolerance. This may have been due to the presence of CA during the temperature establishment time preventing thermotolerance from developing. Heat treatment under anoxic conditions may also reduce the production of heat shock proteins in insects (Thomas and Shellie, 2000) as well as plants.

Slow heating rates (< 8°C/h) and uniform heating appear important in maintaining fruit quality (Klein and Lurie, 1992; Neven and Drake, 2000), thus making the position of insect pests less critical and the feasibility and success of fast-heating air or CA quarantine treatments uncertain. However, Whiting and Hoy's 1998 study predicts variation in mortality responses to high-temperature CA treatments as a result of differences in pest distribution inside the host fruit.

Their results suggest that the few mature larvae situated inside the apples will be exposed to the duration of the delay in heating the inside of the fruit. The few mature larvae situated inside the core will be exposed to treatment with a longer temperature establishment time than those in the more common external sites of establishment (calyx and stem cavities) because of the delay in heat transfer through the fruit flesh (Shellie and Mangan, 1995). If slower establishment of treatment temperature is required to optimize fruit quality or acknowledge differences in insect distribution on the host fruit, then mortality estimates of insect species inside the fruit may be expected to increase by less than or equal to the increase in temperature establishment time (Whiting and Hoy, 1998). Traditional heat treatments naturally contain a CA effect because of internal atmosphere modification. Mitcham and McDonald (1993) demonstrated that CO_2 increased to 13% and O_2 decreased to 6% in mangoes subjected to high-temperature, forced-air treatments. Shellie *et al.* (1997) demonstrated that hot water treatments result in greater modification to internal atmosphere of commodities than hot air treatments.

During hot water heating of grapefruit at 44°C, internal CO_2 increased to 22% and O_2 decreased to 3% over a 3 h period (see Fig. 10.3). This may explain the greater efficacy of hot water over hot air treatments when the heating rate and product temperature are the same. Their data also showed the greater efficacy of hot air with CA (1% O_2 + 20% CO_2) over hot air alone for control of Mexican fruit fly (*A. ludens*).

The efficacy of dual stresses such as heat and CA seems to be based on additive or synergistic effects. The combination of heat and CA (1.2% O_2 at 29°C) often leads to an increase in mortality beyond that predicted by their independent effects (Tabatabai *et al.*, 2000). Based on Sheppard (1996), this is a case of classic or two-factor synergy.

Synergistic effects of stress combinations on mortality have been observed for *E. postvittana* and other insects. It is well known that insects can become acclimated by short-term exposure to temperature extremes (Somme, 1972) or other stresses such as desiccation (Hoffmann, 1990). Therefore, the efficiency of dual stresses may be affected by acclimation (Tabatabai *et al.*, 2000). The synergistic effect of two stresses suggests that a combination of three stresses (heat, CA and cold) may be even more efficient than two stresses.

In codling moth larvae, combinations of three stresses resulted in mortality rates 30% higher than with the most efficient paired combination (Chervin *et al.*, 1998). Chervin *et al.* (1997, 1998) also reported synergistic effects on the mortality of various developmental stages of *E. postvittana* and codling moth after combinations of heat plus CA pulse and cold storage. In addition, Neven (1994) observed some synergistic effects on the mortality of codling moth larvae due to a combination of heat and cold storage (30 min at 45°C killed 70% of fifth-instar larvae, 14 days at 0°C killed about 15% and the combination of heat and cold killed 100%).

10.7 Promising Treatments

Royal Gala and Granny Smith apples tolerated exposure to 1.2% O_2 and 1% CO_2 at 40°C for up to 20 h which provided LT_{99} control of *E. postvittana* and wheatbug (*Nysius huttoni*) (Chervin *et al.*, 1997). A combined treatment of 30 h at 30°C with less than 1% O_2 plus 1 month of cold storage in air at 0°C killed 100% of *E. postvittana* pupae, fifth-instar larvae of codling moth and oriental fruit moth (*G. molesta* (Busck)) and were also tolerated by Packham's Triumph pear fruit (Chervin *et al.*, 1997).

Bartlett pears tolerated treatment with 1% O_2 and 15% CO_2 during forced-air heating at 46°C for 1, 2 and 3 h (Mitcham *et al.*, 1999; Neven, 2004). While ripening was slower after storage, fruit eventually ripened but were more susceptible to skin damage and browning with handling. Codling moth larvae were completely controlled after 2.5 h at 46°C in 1% O_2 + 15% CO_2 (Neven, 2004; Obenland *et al.* 2005). A similar successful protocol was developed for apples by Lisa Neven, USDA ARS in Wapato, Washington, USA, using a heating rate of 12°C/h at an air temperature of 46°C. Fruit are heated to a core temperature of 44.5°C and held at that temperature for 30 min. The total treatment time was about 3 h.

High-temperature CA (1% O_2 + 15% CO_2) treatment protocols (air temperature at 45°C for 45 min or air temperature at 47°C for 25 min) have been developed for control of codling moth (Neven, 2005) and western cherry fruit fly (*Rhagoletis indifferens* Curran) (Neven and Rehfield-Ray, 2006) in sweet cherries and submitted to the Animal and Plant Health Inspection Service (APHIS) for Japanese and Australian markets (Lisa Neven, personal communication).

In addition, two high-temperature CA (1% O_2 + 15% CO_2) treatments for stone fruits have been submitted to APHIS (Neven, 2005) for control of codling moth in shipments to Japan and control of *G. molesta* (Busck) for shipments to Mexico (Lisa Neven, personal communication). The treatments are: (i) heating rate 12°C/h at an air temperature of 46°C to a fruit core temperature of 44.5°C and hold for 30 min; and (ii) heating rate of 24°C/h at an air temperature of 46°C to a core temperature of 44.5°C and hold for 15 min. Thus, it appears as if high-temperature CA treatments will soon be approved by APHIS for commercial use on fruit.

10.8 Summary

The addition of CA to a heat treatment can greatly reduce the exposure time required for complete control of arthropod pests. The shorter treatment can be extremely important in the tolerance of commodities, which is usually the limiting factor in development of quarantine heat treatments. A major benefit of these treatments is the lack of chemical residues on the product, rendering the treatment sustainable for years in the future and also suitable for organic products. After many years of research, a number of high-temperature CA treatments have been submitted to APHIS for approval and we may soon have these treatments available for commercial use.

10.9 References

Benschoter, C.A. (1987) Effects of modified atmospheres and refrigeration temperatures on survival of eggs and larvae of the Caribbean fruit fly (Diptera: Tephritidae) in laboratory diet. *Journal of Economic Entomology* 80, 1223–1224.

Boersig, M.R., Kader, A.A. and Romani, R.J. (1988) Aerobic–anaerobic respiratory transition in pear fruit and cultured pear cells. *Journal of the American Society of Horticultural Science* 113, 869–873.

Carpenter, A.A. (1997) Comparison of the responses of aphids and thrips to controlled atmospheres. *Proceedings of the 6th International Controlled Atmosphere Conference.* Postharvest Horticulture Series, Postharvest Outreach Program, Davis, California, pp. 91–95.

Chavez-Franco, S.H. and Kader, A.A. (1993) Effects of CO_2 on ethylene biosynthesis in 'Bartlett' pears. *Postharvest Biology and Technology* 3, 183–190.

Chervin, C., Brady, C.J., Patterson, B.D. and Faragher, J.D. (1996) Could studies on cell responses to low oxygen levels provide improved options for storage and disinfestation? *Postharvest Biology and Technology* 7, 289–299.

Chervin, C., Kulkarni, S., Kreidl, R., Birrel, F. and Glenn, D. (1997) A high temperature/low oxygen pulse improves cold storage disinfestations. *Postharvest Biology and Technology* 10, 239–245.

Chervin, C., Hamilton, J.A., Kreidl, A. and Kulkarni, S. (1998) Non-chemical disinfestations: combining the combinations. *Acta Horticulturae* 464, 273–278.

Clegg, J.S., Jackson, S.A. and Warner, A.H. (1994) Extensive intracellular translocations of a major protein accompany anoxia in embryos of *Artemia franciscana*. *Experimental Cell Research* 212, 77–83.

De Lima, C.F.P. (1990) Air tight storage: principles and practice. In: Calderon, M. and Barkai-Golan, R. (eds) *Food Preservation by Modified Atmospheres*. CRC Press, Boca Raton, Florida, p. 9.

Dentener, P.R., Lewthwaite, S.E., Rogers, D.J., Miller, M. and Connolly, P.G. (1999) Mortality of Huhu (*Prionoplus reticularis*) subjected to heat and controlled atmosphere treatments. *New Zealand Journal of Forestry Science* 29, 473–483.

Donahaye, J.E., Navarro, S. and Rinder, M. (1994) The influence of temperature on the sensitivity of two nitidulid beetles to low oxygen concentrations. In: Highley, E., Wright, E.J., Banks, H.J. and Champ, B.R. (eds) *Proceedings of the 6th International Working Conference on Stored Product Protection*, 17–23 April, Canberra, Australia. CAB International, Wallingford, UK.

Ellis, R.J. (1991) Chaperone function: cracking the second half of the genetic code. *The Plant Journal* 1, 9–13.

Friedlander, A. (1983) Biochemical reflections on a non-chemical control method. The effect of controlled atmospheres on the biochemical processes in stored product insects. In: Storey, C.L. (ed.) *Proceedings of the 3rd International Working Conference on Stored-Product Entomology*, 23–28 October. Kansas State University, Manhattan, Kansas, pp. 471–480.

Hallman, G.J. (1994) Controlled atmosphere. In: Paul, R.E. and Armstrong, J.W. (eds) *Insect Pests and Fresh Horticultural Products: Treatments and Responses*. CAB International, Wallingford, UK, pp. 121–136.

Hochachka, P.W. (1986) Defense strategies against hypoxia and hypothermia. *Science* 231, 234.

Hoffman, A.A. (1990) Acclimation for desiccation resistance in *Drosophila melanogaster* and the association between acclimation responses and genetic variation. *Journal of Insect Physiology* 36, 885–891.

Hoy, L.E. and Whiting, D.C. (1998) Mortality response of three leafroller (Lepidoptera: Tortricidae) species on kiwi fruit to a high-temperature controlled atmosphere treatment. *New Zealand Journal of Crop and Horticultural Science* 26, 11–15.

Kader, A.A. (1980) Prevention of ripening in fruits by use of controlled atmospheres. *Food Technology* March, 51–54.

Kader, A.A. (1986) Biochemical and physiological basis for effects of controlled and modified atmospheres on fruits and vegetables. *Food Technology* 40, 99–100 and 102–104.

Kader, A.A. (1995) Regulation of fruit physiology by controlled atmospheres. *Acta Horticulturae* 398, 59–70.

Klein, J.D. and Lurie, S. (1992) Pre-storage heat treatment for enhanced postharvest quality: interaction of time and temperature. *HortScience* 27, 326–328.

Lay-Yee, M. and Whiting, D.C. (1996) Response of 'Hayward' kiwi fruit to high-temperature controlled atmosphere treatments for control of two-spotted spider mite (*Tetranychus urticae*). *Postharvest Biology and Technology* 7, 73–81.

Lay-Yee, M., Whiting, D.C. and Rose, K.J. (1997) Response of 'Royal Gala' and 'Granny Smith' apples to high-temperature controlled atmosphere treatments for control of *Epiphyas postvittana* and *Nysius huttoni*. *Postharvest Biology and Technology* 12, 127–136.

Lea, T.J. and Ashley, C.C. (1978) Increase in free Ca^{2+} in muscle after exposure to CO_2. *Nature* 275, 236–238.

Lester, P.J. and Greenwood, D.J. (1997) Pretreatment induced thermotolerance in lightbrown apple moth (Tortricidae: Lepitdoptera) and associated induction of heat shock protein synthesis. *Journal of Economic Entomology* 90, 199–204.

Mbata, G.N. and Phillips, T.W. (2001) Effects of temperature and exposure time on mortality of stored-product insects exposed to low pressure. *Journal of Economic Entomology* 94, 1302–1307.

Mbata, G.N., Hetz, S.K., Reichmuth, C. and Adler, C. (2000) Tolerance of pupae and pharate adults of *Callosobruchus subinno-tatus* Pic (Coleoptera: Bruchidae) to modified atmospheres: a function of metabolic rate. *Journal of Insect Physiology* 46, 145–151.

Mitcham, E.J. and McDonald, R.E. (1993) Respiration rate, internal atmosphere, and ethanol and acetaldehyde accumulation in heat-treated mango fruit. *Postharvest Biology and Technology* 3, 77–86.

Mitcham, E.J., Cristosto, C.H. and Kader, A.A. (1996) Recommendations for maintaining postharvest quality: Bartlett pears. University of California, Postharvest Technology Research and Information Center (available at http://postharvest. ucdavis.edu).

Mitcham, E. J., Zhou, S. and Bikoba, V. (1997) Controlled atmosphere for quarantine control of three pests of table grape. *Journal of Economic Entomology* 90, 1360–1370.

Mitcham, E.J., Nevin, L. and Biasi, B. (1999) Effect of high-temperature controlled-atmosphere treatments for insect control in 'Bartlett' pear fruit. *HortScience* 34, 527.

Mitcham, E.J., Martin, T.A., Zhou, S. and Kader, A.A. (2001) *Potential of CA for Postharvest Arthropod Control in Fresh Horticultural Perishables: an Update of Summary Tables Compiled by Ke and Kader, 1992.* Postharvest Horticulture Series No. 22, Postharvest Technology Research and Information Center, University of California, Davis, California.

Mitcham, E.J., Martin, T. and Zhou, S. (2006) The mode of action of insecticidal controlled atmospheres. *Bulletin of Entomological Research* 96, 213–222.

Neven, L.G. (1994) Combined heat and cold storage effects on mortality of fifth-instar codling moth (Lepidorptera: Torticidae). *Journal of Economic Entomology* 87, 1262–1265.

Neven, L.G. (2004) Disinfestation of fresh horticultural commodities by using hot forced air with controlled atmospheres. In: Dris, R. and Jain, S.M. (eds) *Production Practices and Quality Assessment of Food Crops, Vol. 4: Postharvest Treatment and Technology.* Kluwer Academic Publishers, Netherlands, pp. 297–315.

Neven, L.G. (2005) Combined heat and controlled atmosphere quarantine treatments for control of codling moth, *Cydia pomonella*, in sweet cherries. *Journal of Economic Entomology* 98, 709–715.

Neven, L.G. and Drake, S.R. (1996) Summary of alternative quarantine treatment research in the Pacific Northwest. In: *Proceedings of the 1996 Annual Research Conference on Methyl Bromide Alternatives and Emissions Reductions*, 4–6 November 1996, Orlando, Florida. Methyl Bromide Alternatives Outreach, USEPA and USDA, Fresno, California, pp. 79.1–79.3.

Neven, L.G. and Drake, S.R. (2000) Effects of the rate of heating on apple and pear fruit quality. *Journal of Food Quality* 23, 317–325.

Neven, L.G. and Mitcham, E.J. (1996) CATTS (controlled atmosphere/temperature treatment system): a novel tool for the development of quarantine treatments. *American Entomologist* 42, 56–59.

Neven, L.G. and Rehfield-Ray, L. (2006) Combined heat and controlled atmosphere quarantine treatments for control of western cherry fruit fly in sweet cherries. *Journal of Economic Entomology* 99, 658–663.

Neven, L.G., Haskell, D.W., Guy, C.L., Denslow, N., Klein, P.A., Green, L.G. and Silverman, A. (1992) Association of 70-kilodalton heat-shock cognate proteins with acclimation to cold. *Plant Physiology* 99, 1362–1369.

Nicolas, G. and Sillans, D. (1989) Immediate and latent effects of carbon dioxide on insects. *Annual Review of Entomology* 34, 97–116.

Obenland, D., Neipp, P., Mackey, B. and Neven, L.G. (2005) Peach and nectarine quality following treatment with high temperature forced air combined with controlled atmospheres. *HortScience* 40, 1425–1430.

Paul, R.E. and Armstrong, J.W. (eds) (1994) *Insect Pests and Fresh Horticultural Products: Treatments and Responses*. CAB International, Wallingford, UK.

Schroeder, L.M. and Eidman, H.H. (1986) The effects of pure and blended atmospheric gases on the survival of three bark species. *Journal of Applied Entomology* 101, 353–359.

Shellie, K.C. and Mangan, R.L. (1995) Heating rate and tolerance of naturally degreened 'Dancy' tangerine to high-temperature forced air for fruit fly disinfestations. *HortTechnology* 5, 40–43.

Shellie, K.C., Mangan, R.L. and Ingle, S.J. (1997) Tolerance of grapefruit and Mexican fruit fly larvae to heated, controlled atmospheres. *Postharvest Biology and Technology* 10, 179–186.

Shellie, K.C., Neven, L.G. and Drake, S.R. (2001) Assessing 'Bing' sweet cherry tolerance to a heated controlled atmosphere for insect pest control. *HortTechnology* 11, 308–311.

Sheppard, A.W. (1996) The interaction between natural enemies and interspecific plant competition in the control of invasive pasture weeds. In: Moran, V.C. and Hoffman, J.H. (eds) *Proceedings of the 9th International Symposium on Biological Control of Weeds*. University of Cape Town, Stellenbosh, South Africa, pp. 47–53.

Sisler, E.C. and Wood, C. (1988) Interaction of ethylene and CO_2. *Physiologia Plantarum* 73, 440–444.

Soderstrom, E.L., Brandl, D.G. and Mackey, B.E. (1990) Responses of codling moth (Lepidoptera: Tortricidae) life stages to high carbon dioxide or low oxygen atmospheres. *Journal of Economic Entomology* 83, 472–475.

Soderstrom, E.L., Brandl, D.G. and Mackey, B.E. (1991) Responses of *Cydia pomonella* (L.) (*Lepidoptera: Tortricidae*) adults and eggs to oxygen deficient and carbon dioxide enriched atmospheres. *Journal of Stored Products Research* 27, 95–101.

Soderstrom, E.L., Brandl, D.G. and Mackey, B.E. (1996) High temperature alone and combined with controlled atmospheres for control of diapausing codling moth (Lepidoptera: Tortricidae) in walnuts. *Journal of Economic Entomology* 89, 144–147.

Somme, L. (1972) The effects of acclimation and low temperature on enzyme activities in larvae of *Ephestia kuehniella* (Zell.). *Entomologica Scandinavica* 3, 12–18.

Tabatabai, S., Chervin, C., Hamilton, A. and Hoffmann, A.C. (2000) Sensitivity of pupae of lightbrown apple moth, *Epiphyas postvittana* (Walker) (Lepidoptera: Tortricidae), to combinations of abiotic stresses. *Australian Journal of Entomology* 39, 78–82.

Thomas, D.B. and Shellie, K.C. (2000) Heating rate induces thermotolerance in Mexican fruit fly (Diptera: Tephritidae) larvae, a quarantine pest of citrus and mangoes. *Journal of Economic Entomology* 93, 1373–1379.

Veerman, A. (1977) Photoperiodic termination of diapause in spider mites. *Nature* 266, 526–527.

Vierling, E. (1991) The roles of heat shock proteins in plants. *Annual Review of Plant Physiology and Plant Molecular Biology* 42, 579–531.

Waddell, B.C. and Birtles, D.B. (1992) Disinfestation of nectarines of two-spotted mites (Acari: Tetranychidae). *New Zealand Journal of Crop and Horticultural Science* 20, 229–234.

Wang, C.Y. (1990) Physiological and biochemical effects of controlled atmosphere on fruits and vegetables. In: Calderon, M. and Barkai-Golan, R. (eds) *Food Preservation by Modified Atmospheres*. CRC Press, Boca Raton, Florida, pp. 197–234.

Weichmann, J. (1986) The effects of controlled atmosphere storage on the sensory and nutritional quality of fruits and vegetables. *Horticulture Review* 8, 101–127.

Whiting, D.C. and Hoy, L.E. (1997) High-temperature controlled atmosphere and air treatments to control obscure mealy bug (Hemiptera: Pseudococcidae) on apples. *Journal of Economic Entomology* 90, 546–550.

Whiting, D.C. and Hoy, L.E. (1998) Effect of temperature establishment time on the mortality of *Ephiphyas postvittana* (Lepidoptera: Tortricidae) larvae exposed to a high-temperature controlled atmosphere. *Journal of Economic Entomology* 91, 287–292.

Whiting, D.C. and van den Heuvel, J. (1995)

Oxygen, carbon dioxide, and temperature effects on mortality responses of diapausing *Tetranychus urticae* (Acari: Tetranychidae). *Journal of Economic Entomology* 88, 331–336.

Whiting, D.C., Foster, S.P. and Maindonald, J.H. (1991) Effects of oxygen, carbon dioxide, and temperature on the mortality responses of *Epiphyas postvittana* (Lepidoptera: Tortricidae). *Journal of Economic Entomology* 84, 1544–1549.

Whiting, D.C., Foster, S.P., Van den Heuvel, J. and Maindonald, J.H. (1992a) Comparative mortality responses of four tortricid (Lepidoptera) species to low oxygen-controlled atmosphere. *Journal of Economic Entomology* 85, 2305–2309.

Whiting, D.C., van den Heuvel, J. and Foster, S.P. (1992b) Potential of low oxygen/moderate carbon dioxide atmospheres for postharvest disinfestations of New Zealand apples. *New Zealand Journal of Crop and Horticultural Science* 20, 217–222.

Whiting, D.C., Conner, G.M., van den Heuvel, J. and Maindonald, J.H. (1995) Comparative mortality of six tortricid (Lepidoptera) species to two high-temperature controlled atmospheres and air. *Journal of Economic Entomology* 88, 1365–1370.

Yahia, E.M. (1998) Modified/controlled atmospheres for tropical fruits. *Horticultural Review* 23, 123–183.

Yahia, E.M., Ortega, D., Martinez, A. and Moreno, P. (1999) The mortality of artificially infested third instar larvas of *Anastrepha ludens* and *A. oblique* in mango fruit with insecticidal controlled atmospheres at high temperature. *HortScience* 34, 527.

Zhou, S., Criddle, R.S. and Mitcham, E.J. (2000) Metabolic response of *Platynota stultana* pupae to controlled atmospheres and its relation to insect mortality response. *Journal of Insect Physiology* 46, 1375–1385.

Zhou, S., Criddle, R.S. and Mitcham, E.J. (2001) Metabolic response of *Platynota stultana* pupae under and after extended treatment with elevated CO_2 and reduced O_2 concentrations. *Journal of Insect Physiology* 47, 401–409.

11 The Influence of Heat Shock Proteins on Insect Pests and Fruits in Thermal Treatments

S. Lurie[1] and E. Jang[2]

[1]Department of Postharvest Science, ARO, The Volcani Center, Bet Dagan, Israel; e-mail: slurie43@agri.gov.il; [2]USDA-ARS US Pacific Basin Agricultural Research Center, Hilo, Hawaii, USA; e-mail: jang@pbarc.ars.usda.gov

11.1 Introduction

The response of organisms to high-temperature stress has been widely researched since heat shock response was first observed in *Drosophila* (Ritossa, 1962). The heat shock phenomenon is characterized by dramatic and rapid changes in both transcription and translation of proteins in response to sublethal heat stress. The heat shock response occurs in all organisms, ranging from bacteria and lower eukaryotes to mammals and plants, and is evolutionarily highly conserved. The level of amino acid identity between all prokaryotic and eukaryotic heat shock proteins of 70 kDa size (HSP70), for example, approaches 50% (Gupta and Singh, 1992). This conservation suggests that the response is an essential function for survival of organisms exposed to excessively high, yet sublethal, temperatures.

Heat shock response occurs following an upwards temperature shift of 6–12°C from the normal growth temperature and is characterized by a fast induction of heat shock gene transcription, coupled with a precipitous decline in the transcription of most other genes (Gurley and Key, 1991). The response is further enhanced by the selective translation of heat shock mRNAs (messenger ribonucleic acids) at heat shock temperatures (or rapid turnover of non-heat shock mRNA), resulting in the selective and rapid accumulation of heat shock proteins (HSP) at the elevated temperature. Plants generally have a more complex set of HSP than most other organisms; the complexity and abundance of families of low- or small-molecular weight HSP (sHSP) in plants is particularly unique (Vierling, 1991).

Historically, three observations have suggested that HSP protect cells and organisms from stress. First, their induction has the character of an emergency response, being both rapid and strong. For example, in *Drosophila* there are five copies of the HSP70 gene per haploid genome, and regulation at the levels of transcription, translation, RNA processing and RNA stability works together to produce a greater than thousandfold induction of the protein during high-temperature stress (Lindquist, 1980; Parker and Topol, 1984; Yost and Lindquist, 1986; Zimarino and Wu, 1987; O'Brien and Lis, 1991; Lindquist *et al.*, 1993). The complexity of the regulatory mechanisms for HSP70 suggests that it is vitally important for cells to rapidly accumulate this protein during times of stress.

Secondly, HSP are induced at very different temperatures in different organisms but, in each case, the induction temperature reflects stress conditions for that organism. For example, thermophilic bacteria induce HSP when shifted from 95 to 105°C (Trent *et al.*, 1990; Phipps *et al.*, 1993), while arctic fishes induce them when shifted from 0 to 5–10°C (Parsell and Lindquist, 1993). Organisms induce HSP synthesis when their temperature increases above that which is normal for them, rather than at a universal temperature threshold.

Thirdly, the induction of HSP correlates with the induction of tolerance to extreme heat in a wide variety of cells and organisms (Li and Laszlo, 1985; Sanchez and Lindquist, 1990; Nover, 1991; Fig. 11.1). Thermotolerance is the process whereby an organism is exposed to non-lethal high temperatures and acquires resistance to a normally lethal high temperature due to the production of HSP (Solomon *et al.*, 1991; Vierling, 1991), which also correlates with induced tolerance to other stressful treatments (e.g. high concentrations of ethanol, arsenite or heavy metals) (Lindquist, 1986; Subjeck and Shyy, 1986; Nover, 1991; Sanchez *et al.*, 1992).

Moreover, the correlation is as strong for developmentally regulated inductions as it is for stress-mediated inductions. Germinating wheat seeds constitutively express HSP and have high intrinsic thermotolerance (Abernethy *et al.*, 1989). Conversely, developmental stages in which organisms are unable

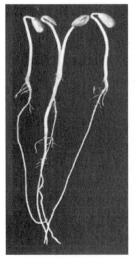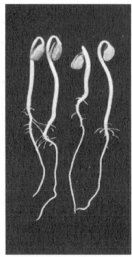

Fig. 11.1. Thermotolerance induced in seedings. Seedlings were grown at 28°C (left panel). Seedlings in centre panel were given a 2-h treatment at 40°C and then 2 h at 45°C. In right panel, seedlings were placed directly at 45°C (from Buchanan *et al.*, 2000).

to express HSP, such as the early embryonic stages of *Drosophila* (Dura, 1981; Graziosi *et al.*, 1983) or plant pollen (Vierling, 1991), are characterized by thermosensitivity.

11.2 Heat Shock Proteins

Several classes of HSP have been described in eukaryotes, including insects and plants (see Table 11.1). They are designated by their approximate molecular weights in kDa as HSP100, HSP90, HSP70, HSP60 and low- or small-molecular weight sHSP (15–45 kDa). Ubiquitin, a small protein involved in ATP (adenosine triphosphate)-dependent, intracellular proteolysis (Hershko and Ciechanover, 1992), is also referred to as an HSP (Lindquist and Craig, 1988; Neumann *et al.*, 1989).

These proteins all fit the criterion that they are heat-induced in a majority of cell types in a wide range of organisms in response to heat stress (Vierling, 1991). sHSP range in size from approximately 15 to 30 kDa and are characterized by a conserved sequence at their C-terminus, related to the α-crystallins of the vertebrate eye lens (Inoglia and Craig, 1982; Sun *et al.*, 2002). In addition to size, the proteins are further characterized by their subcellular localization in the cytosol, mitochondria, chloroplast or nucleus (see Table 11.1).

The importance of HSP extends beyond their role in protection from high-temperature stress. Although HSP were first characterized because their expression increased in response to elevated temperatures, some HSP are found at significant levels in normal, unstressed cells or are produced at particular stages of the cell cycle or at different developmental stages (Lindquist, 1986; Lindquist and Craig, 1988; Morimoto *et al.*, 1990). Additionally, certain cellular proteins are homologous to HSP but do not exhibit increased expression in response to high temperatures (Vierling, 1991).

Different proteins of the HSP70, HSP60 and sHSP are present in various cellular compartments, including semi-autonomous organelles such as mitochondria and chloroplasts (Craig *et al.*, 1989; Engman *et al.*, 1989; Leustek *et al.*, 1989; Mizzen *et al.*, 1989; Marshall *et al.*, 1990). Thus, HSP are members of multi-gene superfamilies in which not all members are regulated by heat.

Table 11.1. Major classes of heat shock proteins (HSP) in eukaryotes.

HSP class	Intracellular location	Size (kDa)
HSP 100	Cytoplasm, chloroplast, mitochondrion	100–114
HSP 80, HSP90	Cytoplasm, endoplasmic reticulum	80–94
HSP 70	Cytoplasm/nucleus, chloroplast, mitochondrion, endoplasmic reticulum	69–71
HSP 60	Chloroplast, mitochondrion	57–60
HSP 40	Cytoplasm, mitochondrion, endoplasmic reticulum	40
Small HSP	Cytoplasm, chloroplast, mitochondrion, endoplasmic reticulum	15–30

Heat is not the only stress treatment that leads to elevated expression of many HSP. Ethanol, arsenite, heavy metals, amino acid analogues, glucose starvation and a number of other abiotic stresses affect the synthesis of some or all HSP in different organisms (Czarnecka *et al.*, 1984, 1988; Nover *et al.*, 1984; Edelman *et al.*, 1988; Winter *et al.*, 1988; Neumann *et al.*, 1989). Consequently, HSP have also been referred to more generally as 'stress proteins'.

A general mechanism of the action of HSP is to aid in the stabilization of proteins in a particular state of folding, so HSPs are termed 'molecular chaperones' (Pelham, 1986; Ellis, 1987; Lindquist and Craig, 1988; Rothman, 1989; Morimoto *et al.*, 1990; Schlesinger, 1990; Nover, 1991). HSP90, HSP70 and HSP60 – with the expenditure of ATP – facilitate a wide diversity of important processes including protein folding, transport of proteins across membranes, assembly of oligomeric proteins and modulation of receptor activities. All of these functions require the alteration or maintenance of specific polypeptide conformations. Although the mechanisms by which sHSP are involved in cell protection are not fully understood, it has been shown that both plant and mammalian sHSP possess non-ATP-dependent, chaperone-like activity *in vitro* (Lee *et al.*, 1995; Lee and Vierling, 2000).

The roles of individual HSP in stress protection are beginning to be dissected. In *Saccharomyces cerevisiae*, for example, some HSP are required for growth at temperatures near the upper end of the normal growth range (HSP70), others for long-term survival at moderately high temperatures (ubiquitin) and still others for tolerance to extreme temperatures (HSP104) (Craig and Jacobsen, 1984; Finley *et al.*, 1987; Sanchez and Lindquist, 1990). Furthermore, different organisms use different HSP to respond to similar levels of stress. For example, tolerance to extreme stress depends largely upon HSP104 in yeast, HSP70 in *Drosophila* and HSP101, HSP70 and HSP27 in mammals (Li and Laszlo, 1985; Landry *et al.*, 1989; Sanchez and Lindquist, 1990; Solomon *et al.*, 1991).

The common signal for HSP induction appears to be the presence of abnormal proteins. Treatments that efficiently induce HSP synthesis *in vivo* also damage or denature proteins. There are at least two ways in which HSP might help cells to cope with stress-induced damage to polypeptides. First, HSP could promote degradation of abnormal proteins. Some HSP have intrinsic proteolyic activities, and others serve auxiliary roles in proteolysis. Secondly, HSP could reactivate stress-damaged proteins by promoting the proper refolding of denatured proteins.

Figure 11.2 shows how the presence of HSP25 can decrease the aggregation of the enzyme citrate synthase when exposed to 45°C, while Fig. 11.3 shows how HSP25 and HSP27, can help the recovery of the enzyme's activity at 22°C following a heat stress (45°C) that caused denaturation and loss of enzyme activity (adapted from Haslbeck and Buchner, 2002).

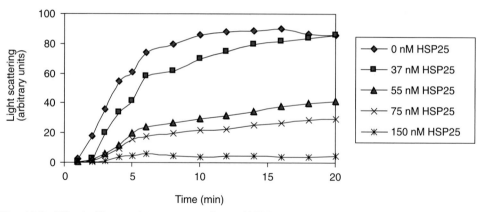

Fig. 11.2. Effect of increasing concentrations of HSP25 in preventing aggregation of citrate synthase (75 nM) during heat treatment at 45°C (adapted from Haslbeck and Buchner, 2002).

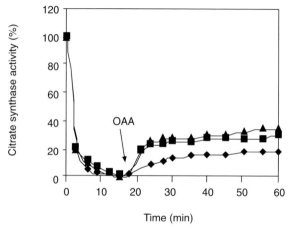

Fig. 11.3. Effect of HSP25 (■), HSP27 (▲) and control (♦) in promoting recovery of enzyme activity of citrate synthase (75 nM) following a heat treatment at 45°C. Substrate (OAA) was added when the solutions were returned to 22°C (adapted from Haslbeck and Buchner, 2002).

11.3 Heat Shock Responses and Heat Shock Proteins in Plant Tissue

Heat shock response in plants is manifested by induction or enhanced synthesis of HSP. In postharvest heat treatments and elevated temperatures in the field, both HSP gene expression and HSP protein accumulation have been demonstrated (Lurie, 1998; Paull and Chen, 2000; Woolf and Ferguson, 2000; Fig. 11.4).

It has been found that the accumulation of HSP in plant tissue induces thermotolerance to higher temperatures, and this has been utilized as a strategy

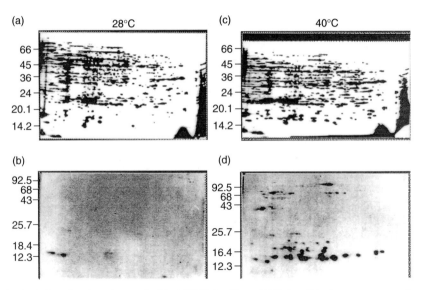

Fig. 11.4. 2-D gels of soybeans held at 28°C (a, b) or 40°C (c, d) in the presence of 3H-leucine for 3 h. (a) and (c) were visualized by silver staining and (b) and (d) by fluorography. The spots in (d) indicate synthesis of HSP (from Buchanan *et al.*, 2000).

for the development of thermal quarantine treatments. Because fruit tissue is often less thermotolerant to high temperatures than that of insects (Couey, 1989), a pretreatment may be given to induce thermotolerance to a higher quarantine-effective temperature. Of course, any heat pretreatment of a crop is also likely to induce HSP production in the insect pest as well (Dean and Atkinson, 1982; Jang, 1992).

Heat pretreatments have been found to increase thermotolerance in many harvested plant commodities. Couey and Hayes (1986) reported a two-stage hot water treatment (30 min at 42°C followed by hot water at 49°C) for papaya disinfestations, which also controlled postharvest fungal diseases (Akamine and Arisumi, 1953). Conditioning in 37 or 39°C before 46 or 47°C hot water disinfestations reduced heat damage on avocados and mangoes (Joyce and Shorter, 1994; Jacobi *et al.*, 1995a, b). Similar benefits of prior temperature conditioning were found for avocados by Woolf and Lay-Yee (1997) and for cucumbers by Chan and Linse (1989). Figure 11.5 shows an avocado in which one half (right) was given a hot water conditioning treatment before high-temperature stress and the other (left) was not. The non-conditioned side developed heat injury.

In some studies, a correlation was found between the induction of thermotolerance and the production of HSP (Woolf *et al.*, 2000) and between the loss of thermotolerance and the disappearance of HSP from both avocados and tomatoes (Sabehat *et al.*, 1996; Woolf *et al.*, 2004). However, it must be stressed that studies of postharvest heat responses are only correlative between HSP and thermotolerance, unlike in some organisms where mutation studies show the necessity of an HSP in conferring thermotolerance.

Fig. 11.5. Avocado, half of which was put in 38°C water for 1 h (left half) before the whole fruit was heated for 3 min at 50°C. The half that was conditioned at 38°C was undamaged by the 50°C treatment, while the untreated half showed peel damage (photograph courtesy of A. Woolf).

There is a clear correlation between HSP and thermotolerance in many organisms, but in plants it has also been established that a heat stress can condition plants to low temperature (Lurie, 1998; Fig. 11.6). The laboratory of Saltveit found that prior high temperature exposure of 2–8 h at 38–42°C hot air affected the chilling sensitivity of tomato discs (Saltveit, 1991), mung bean hypocotyls (Collins *et al.*, 1993) and cucumber cotyledons and seeds (Lafuente *et al.*, 1991; Jennings and Saltveit, 1994).

When a heat treatment of 2–3 days in 38°C air was applied to tomato fruit, their sensitivity to low temperature was reduced and they could be stored for up to 1 month at 2°C without developing chilling injury (Lurie and Klein, 1991; Sabehat *et al.*, 1996; Lurie and Sabehat, 1997). The protection against chilling injury conferred by the preconditioning disappeared if the fruit was held at 20°C before being placed at low temperature (see Table 11.2).

This resistance to low-temperature injury was found to be contingent on the presence of HSP (Lafuente *et al.*, 1991; Sabehat *et al.*, 1996). In a study with avocado discs, maximal HSP production was found after 4 h at 38°C, and heating provided protection from chilling injury (Woolf *et al.*, 1995; Florissen *et al.*, 1996; Table 11.3). This response has been found in numerous other commodities (Table 11.3), including citrus (Wild, 1993), cucumber (McCollum *et al.*, 1995), mango (McCollum *et al.*, 1993), green pepper (Mencarelli *et al.*, 1993), persimmon (Burmeister *et al.*, 1997; Lay-Yee *et al.*, 1997; Woolf *et al.*, 1997) and courgette (Wang, 1994). Such treatments of harvested commodities allow for longer storage without loss of quality, and

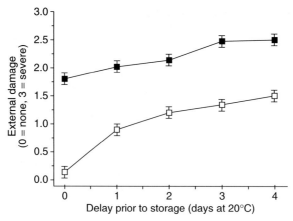

Fig. 11.6. Chilling injury of avocado after storage for 3 weeks at 0°C without any pretreatment (■) or after 6 h at 38°C before storage (□) (from Woolf *et al.*, 2004).

Table 11.2. The effect of high temperature preconditioning and recovery before storage for 3 weeks at 2°C on the development of chilling injury in mature green tomato fruit.

Treatment (day/°C)	Chilling injury (% of fruit)
No pretreatment	86 ± 13.4
4/20	88 ± 14.1
1/38	8 ± 2.1
2/38	4 ± 1.4
2/38 + 2/20	5 ± 2.3
2/38 + 4/20	72 ± 8.6

also the use of cold quarantine in fruits and vegetables that are generally sensitive to chilling injury.

The same pretreatment can induce resistance in a fruit to both high and low temperatures. In general, the best pretreatment of many fruits is at 38°C (see Table 11.3) and, when applied to avocados, decreased damage to further temperature extremes, both high and low, was observed (see Fig. 11.7). This resistance disappeared if the fruit was held at 15°C before the high- or low-temperature exposure (see Fig. 11.8), similar to the response of tomatoes shown in Table 11.2.

11.4 Heat Shock Responses and Heat Shock Proteins in Insects

Drosophila cells are normally grown at 25°C. HSP are induced when the temperature is raised to 29–38°C, but the maximum response in *Drosophila* and other insects (see Fig. 11.9) is observed at 36–38°C. At such temperatures, heat

Table 11.3. The benefits of high-temperature preconditioning before storage on preventing chilling injury in different commodities.

Commodity	Phenomenon/Appearance	Regime	Temperature (°C)/Time
Apple	Scald	HAT	38/4 days or 42/2 days
Avocado	Skin browning, internal browning, pitting	HAT then HWT	38/3–10 h then 40/30 min
		HWT	38/60 min
Cactus pear	Rind pitting, brown staining	HAT or HWT	38/24 h or 55/5 min
Citrus	Rind pitting	HAT	34–36/48–72 h
		HWT	50–54/3 min; 53/2–3 min
		HWB	59–62/15–30 s
Cucumber	Pitting	HWT	42/30 min
Green pepper	Pitting	HAT	40/20 h
Mango	Pitting	HAT	38/2 days
Persimmon	Gel formation	HWT	47/90–120 min; 50/30–45 min
		HAT	52/20–30 min
Tomato	Pitting	HAT	38/2–3 days
		HWT	48/2 min
			42/60 min
Courgette	Pitting	HWT	42/30 min

HAT, hot forced-air treatment; HWT, hot water treatment; HWB, hot spray and brush treatment.

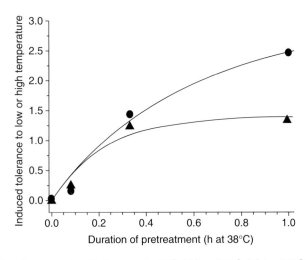

Fig. 11.7. Induction in avocado of tolerance to 0°C (▲) or 50°C (●) by 38°C hot water pretreatments (from Woolf *et al.*, 2004).

shock messages are produced within 4 min. Within 1 h, several thousand transcripts are present per cell (Lindquist, 1980). At the same time, both the transcription of previously active genes (Parsell and Lindquist, 1993) and the translation of pre-existing messages are repressed (Lindquist, 1986). Heat shock messages appear in the cytoplasm within a few minutes of temperature elevation and are immediately translated with very high efficiency (Lindquist, 1980).

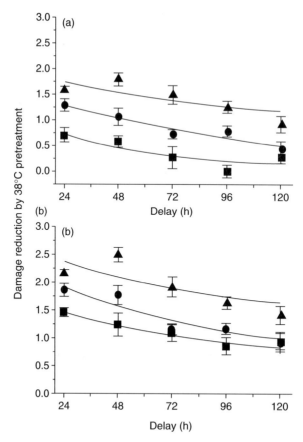

Fig. 11.8. Damage reduction in fruit by 38°C hot water pretreatment for 5 (■), 20 (●) or 60 min (▲) and then holding at 15°C for 24–120 h before placing at 0°C (a) or 50°C (b). A higher value indicates less damage (from Woolf *et al.*, 2004).

As long as the cells are maintained at high temperatures, HSP continue to be the primary products of protein synthesis. When cells are returned to normal temperatures of around 20°C, normal protein synthesis gradually resumes, the timing being a reflection of the severity of the preceding heat shock (Lindquist and DiDomenico, 1985). The HSP may disappear in a few hours at ambient temperature (see Fig. 11.10).

In *Drosophila*, the majority of cells respond to heat in a similar fashion. Imaginal wing discs, the brain, Malpighian tubules, salivary glands and tissue culture cells respond identically (Lindquist, 1986). However, the developmental stage may have an important effect on the response. The ability to respond to high temperature is absent in early embryos. Competence is achieved at the time of cellular blastoderm formation (Dura, 1981; Graziosi *et al.*, 1983).

The development of thermotolerance in *Drosophila* is tied to the production of HSP70, which is virtually absent from non-stressed cells (Velazquez *et al.*, 1983), but may account for the bulk of protein synthesis minutes after exposure to high temperature (Palter *et al.*, 1986). Too little

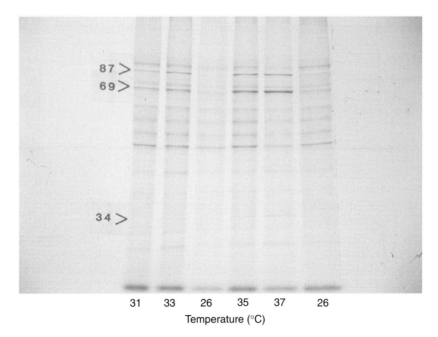

Fig. 11.9. 35S methionine labelling of heat shock proteins from cultured cells (TFVRL-CC1) of the Mediterranean fruit fly, *Ceratitis capitata*, after PAGE electrophoresis and autoradiography. Values on the left represent approximate MW of polypeptides separated. Temperatures at the bottom are exposure temperatures of cells (photograph courtesy of E. Jang).

HSP70 expression, induced either by shortening the stress treatments or lowering the treatment temperature, restricts the level of thermotolerance that develops (Krebs and Feder, 1998). Too much HSP70 can also be a problem. *Drosophila melanogaster* lines that carry many extra HSP70 copies tolerate stress less well than do control lines with normal copy numbers under pretreatment conditions that maximize HSP70 expression (Krebs and Feder, 1997). However, engineering lines to overproduce HSP70 can increase thermotolerance in some stages of development (Feder *et al.*, 1996).

It is unclear whether the responses that are found in *D. melanogaster* can be generally applied to other flies or other insects or whether individuals from hot environments will express more HSP70 than those from cooler environments. Studies suggest that this pattern exists for Chrysmelid beetles, as individuals from low altitudes express more HSP70 than do those from high altitudes (Dahlhoff and Rank, 1998). However, HSP may impose fitness costs on growth, particularly at non-lethal temperatures (Dorner *et al.*, 1992; Krebs and Feder, 1997). Potentially, insects from warmer climates may have tighter controls on HSP production in order to limit unnecessary production. *Drosophila melanogaster* maintained for years at 28°C expressed less HSP70 than comparable lines held at 18 or 25°C (Bettencourt *et al.*, 1999).

The environmental effect on HSP70 and thermotolerance was examined in three species of *Drosophila*: two common species, *D. melanogaster* and *D.*

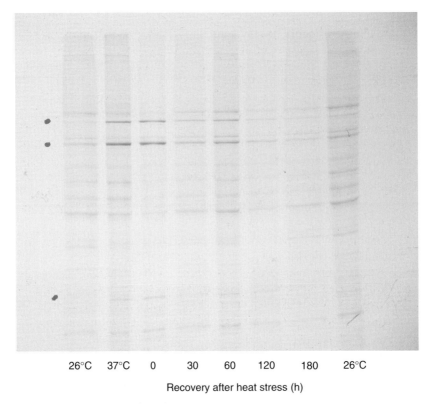

26°C   37°C   0   30   60   120   180   26°C

Recovery after heat stress (h)

Fig. 11.10. Time course of recovery after heat shock at 37°C for 1 h to normal protein synthesis of Mediterranean fruit fly, *Ceratitis capitata*. Cells were pulse labelled with 35S methionine at the indicated times after heat shock for 60 min at 26°C and exposed to autoradiography. Marks on the left indicate bands that decreased or disappeared during recovery from heat stress (photograph courtesy of E. Jang).

simulans, and one distantly related species, *D. mojavensis*, which inhabits the Sonoran desert in the USA and Mexico (Krebs, 1999). Larvae of *D. melanogaster* and *D. simulans* responded similarly to heat by expressing HSP70 maximally at 36–37°C, with a 2°C increase in heat tolerance.

The adult flies were less tolerant to heat than the larvae, but a close link between peak HSP70 expression and maximal induction of thermotolerance was found in *D. melanogaster*, but not in the other two species. *Drosophila mojavensis* tolerated higher temperatures than did either *D. melanogaster* or *D. simulans*. However, there was a poor relation between peak HSP70 production in this fly and survival after a further heat shock. Of the three species examined, only *D. melanogaster* showed a close relationship between HSP70 expression and an increase in thermotolerance in both larvae and adults (Krebs, 1999).

Similarly to the findings with *Drosophila*, thermotolerance of insect pests after exposure to non-lethal temperatures has been correlated with the induction of HSP (Dahlgaard *et al.*, 1998; Neven, 2000). With regard to quarantine treatments, research into inducible insect HSP is useful because temperature fluctuations in the field or during postharvest can cause a heat shock response that could subsequently compromise the efficacy of thermal treatments (Hallman and Mangan, 1997).

Increase in heat resistance occurs in: (i) the Mediterranean fruit fly, *Ceratitis capitata* (Wiedemann) [Diptera: Tephritidae] following exposure to sublethal temperatures between 32 and 42°C (Jang, 1992; Feder *et al.*, 1996); (ii) the Queensland fruit fly, *Bactrocera tryoni* (Froggatt), [Diptera: Tephritidae] (Beckett and Evans, 1997; Waddell *et al.*, 2000); (iii) the flesh fly, *Sarcophaga crassipalpis* (Macquart) [Diptera: Sarcophagidae] (Yocum and Denlinger, 1992); and (iv) the light brown apple moth, *Epiphyas postvittana* (Walker) [Lepidoptera: Tortricidae] (Beckett and Evans, 1997; Lester and Greenwood, 1997).

Hallman (1994) observed that the third-instars of the Caribbean fruit fly, *Anastrepha suspensa* (Loew) [Diptera: Tephritidae], reared at 30°C were significantly more heat tolerant than those reared at 20°C. Thomas and Shellie (2000) reported induction of thermotolerance in third-instar Mexican fruit fly, *Anastrepha ludens* (Loew) [Diptera: Tephritidae], after exposure to 44°C at a slow heating rate of 120 min. Thermal death responses in codling moth larvae to high temperature treatments without preconditioning have been reported (Yokoyama *et al.*, 1991; Neven and Rehfield, 1995; Neven, 1998; Wang *et al.*, 2002, 2004). Figure 11.11 shows how the induction of thermotolerance increases the time of exposure needed to kill insects such as codling moth.

In only a few cases has thermotolerance been examined with regard to production of HSP: (i) light brown apple moth, *E. postvittana* (Lester and Greenwood, 1997); (ii) Colorado potato beetle, *Leptinotarsa decemlineata* (Say) [Coleoptera: Chysomalidae] (Yocum, 2001); (iii) African migratory locust,

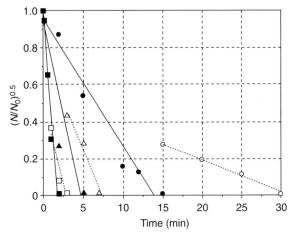

Fig. 11.11. Thermal mortality curves of fifth-instar *Cydia pomonella* with (empty symbols) or without (solid symbols) pretreatment at 35°C for 6 h before exposure to 48 (circles), 50 (triangles) or 52°C (squares) (from Yin *et al.*, 2006a).

Locusta migratoria L. [Orthoptera: Acrididae] (Qin *et al.*, 2003); and (iv) codling moth, *Cydia pomonella* [Lepidoptera: Tortricidae] (Yin *et al.*, 2006a, b).

Figure 11.12 indicates that the presence of HSP70 protein induced by heat conditioning pretreatment in codling moth is related to the survival of the insect after a quarantine heat treatment. Also, for the flesh fly, *Sarcophaga crassipalpis*, a 2 h treatment at 40°C conferred thermal tolerance to the normally lethal temperature of 45°C (Yocum and Denlinger, 1992). The thermal tolerance disappeared over time, as did the accumulation of HSP70 in the insect (see Fig. 11.13).

11.5 Discussion

It is important to recognize that many other factors are also likely to play a role in high-temperature stress tolerance in insects and plants. Heat shock has been found to result in the alteration of transcription of hundreds of genes in addition to HSP (Rizhsky *et al.*, 2002), and the accumulation of more than 100 different metabolites (Kaplan *et al.*, 2004).

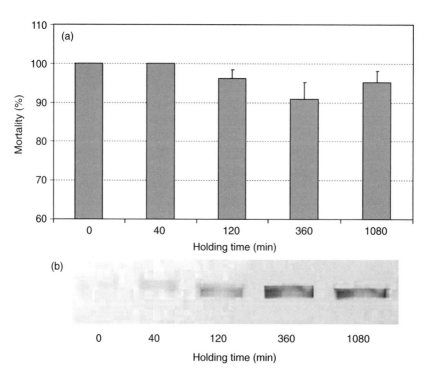

Fig. 11.12. (a) Mortality of fifth-instar codling moth (*Cydia pomonella*) at 50°C for 5 min, following preconditioning at 38°C for different times; (b) Western blot of insect protein using an antibody to HSP70. The control insects were held at 22°C for 18 h (0 min preconditioning) before heating. Preconditioned insects were held at 35°C for 40, 120, 360 and 1080 min and then heat treated (from Yin *et al.*, 2006b).

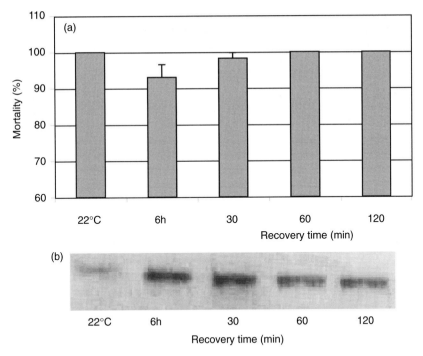

Fig. 11.13. (a) Recovery of high-temperature sensitivity of fifth-instar codling moths (*Cydia pomonella*) after preconditioning. The insects were pretreated for 6 h at 35°C and then held for up to 120 min at 22°C before exposure to 50°C for 5 min; (b) Western blot of insect protein with an HSP70 antibody showing decrease in HSP70 during the recovery period at 22°C (from Yin *et al.*, 2006b).

Trehalose, for example, is a disaccharide whose accumulation in many systems closely parallels acquisition of a stress-tolerant state (Colaco *et al.*, 1992). It is possible that trehalose directly promotes the reactivation of stress-damaged proteins. *In vitro* studies show that labile enzymes dried in the presence of trehalose can be reconstituted after prolonged storage with no loss of activity and can withstand exposure to extreme heat (Colaco *et al.*, 1992).

Other small molecules, like glycerol, betaine and proline, accumulate in various organisms under stress-tolerant conditions and may play a similar role. It is not yet clear whether these small molecules stabilize protein structures by interacting with them directly or by affecting the solvent properties of nearby water molecules (Yancey *et al.*, 1982). Whatever the mechanism, it is likely that small organic molecules and inorganic ions play an important role in protecting cellular proteins that become denatured during conditions of stress.

What is also often ignored during discussion of heat stress is another consequence of HSP induction, namely the diversion of normal protein synthesis; the heat response involves a substantial shift in protein synthesis away from normal turnover to that of HSP synthesis. This may have beneficial consequences in plant tissues. For example, in cut lettuce phenylalanine ammonia-lyase (PAL) and polyphenol oxidase (PPO) enzymes that cause browning are inhibited by prior heat treatment (Saltveit, 2000), indicating that the wounding response is overridden by the heat stress response. Thus, the dynamics of heat shock and HSP production may determine some physiological responses to heat.

Also, in harvested commodities, a wide range of fruit ripening processes are affected by heat, such as ethylene synthesis, respiration, fruit softening, cell wall metabolism, pigment metabolism and volatile production (Lurie, 1998; Paull and Chen, 2000; Chapter 4, this volume). These responses may have the consequence of slowing ripening and thereby extending the storage and shelf life of certain commodities.

It has been suggested that treatments to control insect pests should utilize the production of HSP to model insect responses. As discussed above with different species of *Drosophila*, this approach does not look promising. Although the accumulation of HSP70 followed thermoresistance to high temperatures in *D. melanogaster* quite well, it was less indicative of thermotolerance in *D. simulans* and *D. mojavensis*. On the other hand, in a recent study of codling moth, decrease in HSP70 was linked to the disappearance of thermotolerance induced by a 35°C pretreatment and the sensitivity of the insect to a 50°C quarantine treatment (Yin *et al.*, 2006b; Fig. 11.12).

Another problem with this approach is that the effective temperatures for pretreatments of plant tissue are between 38 and 42°C. Above 43°C, the accumulation of HSP decreases drastically in plants (Ferguson *et al.*, 1994). However, HSP are induced at lower temperatures that do not as successfully induce thermotolerance. This raises the question of whether the critical aspect of temperature response is the actual threshold temperature or the shift from normal growth temperature to a higher temperature. There may be a dose response involving the amount of heat *versus* time of treatment, as has been suggested by Woolf *et al.* (1995, 2000). It is unclear whether this relates to the production of HSP or to other factors as well.

In conclusion, the response of insects and plant tissues to elevated temperatures involves the accumulation of HSP, which can help protect organisms against a further elevation in temperature that would generally be harmful and, in the case of plants, also against a low temperature that may cause chilling injury. In some cases, the presence or persistence of HSP can indicate the resistance of the tissue to a further stress, but it is unclear whether HSP alone can be used as markers to determine whether an organism will be sensitive or resistant to a high-temperature stress. Further work in this area is needed, particularly on quarantine pests, before general conclusions can be drawn.

11.6 References

Abernethy, R.H., Thiel, D.S., Petersen, N.S. and Helm, K. (1989) Thermotolerance is developmentally dependent in germinating wheat seeds. *Plant Physiology* 89, 569–579.

Akamine, E.K. and Arisumi, T. (1953) Control of postharvest storage decay of fruits of papaya (*Carica papaya* L.) with special reference to the effect of hot water. *Proceedings of the American Society of Horticultural Science* 61, 270–274.

Beckett, S.J. and Evans, D.E. (1997) The effects of thermal acclimation on immature mortality in the Queensland fruit fly *Bactrocera tryoni* and the light brown apple moth *Epiphyas postvittana* at a lethal temperature. *Entomologia Experimentalis et Applicata* 82, 45–51.

Bettencourt, B.R., Feder, M.E. and Cavicchi, S. (1999) Experimental evolution of HSP70 expression and thermotolerance in *Drosophila melanogaster*. *Evolution* 53, 484–492.

Buchanan, B., Gruissen, W. and Jones, R.L. (2000) *Biochemistry and Molecular Biology of Plants*. American Society of Plant Physiologists, Rockville, Maryland, pp. 1158–1204.

Burmeister, D., Bal, S., Green, S. and Woolf, A.B. (1997) Interactions of hot water treatments and controlled atmosphere storage on quality of 'Fuyu' persimmons. *Postharvest Biology and Technology* 12, 71–82.

Chan, H.T. and Linse, E. (1989) Conditioning cucumbers for quarantine heat treatments. *HortScience* 24, 985–989.

Colaco, C., Sen, S., Thangavelu, M., Pinder, S. and Roser, B. (1992) Extraordinary stability of enzymes dried in thehalose: simplified molecular biology. *Biotechnology* 10, 1007–1011.

Collins, G.G., Nie, X. and Saltveit, M.E. (1993) Heat shock increases chilling tolerance of mung bean hypocotyls tissue. *Physiologia Plantarium* 89, 117–124.

Couey, H.M. (1989) Heat treatment for control of postharvest diseases and insect pests of fruits. *HortScience* 24, 198–201.

Couey, H.M. and Hayes, C.F. (1986) A quarantine system for papayas using fruit selection and a two stage hot water treatment. *Journal of Economic Entomology* 79, 1307–1314.

Craig, E.A. and Jacobsen, K. (1984) Mutations of the heat inducible 70 kilodalton genes of yeast confer temperature sensitive growth. *Cell* 38, 841–849.

Craig, E.A., Kramer, J., Shilling, J., Werner-Washburne, M. and Holmes, S. (1989) SSCI, an essential member of the yeast HSP70 multigene family, enclodes mitochondrial proteins. *Molecular Cell Biology* 9, 3000–3008.

Czarnecka, E., Edelman, L., Schoffl, F. and Key, J.L. (1984) Comparative analysis of physical stress responses in soybean seedlings using cloned heat shock cDNAs. *Plant Molecular Biology* 3, 45–58.

Czarnecka, E., Nagao, R.T., Key, J.L. and Gurley, W.B. (1988) Characterization of Gmhsp26-A, a stress gene encoding a divergent heat shock protein of soybean: heavy metal induced inhibition of intron processing. *Molecular Cell Biology* 8, 1113–1122.

Dahlgaard, J., Loeschcke, V., Michalak, P. and Justesen, J. (1998) Induced thermotolerance and associated expression of the heat-shock protein Hsp70 in adult *Drosophila melanogaster*. *Functional Ecology* 12, 786–793.

Dahlhoff, E.P. and Rank, N.E. (1998) Altitudinal variation in body temperature and heat shock protein expression in a montane leaf beetle. *American Zoologist* 38, 115A.

Dean, R.L. and Atkinson, B.G. (1982) The acquisition of thermal tolerance in larvae of *Calpodes ethlius* (Lepidoptera) and the *in situ* and *in vitro* synthesis of heat shock proteins. *Canadian Journal of Biochemistry and Cell Biology* 61, 472–479.

Dorner, A.J., Wasley, A.C. and Kaufman, R.J. (1992) Overexpression of GRP78 mitigates stress induction of glucose regulated proteins and blocks secretion of selective proteins in Chinese hamster ovary cells. *EMBO Journal* 11, 1563–1571.

Dura, J.M. (1981) Stage dependent synthesis of heat shock induced proteins in early embryos of *Drosophila melanogaster*. *Molecular and General Genetics* 184, 381–385.

Edelman, L., Czarnecka, E. and Key, J.L. (1988) Induction and accumulation of heat shock specific poly (A+) RNAs and proteins in soybean seedlings during arsenite and cadmium treatments. *Plant Physiology* 86, 1048–1056.

Ellis, J. (1987) Proteins as molecular chaperones. *Nature* 328, 378–379.

Engman, D.M., Kirchhoff, L.V. and Donelson, J.E. (1989) Molecular cloning of *mtp70*, a mitochondrial member of the *hsp70* family. *Molecular Cellular Biology* 9, 5163–5168.

Feder, M.E., Cartano, N.V., Milos, I., Krebs, R.A. and Lindquist, S. (1996) Effect of engineering Hsp70 copy number on Hsp70 expression and tolerance of ecologically relevant heat shock in larvae and pupae of *Drosophila melanogaster*. *Journal of Experimental Biology* 199, 1837–1844.

Ferguson, I.B., Lurie, S. and Bowen, J.H. (1994) Protein synthesis and breakdown during heat shock of cultured pear cells. *Plant Physiology* 104, 1429–1437.

Finley, D., Ozkaynak, E. and Varshavsky, A. (1987) The yeast polyubiquitin gene is essential for resistance to high temperatures, starvation, and other stresses. *Cell* 37, 1035–1046.

Florissen, P., Ekman, J.S., Blumenthal, C., McGlasson, W.B., Conroy, J. and Holford, P. (1996) The effects of short heat treatments on the induction of chilling injury in avocado fruit (*Persea americana* Mill). *Postharvest Biology and Technology* 8, 129–141.

Graziosi, G., Cristinii, F.D., Marcotullio, A.D., Marzari, R. and Micali, F. (1983) Morphological and molecular modifications induced by heat shock in *Drosophila melanogaster* embryos. *Journal of Embryology and Experimental Morphology* 77, 167–182.

Gupta, R.S. and Singh, B. (1992) Cloning of the HSP70 gene from *Halobacterium marismortui*: relatedness of archae-bacterial HSP70 to its eubacterial homologs and a model for the evolution of the HSP70 gene. *Journal of Bacteriology* 174, 4594–4605.

Gurley, W.B. and Key, J.L. (1991) Transcription regulation of the heat shock response: a plant perspective. *Biochemistry* 30, 1–12.

Hallman, G.J. (1994) Mortality of third-instar Caribbean fruit fly (Diptera: Tephritidae) reared at three temperatures and exposed to hot water immersion or cold storage. *Journal of Economic Entomology* 87, 405–408.

Hallman, G.J. and Mangan, R.L. (1997) Concerns with temperature quarantine treatment research. *Proceedings of the Annual International Research Conference on Methyl Bromide Alternatives and Emission Reduction*, 10–11 November, Orlando, Florida, pp. 79.1–79.4.

Haslbeck, M. and Buchner, J. (2002) Chaperone function of sHSPs. In: Arrigo, A.P. and Müller, W.E.G. (eds) *Progress in Molecular and Subcellular Biology*, Vol. 28. Springer Verlag, Berlin, pp. 37–59.

Hershko, A. and Ciechanover, A. (1992) The ubiquitin system for protein degradation. *Annual Review of Biochemistry* 61, 761–807.

Inoglia, T. and Craig, E. (1982) Four small *Drosophila* heat shock proteins are related to each other and to mammalian α-crystallin. *Proceedings of the National Academy of Sciences of the USA of America* 79, 2360–2364.

Jacobi, K.K., Giles, J., MacRae, E. and Wegrzyn, T. (1995a) Conditioning 'Kensington' mango with hot air alleviates hot water disinfestations injuries. *HortScience* 30, 562–565.

Jacobi, K.K., Wong, L.S. and Giles, J.E. (1995b) Effects of fruit maturity on quality and physiology of high humidity hot air treated 'Kensington' mango (*Mangifera indica* Linn.). *Postharvest Biology and Technology* 5, 149–159.

Jang, E.B. (1992) Heat shock proteins and thermotolerance in a cultured cell line from the Mediterranean fruit fly, *Ceratitis capitata*. *Archives of Insect Biochemistry and Physiology* 19, 93–103.

Jennings, P. and Saltveit, M.E. (1994) Temperature and chemical shocks induce chilling tolerance in germinating *Cucumis sativus* (cv. Poinsett 76) seeds. *Physiologia Plantarum* 91, 703–707.

Joyce, D.S. and Shorter, A.J. (1994) High temperature conditioning reduces hot water treatment injury of 'Kensington Pride' mango fruit. *HortScience* 28, 1047–1051.

Kaplan, F., Kopka, J., Haskell, D.W., Wei Zhao, K., Schiller, C., Gratzke, N., Sung, D.Y. and Guy, C.L. (2004) Exploring the temperature stress metabolome of *Arabidopsis*. *Plant Physiology* 136, 4159–4168.

Krebs, R.A. (1999) A comparison of Hsp70 expression and thermotolerance in adults and larvae of three *Drosophila* species. *Cell Stress and Chaperones* 4, 243–249.

Krebs, R.A. and Feder, M.E. (1997) Deleterious consequences of Hsp70 overexpression in *Drosophila melanogaster* larvae. *Cell Stress and Chaperones* 2, 60–71.

Krebs, R.A. and Feder, M.E. (1998) Heritability of expression of the 70kD heat shock protein in *Drosophila melanogaster* and its relevance to the evolution of thermotolerance. *Evolution* 52, 841–847.

Lafuente, M.T., Belver, A., Guye, M.G. and Saltveit, M.E. (1991) Effect of temperature conditioning on chilling injury of cucumber cotyledons. *Plant Physiology* 95, 443–449.

Landry, J., Cretien, P., Lambert, H., Hickey, E. and Weber, L.A. (1989) Heat shock resistance conferred by expression of the human HSP27 gene in rodent cells. *Journal of Cellular Biology* 109, 7–15.

Lay-Yee, M., Ball, S., Forbes, S.K. and Woolf, A.B. (1997) Hot water treatment for insect disinfestations and reduction of chilling sensitivity of 'Fuyu' persimmon. *Postharvest Biology and Technology* 10, 81–87.

Lee, G.J. and Vierling, E. (2000) A small heat shock protein cooperates with heat shock protein 70 systems to reactivate a heat-denatured protein. *Plant Physiology* 122, 189–198.

Lee, G.J., Pokala, N. and Vierling, E. (1995) Structure and *in vitro* molecular chaper-one activity of cytosolic small heat-shock protein from pea. *Journal of Biological Chemistry* 270, 10432–10438.

Lester, P.J. and Greenwood, D.R. (1997) Pretreatment induced thermotolerance in lightbrown apple moth (Lepidoptera: Tortricidae) and associated induction of heat shock protein synthesis. *Journal of Economic Entomology* 90, 199–204.

Leustek, T., Dalie, B., Smir-Shapira, D., Brot, N. and Weissbach, H. (1989) A member of the Hsp70 family is localized in mitochondria and resembles *Escherichia coli* DnaK. *Proceedings of the National Academy of Sciences of the USA of America* 86, 7805–7808.

Li, G.C. and Laszlo, A. (1985) Thermotolerance in mammalian cells: a possible role for heat shock proteins In: Atkinson, B.G. and Walden, D.B. (eds) *Changes in Eukaryotic Gene Expression in Response to Environmental Stress*. Academic Press, New York, pp. 227–254.

Lindquist, S. (1980) Translational efficiency of heat induced messages in *Drosphila melanogaster* cells. *Journal of Molecular Biology* 137, 151–158.

Lindquist, S. (1986) The heat shock response. *Annual Review of Biochemistry* 55, 1151–1191.

Lindquist, S. and Craig, E.A. (1988) The heat shock proteins. *Annual Review of Genetics* 22, 631–677.

Lindquist, S. and DiDomenico, B. (1985) Coordinate and noncoordinate gene expression during heat shock: a model for regulation. In: Atkinson, B.G. and Walden, D.B. (eds) *Changes in Eukaryotic Gene Expression in Response to Environmental Stress*. Academic Press, New York, pp. 71–89.

Lindquist, S., Parsell, D.A., Sanchez, Y., Taulien, J. and Craig, E.A. (1993) Heat shock proteins in stress tolerance. *Journal of The University of Occupational and Environmental Health* 15, 1–9.

Lurie, S. (1998) Postharvest heat treatments. *Postharvest Biology and Technology* 14, 257–269.

Lurie, S. and Klein, J.D. (1991) Acquisition of low temperature tolerance in tomatoes by exposure to high temperature stress. *Journal of the American Society of Horticultural Science* 116, 1007–1012.

Lurie, S. and Sabehat, A. (1997) Prestorage temperature manipulations to reduce chilling injury in tomatoes. *Postharvest Biology and Technology* 11, 57–62.

Marshall, J.S., DeRocher, A.E., Keegstra, K. and Vierling, E. (1990) Identification of heat shock protein hsp70 homologues in chloroplasts. *Proceedings of the National Academy of Sciences* 87, 374–378.

McCollum, T.G., D'Aquino, S. and McDonald, R.E. (1993) Heat treatment inhibits mango chilling injury. *HortScience* 28, 197–198.

McCollum, T.G., Doostdar, H., Mayter, R.T. and McDonald, R.E. (1995) Immersion of cucumber fruit in heated water alters chilling-induced physiological changes. *Postharvest Biology and Technology* 6, 55–64.

Mencarelli, F., Ceccantoni, B., Bolini, A. and Anelli, G. (1993) Influence of heat treatment on the physiological response of sweet pepper kept at chilling temperature. *Acta Horticulturae* 343, 238–243.

Mizzen, L.A., Chang, C., Garrels, J.I. and Welch, W.J. (1989) Identification, characterization, and purification of two mammalian stress proteins present in mitochondria, grp75, a member of the hsp70 family and hsp58, a homolog of the bacterial groEL protein. *Journal of Biological Chemistry* 264, 20664–20675.

Morimoto, R.I., Tissieres, A. and Georgopoulos, C. (eds) (1990) *Stress Proteins in Biology and Medicine*. Cold Spring Harbor Laboratory Press, New York.

Neumann, D., Nover, L., Parthier, H., Rieger, R. and Scharf, K.D. (1989) Heat shock and other stress response systems of plants. *Biologisches Zentralblatt* 108, 1–156.

Neven, L.G. (1998) Effects of heating rate on the mortality of fifth-instar codling moth (Lepidoptera: Tortricidae). *Journal of Economic Entomology* 91, 297–301.

Neven, L.G. (2000) Physiological responses of insects to heat. *Postharvest Biology and Technology* 21, 103–111.

Neven, L.G. and Rehfield, L.M. (1995) Comparison of prestorage heat treatments on fifth-instar codling moth (Lepidoptera: Tortricidae) mortality. *Journal of Economic Entomology* 88, 1371–1375.

Nover, L. (1991) *Heat Shock Response*. CRC Press, Boca Raton, Florida.

Nover, L., Hellmund, D., Neumann, D., Scharf, K.D. and Serfling, E. (1984) The heat shock response of eukaryotic cells. *Biologisches Zentralblatt* 103, 357–435.

O'Brien, T. and Lis, J.T. (1991) RNA polymerase II pauses at the 5' end of the transcriptionally induced *Drosophila* hsp70 gene. *Molecular and Cellular Biology* 11, 5285–5290.

Palter, K.B., Watanabe, M. Stinson, L., Mahowald, A.P. and Craig, E.A. (1986) Expression and localization of *Drosophila melanogaster* hsp70 cognate proteins. *Molecular and Cellular Biology* 6, 1187–1203.

Parker, C.S. and Topol, J. (1984) A *Drosophila* RNA polymerase II transcription factor binds to the regulatory site of an hsp 70 gene. *Cell* 37, 273–283.

Parsell, D.A. and Lindquist, S. (1993) The function of heat shock proteins in stress tolerance: degradation and reactivation of damaged proteins. *Annual Review of Genetics* 27, 437–496.

Paull, R.E. and Chen, N. (1990) Heat shock response in field grown ripening papaya fruit. *Journal of American Society of Horticultural Science* 115, 623–631.

Paull, R.E. and Chen, N. (2000) Heat treatment and fruit ripening. *Postharvest Biology and Technology* 21, 21–38.

Pelham, H.R. (1986) Speculations on the functions of the major heat shock and glucose regulated proteins. *Cell* 46, 959–961.

Phipps, B.M., Typke, D., Heger, R., Volker, S. and Hoffmann, A. (1993) Structure of a molecular chaperone from a thermophilic archaebacterium. *Nature* 361, 475–477.

Qin, W., Tyshenko, M.G., Wu, B.S., Walker, V.K. and Robertson, R.M. (2003) Cloning and characterization of a member of the *hsp70* gene family from *Locusta migratoria*, a highly thermotolerant insect. *Cell Stress and Chaperones* 8, 144–152.

Ritossa, F. (1962) The response of *Drosophila melanogaster* to elevated temperature. *Experientia* 18, 571–573.

Rizhsky, L., Liang, H. and Mittler, R. (2002) The combined effect of drought stress and heat shock on gene expression in tobacco. *Plant Physiology* 130, 1043–1151.

Rothman, J.E. (1989) Polypeptide chain binding proteins: catalysis of protein folding and related processes in cells. *Cell* 59, 591–601.

Sabehat, A., Weiss, D. and Lurie, S. (1996) The correlation between heat shock protein accumulation and persistence and chilling tolerance in tomato fruit. *Plant Physiology* 110, 531–537.

Saltveit, M. (1991) Prior temperature exposure affects subsequent chilling injury. *Physiologia Plantarium* 82, 529–536.

Saltveit, M. (2000) Heat shock prevents tissue browning by redirecting stress induced protein synthesis. *Postharvest Biology and Technology* 21, 61–70.

Sanchez, Y. and Lindquist, S. (1990) HSP104 is required for induced thermotolerance. *Science* 248, 1112–1115.

Sanchez, Y., Taulien, J., Borkovich, K.A. and Lindquist, S. (1992) Hesp104 is required for tolerance to many forms of stress. *EMBO Journal* 11, 2357–2364.

Schirra, M. and Mulas, M. (1995) Influence of postharvest hot water dip and imazalil fungicide treatments on cold stored Di Massa lemons. *Advances in Horticultural Science* 1, 43–46.

Schlesinger, M.J. (1990) Heat shock proteins. *Journal of Biological Chemistry* 265, 12111–12114.

Solomon, J.M., Rossi, J.M., Golic, K., McGarry, T. and Lindquist, S. (1991) Changes in Hsp70 after thermotolerance and heat shock regulation in *Drosophila*. *New Biology* 3, 1106–1120.

Subjeck, J.R. and Shyy, T.T. (1986) Stress protein systems of mammalian cells. *American Journal of Physiology* 250, C1–C17.

Sun, W., van Montagu, M. and Verbruggen, N. (2002) Small heat shock proteins and stress tolerance in plants. *Biochimica Biophysica Acta* 1577, 1–9.

Thomas, D.B. and Shellie, K.C. (2000) Heating rate and induced thermotolerance in Mexican fruit fly larvae, a quarantine pest of citrus and mangoes. *Stored Product and Quarantine Entomology* 93, 1373–1379.

Trent, J.D., Osipiuk, J. and Pinkau, T. (1990) Acquired thermotolerance and heat shock in the extremely thermophilic archaebacterium *Sulfolobus* sp. Strain B12. *Journal of Bacteriology* 172, 1478–1484.

Velazquez, J.M., Sonoda, S., Bugaisk, G. and Lindquist, S. (1983) Is the major *Drosophila* heat shock protein present in cells that have not been heat shocked? *Journal of Cellular Biology* 96, 286–290.

Vierling, E. (1991) The roles of heat shock proteins in plants. *Annual Review of Plant Physiology and Plant Molecular Biology* 42, 579–620.

Waddell, B.C., Jones, V.M., Petry, R.J., Sales, F., Paulaud, D., Maindonald, J.H. and Laidlaw, W.G. (2000) Thermal conditioning in Bactrocera tryoni eggs following hot-water immersion. *Postharvest Biology and Technology* 21, 113–128.

Wang, C.Y. (1994) Combined treatment of heat shock and low temperature conditioning reduces chilling injury in zucchini squash. *Postharvest Biology and Technology* 4, 65–73.

Wang, S., Ikediala, J.N., Tang, J. and Hansen, J.D. (2002) Thermal death kinetics and heating rate effects for fifth-instar *Cydia pomonella* (L.). *Journal of Stored Products Research* 38, 441–453.

Wang, S., Yin, X., Tang, J. and Hansen, J.D. (2004) Thermal resistance of different life stages of codling moth (Lepidoptera: Tortricidae). *Journal of Stored Products Research* 40, 565–574.

Wild, B.L. (1993) Reduction of chilling injury in grapefruit and oranges stored at 1°C by prestorage hot dip treatments, curing, and wax application. *Australian Journal of Experimental Agriculture* 33, 495–498.

Winter, J., Wright, R., Duck, N., Gasser, C., Fraley, R. and Shah, D. (1988) The inhibition of petunia hsp70 mRNA processing during CdCl2 stress. *Molecular and General Genetics* 211, 315–319.

Woolf, A.B. and Ferguson, I.B. (2000) Post-harvest responses to high fruit temperature in the field. *Postharvest Biology and Technology* 21, 7–20.

Woolf, A.B. and Lay-Yee, M. (1997) Pretreatments at 38°C of 'Hass' avocado confer thermotolerance to 50°C hot water treatments. *HortScience* 32, 705–708.

Woolf, A.B., Watkins, C.B., Bowen, J.H., Lay Yee, M., Maindonald, J.H. and Ferguson, I.B. (1995) Reducing external chilling injury in stored 'Hass' avocados with dry heat treatments. *Journal of the American Society of Horticultural Science* 120, 1050–1056.

Woolf, A.B., Ball, S., Spooner, K.J., Lay Yee, M., Ferguson, I.B., Watkins, C.B., Gunson, A. and Forbes, S.K. (1997) Reduction of chilling injury in the sweet persimmon 'Fuyu' during storage by dry air heat treatments. *Postharvest Biology and Technology* 11, 155–164.

Woolf, A.B., Weksler, A., Prusky, D. and Lurie, S. (2000) Direct sunlight influences postharvest temperature responses and ripening of five avocado cultivars. *Journal of the American Society of Horticultural Science* 125, 370–376.

Woolf, A.B., Bowen, J.H., Ball, S., Durand, S., Laidlaw, W.G. and Ferguson, I.B. (2004) A delay between a 38C pretreatment and damaging high and low temperature treatments influences pretreatment efficacy in 'Hass' avocados. *Postharvest Biology and Technology* 34, 143–153.

Yancey, P.H., Clark, M.E., Hand, S.C., Bowlus, R.D. and Somero, G.N. (1982) Living with water stress: evolution of osmolyte systems. *Science* 217, 1214–1222.

Yin, X., Wang, S., Tang, J. and Hansen, J.D. (2006a) Thermal resistance of fifth-instar *Cydia pomonella* (L.) (Lepidoptera: Tortricidae) as affected by pretreatment conditioning. *Journal of Stored Products Research* 42, 75–85.

Yin, X., Wang, S., Tang, J., Lurie, S. and Hansen, J.D. (2006b) Thermal preconditioning of fifth-instar *Cydia pomonella* (Lepidoptera: Toretricidae) affects HSP70 accumulation and insect mortality. *Physiological Entomology* 31, 241–247.

Yocum, G.D. (2001) Differential expression of two HSP 70 transcripts in response to cold shock, thermoperiod, and adult diapause in the Colorado potato beetle. *Journal of Insect Physiology* 47, 1139–1145.

Yocum, G.D. and Denlinger, D.L. (1992) Prolonged thermotolerance in the flesh fly, *Sarcophaga crassipalpis*, does not require continuous expression or persistence of the 72 kDa heat shock protein. *Physiological Entomology* 19, 152–158.

Yokoyama, V.Y., Miller, G.T. and Dowell, R.V. (1991) Response of codling moth (Lepidoptera: Tortricidae) to high temperature, a potential quarantine treatment for exported commodities. *Journal of Economic Entomology* 84, 528–531.

Yost, H.J. and Lindquist, S. (1986) RNA splicing is interrupted by heat shock and is rescued by heat shock protein synthesis. *Cell* 45, 185–193.

Zimarino V. and Wu, C. (1987) Induction of sequence specific binding of *Drosophila* heat shock activator protein without protein synthesis. *Nature* 327, 727–730.

12 Thermal Treatment Protocol Development and Scale-up

J. Tang,[1]* S. Wang,[1]** and J.W. Armstrong[2]

[1]Department of Biological Systems Engineering, Washington State University, Pullman, Washington, USA; e-mails: *jtang@wsu.edu; **shaojin_wang@wsu.edu; [2]USDA-ARS, US Pacific Basin Agricultural Research Center, Hilo, Hawaii, USA; e-mail: jarmstrong@pbarc.ars.usda.gov

12.1 Introduction

Heat (thermal) treatments to control insect pests in agricultural commodities must satisfy three criteria to be commercially successful: (i) provide adequate insect mortality to meet quarantine or phytosanitary requirements; (ii) cannot adversely affect commodity quality; and (iii) be technically and economically feasible to use in commercial operations. This chapter discusses a systematic approach to developing heat treatments and provides examples of quarantine treatment development using radio frequency (RF) energy to control target pests in agricultural commodities.

12.2 Strategies for Thermal Treatment Development

Important Scientific Information

When developing effective treatment protocols, it is essential to know the thermotolerance characteristics of both the target insects and commodities over a relatively wide range of time–temperature combinations. Methods to obtain insect thermal mortality data are discussed in Chapter 5, this volume. From these data, it is possible to define the treatment time and temperature parameters that will cause 100% mortality to a targeted insect population (see Fig. 12.1) and also maintain commodity quality and shelf life. The upper temperature and time limits of this range, which we will call the 'quality curve,' are illustrated in Fig. 12.1.

All quality attributes of a commodity must be considered in the development of this curve. For example, stem colour and attachment, fruit appearance and firmness, total solid soluble content and titratable acidity are

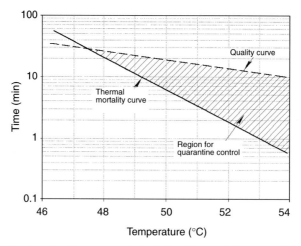

Fig. 12.1. Schematic showing the concept of relative thermotolerance of a target pest and the commodity it infests (from Tang *et al.*, 2000).

the major quality attributes that must be factored into the quality curve when developing a treatment for cherries (Drake *et al.*, 2005). Each quality attribute may react differently to different temperature ranges and, if any attribute becomes unacceptable to the consumer, the treated commodity loses its market value. In quarantine heat treatment research, it is more complicated and time-consuming to develop a quality thermotolerance curve than a mortality curve for an insect, because many quality attributes must be preserved, and the acceptability of relatively minor changes are often difficult to evaluate.

Theoretically, fresh commodities are less heat-sensitive than insects because quality changes in commodities are associated with smaller activation energy (e.g. E_a = 100 kJ/mole for texture softening, Rao and Lund, 1986; Taoukis *et al.*, 1997) than that required to obtain insect mortality (E_a = 400–956 kJ/mole, Wang *et al.*, 2002a, b). That is, the slope of the quality thermotolerance curve (see Fig. 12.1) is usually smaller than that of the insect mortality curve (Tang *et al.*, 2000). The positions and shapes of quality and insect thermotolerance curves are commodity- and insect-dependent. The overlap between the area below the quality curve and the area above the insect mortality curve (Fig. 12.1) identifies a range of temperatures and exposure times at which a thermal treatment provides complete control of the target insects without damaging the commodity.

Separate studies on commodity and insect thermotolerance are performed to identify treatment time and temperature parameters in the development of candidate heat treatments. Quality curves for some commodities may remain below the mortality curves of the target insect regardless of temperature. In such cases, heat treatments that control the target insects would cause unacceptable damage to the commodity, regardless of application method. Developing thermotolerance curves for both target insects and commodities is an important first step that will accelerate the development of quarantine heat treatments and avoid unnecessary and unsuccessful efforts in the process.

Systematic Treatment Development

Systematic development of effective thermal treatments should consist of the steps illustrated in Fig. 12.2:

1. Developing baseline data – including insect and commodity thermotolerance curves – and selecting the most appropriate technology for application of the thermal treatment for insect control.

2. Developing a treatment protocol using small- or pilot-scale heating equipment to prove that the treatment works and to identify the optimum treatment parameters.

3. Confirming treatment efficacy by testing it against the commodity infested with live insects and evaluating the quality of the treated commodity.

4. Engineering the treatment to commercial scale and performing confirmatory tests with the commercial equipment. Confirmatory tests provide direct evidence to regulatory agencies that the treatment and equipment will operate to specification and will ensure quarantine security against potential infestations.

Pilot-scale testing (Step 2) is usually performed in laboratories with relatively small samples in order to study different combinations of treatment conditions economically and in a timely manner. Direct temperature measurement with thermocouples, fibre-optic sensors or infrared temperature imaging systems is used to determine heating rates and uniformity under controlled laboratory conditions. Optimal treatment parameters are then selected to ensure that all temperatures are as uniform as possible throughout the load of treated commodity (whether in single layers or bulk bins), and that the temperatures fall within the range that does not damage the commodity (see Fig. 12.1).

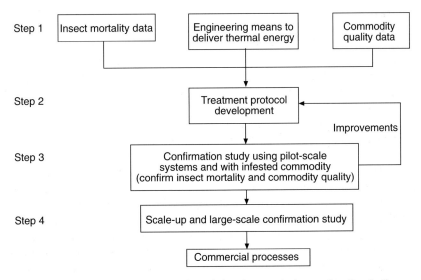

Fig. 12.2. Flowchart showing the research and development phases leading to the commercialization of a quarantine treatment.

Frequently, adequate information to develop the quality curve (see Fig. 12.1) does not exist, and pilot-scale testing must be conducted using a wide range of treatment times and temperatures to determine quality differences between commodities. Optimal treatment parameters are selected from evaluations that determine the best commodity quality while ensuring quarantine security against the target pest.

After the time and temperature parameters for a candidate treatment have been selected, tests are conducted (Step 3) using commodities infested with live insects to confirm treatment efficacy (see Fig. 12.2). Quality studies are carried out using uninfested commodities with the same treatment parameters used in the efficacy tests to avoid any potential damage caused by the target pest interfering with the quality data. A number of efficacy and quality tests may be necessary between treatment protocol development in Step 2 and confirmation testing in Step 3 to optimize the parameters that provide control of the targeted insect without causing damage to the commodity.

Engineering a quarantine treatment from pilot-scale tests to commercial application (Step 4, Fig. 12.2) requires the development of equipment that will reproduce a given treatment consistently, can be incorporated into the postharvest process, is easy to use, and is economically viable. Confirmatory tests consisting of a series of commercial treatments using infested commodities are required to demonstrate that the equipment will produce consistent treatments that provide quarantine security.

Correctly engineered treatment equipment that has undergone proper testing beforehand should not result in survival of target insects in the confirmatory tests, as this would result in costly delays while the causes of the test failure are determined, the commercial equipment is adjusted accordingly or corrections are made to the treatment parameters. Regulatory agencies will not approve equipment until confirmatory testing is successfully completed.

Once the equipment is certified, a compliance agreement between the industry, treatment facility, exporter or other party responsible for performing the quarantine treatment and the regulatory agency is established. Compliance agreements are legally binding contracts developed by regulatory agencies that describe in detail all aspects of the required treatment parameters, equipment operations, commodity inspection, handling, storage and shipping procedures, as well as other operational aspects that ensure quarantine security. In some cases, each treatment must be validated on-site by regulatory personnel. Violating a compliance agreement can result in closure of a treatment facility, forfeiture of export rights and severe fines. If a regulatory agency requires that a quarantine treatment ensures product quality as well as quarantine security, quality tests with uninfested fruit become part of the confirmatory test process.

Commercializing a quarantine treatment also includes integrating it into existing postharvest operations to minimize additional handling and reduce capital investment. Here, engineering and economics combine to evaluate and determine the economic feasibility of the proposed commercial treatment. These studies include: (i) critical pretreatment (e.g. field-to-packing house time intervals, initial commodity temperatures, pretreatment storage conditions and handling procedures); (ii) treatment; and (iii) post-treatment parameters (e.g.

commodity throughput, treatment duration, treatment cost, post-treatment cooling, handling, processing, storage and shipping).

Ultimately, economic feasibility is determined by whether the quarantine treatment can be carried out with minimum economic impact on the overall field-to-market activities and profit margin after treatment. The development and application of conventional thermal treatments currently used by export industries throughout the world are discussed in Chapter 13, this volume. In the following sections we look at the actual development of a novel quarantine treatment for walnuts and fruits based on RF energy.

12.3 Systematic Development of RF Treatment for In-shell Walnuts

In walnuts, RF energy generates heat by agitating bound water molecules in the nutmeat. In insects, the same RF energy generates heat through ionic conduction and agitation of the free water molecules in the insect body (Wang et al., 2003b). As a result, more thermal energy is converted in the insects than in the walnuts when exposed to an RF field (Wang et al., 2003a).

Codling moth (Cydia pomonella L.) and navel orangeworm (Amyelois transitella Walker) are major field pests of walnuts during postharvest storage before processing. Codling moth is regulated by Japan and Korea as a quarantine pest, while the presence of navel orangeworm is a phytosanitary concern for Australian and European markets and can result in regulatory action. Indianmeal moth (Plodia interpunctella Hübner) and red flour beetle (Tribolium castaneum Herbst) are also commonly found in stored walnuts, and are often responsible for consumer complaints (Johnson et al., 2004).

To accelerate RF treatment development against all these insect pests, we wanted to use the most thermotolerant insect species and the most thermotolerant life stage of that species. We found that the fifth-instar navel orangeworm was the most heat-resistant life stage and species of the four pest species at temperatures > 50°C (Wang et al., 2002c; Johnson et al., 2004; Fig. 12.3). Therefore, any treatment parameters that control fifth-instar navel orangeworm will also control all other life stages of codling moth, Indianmeal moth, and red flour beetle.

Pilot-scale Protocol Development

In the pilot-scale phase of developing an RF treatment protocol, a walnut kernel temperature–time profile was determined first with selected treatment parameters. Since the walnut kernel temperature is linearly proportional to the heating time in RF systems (Wang et al., 2001a, b), increasing treatment time at a given heating rate results in increased walnut kernel temperatures, thereby increasing insect mortality.

The first laboratory-scale RF treatment to control codling moth in walnuts

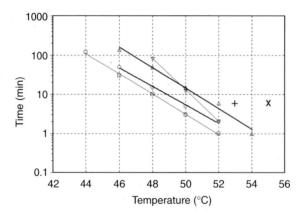

Fig. 12.3. Selected time–temperature combinations to control codling moth (+) and navel orangeworm (x) in walnuts, based on experimentally determined minimum time–temperatures for complete kill of 600 fifth-instar Indianmeal moths (□), codling moths (○), navel orangeworms (Δ) and red flour beetle (∇) after heating at 18°C/min in a heating block system (from Wang *et al.*, 2002c; Johnson *et al.*, 2004).

was developed in 2001 to meet the phytosanitary regulations of Japan and South Korea (Wang *et al.*, 2001a). Based on the thermal death time (TDT) curve of codling moth, heating times of 1, 2 and 3 min were selected to obtain a range of mortality levels after 5 min of holding time at the final core temperature. Candidate RF treatments that provided 100% mortality of codling moth were then tested against uninfested walnuts to evaluate commodity quality.

A 6 kW, 27 MHz pilot-scale RF system set to provide 0.4 kW power was used to heat walnuts in the pilot-scale tests. Twenty in-shell walnuts were placed in a closed plastic box between the electrodes. The plastic box did not heat during the RF treatment. A type-T thermocouple was inserted through a pre-drilled hole in each shell to measure kernel temperatures after RF treatments. Test results showed that kernel temperatures were 31.0, 42.6 and 53.3°C after 1, 2 and 3 min of RF heating, respectively. The protocol point after the 3 min of RF treatment with 5 min holding is shown in Fig. 12.3, which resulted in 100% mortality of third- and fourth-instar codling moth (see Table 12.1).

The next phase of the research was to determine whether the RF treatment parameters that killed the target pests damaged walnut quality. Oxidation of the 54% unsaturated fatty acids (FA) in walnut kernels makes them one of the most heat-sensitive, low-moisture commodities. The two most important quality factors of walnut oil, peroxide value (PV) and FA, were analysed after RF treatment to 53.3°C kernel temperature and 5 min holding at that temperature, followed by an accelerated storage time.

Accelerated storage processes consist of holding the walnuts at a higher temperature for shorter times that is equivalent to holding the walnuts at normal colder storage conditions for longer times (Wang *et al.*, 2006b). For treated walnuts, accelerated storage times of 10, 20 and 30 days at 35°C were followed to simulate 2 years' storage at 4°C. The test results showed that

Table 12.1. Mortality of third- and fourth-instar codling moth in walnuts after radio frequency (27 MHz) heating to three different final product temperatures and 5 min holding (adapted from Wang *et al.*, 2001a).

Heating time (min)	Product temperature (°C)	Recovered (n)	Alive (n)	Dead (n)	Repeat number (n)	Mortality ± SD (%)
0	Control	17	17	0	2	0
1	31.0	61	32	29	2	47.5 ± 11
2	42.6	84	18	66	5	78.6 ± 6
3	53.3	70	0	70	4	100 ± 0

neither PV nor FA values were affected by the RF treatment and storage parameters (see Table 12.2), indicating an efficacious treatment against codling moth that does not adversely affect walnut quality.

A similar treatment protocol was developed in 2002 using the same RF system with an auxiliary hot air system that helped to heat and maintain walnut surface temperature during holding time. This candidate treatment was used to control fifth-instar navel orangeworm in walnuts to meet Australian and European phytosanitary regulations (Wang *et al.*, 2002c). Sixty walnuts were used from the same batch of samples for infestation and quality tests. Because navel orangeworm was more heat-resistant than codling moth (Fig. 12.3), treatment parameters of 55°C final temperature and 5 or 10 min holding time were selected to obtain complete control of fifth-instar navel orangeworm in walnuts. Results showed that RF heating to 55°C and holding in hot air for 5 min killed all navel orangeworm with no adverse effects on walnut quality (Wang *et al.*, 2002c). Because fifth-instar navel orangeworm is the most heat-tolerant of all the target species, the treatment also would be effective against codling moth, Indianmeal moth and red flour beetle.

Heating Uniformity

Heating uniformity, the distribution of hot or cold spots throughout a treated load, is extremely important when developing a commercial RF treatment. Hot spots can lead to loss of quality, while cold spots can lead to survival of the

Table 12.2. Quality characteristics (means ± SD over two replicates) of in-shell walnuts treated by radio frequency (RF) energy at 3 min with a final temperature of 53.3°C for 5 min (from Wang *et al.*, 2001a).

Storage time at 35°C (days)	Peroxide value (PV) (mEq/kg)		Fatty acid (%)	
	Control	RF treated	Control	RF treated
0	0.26 ± 0.04	0.28 ± 0.04	0.08 ± 0.01	0.08 ± 0.01
10	0.49 ± 0.05	0.51 ± 0.05	0.08 ± 0.01	0.09 ± 0.01
20	0.93 ± 0.05	0.98 ± 0.05	0.10 ± 0.01	0.08 ± 0.01
30	1.04 ± 0.05	1.17 ± 0.05	0.11 ± 0.01	0.10 ± 0.01

target pest. Non-uniform heating was found in walnuts during RF treatment because of the individual size and shape of each walnut and the location of walnuts in a non-uniform electromagnetic field (Wang *et al.*, 2005). The effect of walnut orientation (the direction the walnut faces in relation to the other walnuts in the treatment) on temperature is shown in Fig. 12.4.

After 6 min of RF heating, the four vertically oriented walnuts were hotter (61.3 ± 1.9°C) compared with the four horizontally oriented walnuts (53.9 ± 1.1°C) (Wang *et al.*, 2006b). This temperature difference (see Fig. 12.4) can be reduced by blending and rotating (i.e. mixing) the walnuts during RF treatment. Mixing is generally achieved by stirring manually or tumbling mechanically to obtain thorough changes of walnut orientations and positions.

Although treatment uniformity depends on the amount of mixing in a given RF treatment, mixing increases equipment and operational costs. Therefore, the minimum amount of mixing required to achieve heating uniformity must be identified to reduce costs related to mixing time. A mathematical model was developed to determine the amount of mixing needed to achieve treatment uniformity at a selected average final temperature in treated products (Wang *et al.*, 2005). The amount of mixing operations between consecutive RF heating steps is referred to as the 'mixing number'.

Equation 12.1 was based on the following assumptions:

1. An RF treatment consists of $n + 1$ RF heating units, with one mixing operation taking place between two consecutive RF units.
2. No mixing occurs within each RF unit.
3. Walnut temperatures increase linearly with RF heating time.
4. Product temperature follows a normal distribution from colder to hotter walnuts in a treated load.
5. Heat losses during mixing are negligible.
6. Mixing randomly changes the position and orientation of each walnut in the treated load.

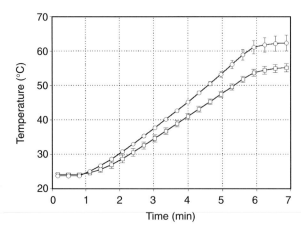

Fig. 12.4. Typical temperature–time heating profile (means ± SD) of four horizontally (□) and vertically (○) oriented walnuts in the centre of a single layer of walnuts subjected to radio frequency (RF) heating (from Wang *et al.*, 2006b).

7. The mathematical model was based on the lower temperature limit for insect control (Wang *et al.*, 2005).

The minimum number (n) of mixings can be expressed as

$$n = \frac{\left(\mu_T - \mu_0\right)^2 \cdot \lambda^2}{\left(\dfrac{L - \mu_T}{z_p}\right)^2 - \sigma_0^{\,2}} - 1 \qquad (12.1)$$

where μ_0 and σ_0 represent the mean and standard deviation (°C), respectively, of the initial product temperature, which can be measured before the experiments; and μ_T and L are the desired mean and minimum temperature (°C), respectively, which can be determined by the TDT curve of the targeted insect. λ is the uniformity index; and normal score z_p is determined by probability P based on the desired level of insect mortality (e.g. $z_p = -4.0$ when $P = 0.000032$).

λ is determined by a series of experiments covering RF operational conditions, walnut samples and specific container sizes. It reflects the heating uniformity in a given RF unit and depends on the design of RF applicators and interactions between the RF unit and product. It can be derived from product temperature measurements during RF treatments using the following equation:

$$\lambda = \frac{\Delta \sigma}{\Delta \mu} \qquad (12.2)$$

where $\Delta \sigma$ is the rise in standard deviation of product temperature (°C) and $\Delta \mu$ is the rise in mean product temperature (°C) over the treatment time. By definition, when mean temperature increases by one degree in RF heating, the standard deviation will increase by λ degrees. The smaller the λ value, the more uniform heating in the product.

Figure 12.5 shows the predicted number of mixings using Eq. 12.1 for mean walnut temperatures of 54–70°C under several combinations of P, L and $\lambda = 0.165$ for a container with 4.5 kg of in-shell walnuts in a 12 kW RF unit. The number of mixings decreased with increasing L and mean temperature and increasing P. For example, if the desired treatment requires that < 0.1% ($P = 0.001$) of the walnuts be < 48°C, a treatment can achieve a mean temperature of 66°C with one mixing.

A three-step RF heating with two intermittent mixings would further reduce the required mean product temperature to 61°C. It is clear from this example that increasing the mixing number reduces the use of RF energy and adverse thermal impact on product quality, but increasing mixing times inevitably increases capital costs and reduces throughputs. Therefore, a balance between capital investment of RF units, energy use, throughput and product quality must be achieved before commercialization of the RF treatment takes place.

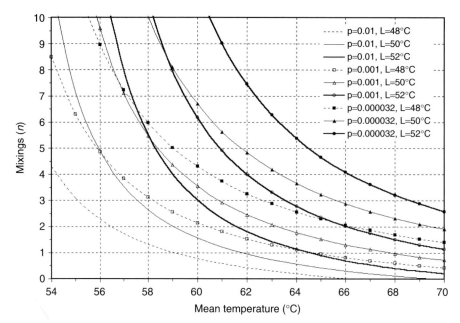

Fig. 12.5. Minimum number (*n*) of mixings predicted by a mathematical model as a function of the mean temperature with different probability levels (*P*) for walnut temperatures below the lower limit (*L*) (from Wang *et al.*, 2005).

Commercialization

Engineering and economic studies to determine whether a candidate quarantine treatment is commercially viable were discussed above in reference to pre- and post-treatment operations. Here, we will discuss these requirements as applied to walnuts. Typical walnut operations (in consecutive order) include harvesting, hulling, cleaning, field drying, fumigation, sizing, washing, bleaching, static air drying and packaging (see Fig. 12.6).

The best points for inserting an RF treatment are either (i) before washing and bleaching; or (ii) after static air drying. The first option has the advantage of preventing pests from entering into the processing facility. In this case, RF treatments should be designed to treat walnuts with a relatively wide range of moisture contents. The second option takes advantage of uniform moisture content and the residual heat from the static air dryers to reduce RF energy because of increased walnut temperatures (37–41°C) after hot-air drying.

An industrial 27.12 MHz, 25 kW RF system with two pairs of identical electrodes was used in commercialization studies for pest control in in-shell walnuts (see Fig. 12.7). Coupling of RF power into the load was adjusted by changing the gap between the electrodes, and throughput was adjusted by changing conveyor belt speeds. The RF system was equipped with an auxiliary air heating system to heat and dry the walnut surfaces.

The first step in commercializing the RF treatment was to determine heating uniformity. The critical factors considered in this study included initial

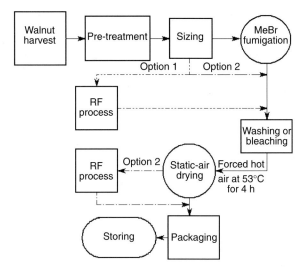

Fig. 12.6. Flowchart for deciding which of two radio frequency (RF) treatment options to use as a replacement for fumigation to control pests in walnuts. MeBr, methyl bromide.

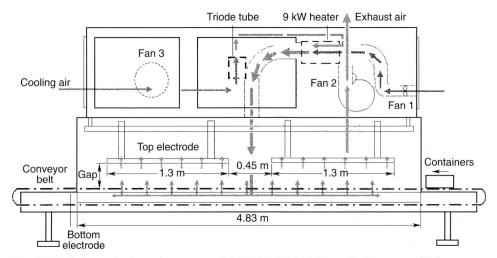

Fig. 12.7. Schematic view of a commercial 25 kW, 27.12 MHz radio frequency (RF) system with plate electrodes and a hot-air heating system (from Wang *et al.*, 2007).

temperatures (similar to the ambient walnut temperatures before treatment) and moisture content (5–8% after pre-dehydration). Tests were conducted to determine the heating uniformity index for a single perforated plastic container and a continuous process (17 identical containers) using stored walnuts from the previous season (11 kg per container).

Walnut surface temperatures were measured by a thermal imaging camera before and after RF heating. In the commercial RF system, the uniformity index value (λ in Eq. 12.2) was determined by thermal imaging to be 0.103 for in-shell walnuts in stationary containers and 0.061 (see Table 12.3) when heated on a moving belt with circulating air at 60°C added to maintain surface temperature and a thorough mixing in the middle of the RF heating time.

Mixing was achieved by a single pass through a riffle-type sample splitter and dividing the treated nuts into two representative samples that were then added back into the treatment container. The mixing process showed that nuts were redistributed evenly throughout the sample after mixing. Using the uniformity index λ value of 0.061, the results showed that when the mean walnut shell temperature reached 60°C, 99.9968% of the walnut kernels would be at a temperature of at least 52°C with one mixing. Because 5 min at 52°C killed all fifth-instar navel orangeworm (Wang *et al.*, 2002b), a 60°C walnut surface temperature was selected for commercialization studies.

Table 12.4 shows the insect mortality and temperatures in unwashed and air-dried in-shell walnuts after RF treatments, along with insect mortality in controls. The RF treatments were continuous at a belt speed of 57 m/h and electrode gap of 280 mm. Minor differences in final temperatures were due primarily to normal variability in walnut moisture content. The final average shell and kernel temperatures were at least 55°C after the RF treatment (see Table 12.4). All the RF treatments resulted in 100% mortality, compared to < 1.1% in untreated control larvae.

The efficacy results agreed with the TDT curves of fifth-instar navel orangeworm obtained by the heating block system (Wang *et al.*, 2002b). The TDT curve showed that holding times of 5 min at 52°C or 1 min at 54°C should result in 100% mortality (see Fig. 12.8), which concurred with the pilot-scale protocol development for walnuts (Wang *et al.*, 2001a, 2002c) discussed above regarding product quality *versus* insect mortality.

In-shell walnuts also were treated to identify any adverse effects on quality. Development of rancidity values based on PV, FA and kernel colour (*L* values) were used as quality indicators. No statistically significant quality differences were observed between control and RF-treated samples (see Table 12.5).

Table 12.3. Heating uniformity index (means ± SD over two replicates) of stored in-shell walnuts after radio frequency (RF) treatments under different operational conditions.

| Containers | Treatments[a] | | | Uniformity index (λ) |
	Moving	Single mixing	Hot air	
Single	No	No	No	0.103 ± 0.017
Single	Yes	No	No	0.087 ± 0.005
Single	Yes	No	Yes	0.083 ± 0.012
Single	Yes	Yes	Yes	0.061 ± 0.012
Continuous	Yes	Yes	Yes	0.062 ± 0.012

[a] The treatments were conducted with: (i) a single or continuous process (17 containers); (ii) stationary (no belt movement) or movement at the maximum conveyor belt speed of 57 m/h; (iii) with or without 60°C hot air; and (iv) with or without a single mixing between two RF exposures.

Table 12.4. Average nut kernel temperature, nut surface temperature and insect mortality rate in radio frequency (RF)-treated in-shell walnuts.

		Average temperature (°C)		
Replicate	Treatment	Nut kernel	Nut surface	Insect mortality (%)
Unwashed walnuts				
1	Control	26.0	27.8	0
	RF	58.5	63.8	100
2	Control	23.7	22.7	1.1
	RF	57.9	62.6	100
3	Control	24.3	23.9	0
	RF	55.9	59.2	100
Washed and air-dried walnuts				
1	Control	28.6	27.5	1.0
	RF	59.7	66.5	100
2	Control	27.8	28.3	0
	RF	57.6	60.4	100

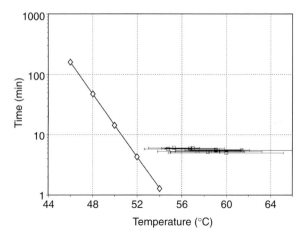

Fig. 12.8. Variations in final walnut kernel temperatures (means ± SD) after commercial radio frequency (RF) treatments that provided complete control (□) of fifth-instar navel orangeworm related to the thermal death time (TDT) curve (◇) obtained in a heating block study (from Wang *et al.*, 2002b).

The final PV, FA and *L* values during accelerated storage of up to 20 days (simulating 2 years' storage at 4°C) were all below the threshold values (PV < 1.0 mEq/kg; FA < 0.6%; *L* ⩾ 40) that are the industry's quality standards for walnuts. The quality evaluations confirmed earlier pilot-scale results in which RF heating did not increase rancidity after 10 and 20 days of storage when final walnut temperatures were raised to 53°C for 5 min (Wang *et al.*, 2001a), 55°C for 10 min (Wang *et al.*, 2002c) and 75°C for 5 min (Mitcham *et al.*, 2004).

Calculations of electric power-to-thermal energy conversion based on

Table 12.5. Storage quality characteristics of in-shell walnuts with a commercial radio frequency (RF) system.

Storage time at 35°C (days)	Peroxide value (PV)[a] (mEq/kg)		Fatty acid (FA)[a] (%)		Kernel colour (L value)	
	Control	RF treated	Control	RF treated	Control	RF treated
Unwashed walnuts						
0	0.04 ± 0.03	0.18 ± 0.29	0.15 ± 0.07	0.19 ± 0.06	–	–
10	0.32 ± 0.21	0.21 ± 0.03	0.23 ± 0.07	0.24 ± 0.04	47.46 ± 1.04	46.55 ± 1.29
20	0.71 ± 0.41	0.86 ± 0.26	0.20 ± 0.02	0.20 ± 0.05	46.80 ± 0.97	45.76 ± 1.00
Hot air-dried walnuts						
0	0.15 ± 0.11	0.09 ± 0.11	0.42 ± 0.36	0.17 ± 0.12	–	–
10	0.41 ± 0.13[ab]	0.75 ± 0.08[b]	0.17 ± 0.07	0.26 ± 0.07	46.30 ± 3.14	46.77 ± 2.77
20	0.74 ± 0.30	0.82 ± 0.28	0.23 ± 0.02	0.22 ± 0.08	44.98 ± 2.46	44.94 ± 2.42

[a] Accepted PV and FA values for good quality < 1.0 mEq/kg and 0.6%, respectively
[b] Different letters indicate that means are significantly different ($P < 0.05$) between the control and RF-treated walnuts.

measured product temperature and electronic power revealed a 79.5% average RF energy efficiency. The two 25 kW RF units in series were operated at 19.2 kW, leading to a throughput of 1561.7 kg/hr. The treatment capacity can be increased by employing multiple 25 kW systems or higher power systems arranged in series. While there would be a substantial capital investment for RF systems, the electrical costs for the walnut RF treatment are comparable to the current cost of methyl bromide fumigant (Aegerter and Folwell, 2001), and would decrease as the fumigant becomes increasingly expensive upon phase-out. Therefore, RF treatment of walnuts is both technically feasible and economically viable.

We discussed previously the effects of walnut size and heterogeneity structure (i.e., fibrous shell and high oil content kernel) on the non-uniformity of RF heating. However, smaller agricultural commodities may heat more uniformly. For example, our studies using the same RF unit revealed that the RF heating uniformity index λ in Eq. 12.2 was 0.032 for lentils and 0.048 for soybeans, much lower than the 0.165 uniformity index for in-shell walnuts. In addition, Tang and Sokhansanj (1993) reported that lentils at < 18% moisture content (on a wet basis) experienced no significant drop in germination rates (an index of thermal damage) after exposure at up to 70°C for 1 h, which would be more than adequate to control insects in lentils.

12.4 Developing RF Treatments for Fresh Fruits

Fresh fruits are more sensitive to heat compared with low-moisture agricultural commodities such as walnuts and lentils. Figure 12.9 illustrates the tolerances for 'Bing' cherries in relation to control of fifth-instar codling moth (Feng *et al.*,

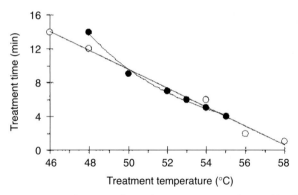

Fig. 12.9. Maximum exposure time at selected temperatures tolerated by 'Bing' cherries (○) compared with the thermal death time curve for 100% mortality (●) of fifth-instar codling moth (from Feng *et al.*, 2004).

2004). Because the overlap of exposure times and treatment temperatures between controlling target insects and maintaining product quality is so small, the fruits must be heated very uniformly to avoid either insect survival or thermal damage to product quality. Heating uniformity cannot be assisted by hot air heating because it would cause damage to the fruit surfaces and dehydration. Therefore, studies were conducted to explore the possibilities of combining RF heating with water heating to improve heating uniformity in heat-sensitive fresh fruits, such as cherries, apples and oranges by using small-scale pilot systems.

A 'fruit mover' was designed for RF heating of fresh fruit (Birla *et al.*, 2004) to simulate continuous flow-through (see Fig. 12.10) and provide 5–8°C/min heating in a 12 kW RF system. The fruit mover prevented fruit from remaining in the corners of the treatment unit and heating non-uniformly. The heating uniformity of apples, cherries and oranges was significantly improved by the addition of the fruit mover (Birla *et al.*, 2004).

Experiments were conducted with a 6 kW, 27 MHz pilot-scale RF system to develop a treatment for control of codling moth in cherries. Treatments consisted of preheating the cherries with hot water at 38°C and RF heating to final pulp temperatures of 50, 52, 53 or 54°C. Heating infested fruit in the RF unit to 50°C and holding for 5 or 6 min, 52°C for 4 min or 53 and 54°C without holding resulted in 100% larval mortality (Mitcham *et al.*, 2005). The treatments had no significant effects on cherry colour, decay, or shrivel, and only a slight (not commercially significant) effect on browning. Regardless of shipment conditions, stem browning and berry pitting were observed in RF-treated cherries. Cherry quality was most affected after storage that simulated sea shipment (0°C for 2 weeks).

Experiments were also conducted to explore the possibility of developing RF treatment to control codling moth in apples (Wang *et al.*, 2006a). Treatment conditions included water preheating at 45°C for 30 min, RF heating to 48°C, holding in 48°C water for 5, 10, 15 and 20 min and hydrocooling in ice water for 30 min (see Fig. 12.11).

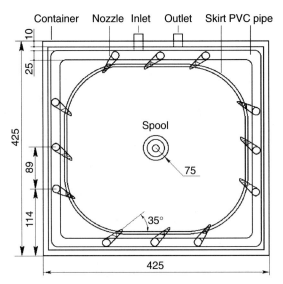

Fig. 12.10. Diagram (overhead view) of the fruit mover (all dimensions in mm) (from Birla *et al.*, 2004).

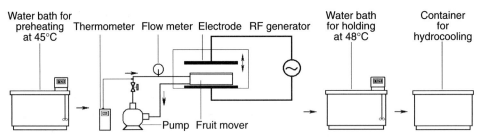

Fig. 12.11. Flowchart for the water-assisted radio frequency (RF) treatment process (from Wang *et al.*, 2006a).

The hydrocooled apples were packed again in ventilated containers and stored at 4°C for 0, 7 and 30 days. The results showed that the hot water-assisted RF treatments provided relatively uniform heating in apples (Wang *et al.*, 2006a). However, at least 15 min of holding at 48°C in a water bath was required to ensure that all larvae were dead (see Table 12.6). There was no significant effect of RF heat treatments on titratable acidity or soluble or solid content of apples even after 30 day storage at 4°C. But other fruit quality indices, such as appearance and flavour, were adversely affected by the RF treatments.

Birla *et al.* (2005) studied the potential for using hot water-assisted RF treatments to control fruit flies in oranges. Treatment time–temperature combinations were selected based on Mediterranean fruit fly thermotolerances developed by Gazit *et al.* (2004), showing that 100% mortality could be achieved by exposing infested fruit to 48°C for 15 min, 50°C for 4 min or 52°C for 1 min. RF treatment temperatures of 48, 50 and 52°C with different holding times were chosen (see Fig. 12.12).

Table 12.6. Total number of live and dead fifth-instar codling moths recovered from 'Red Delicious' apples, with the mortality (means ± SD, %) for the controls at room air and water temperatures and three treatments with the same water preheating (45°C for 30 min), radio frequency (RF) heating (27 MHz and 6 kW for 1.25 min) and hydrocooling for 30 min (from Wang *et al.*, 2006a).

Temperature + holding time	Alive (*n*)	Dead (*n*)	Mortality (%)[a]
Control at indoor ambience	99	6	5.7 ± 2.9
Control at room water for 80 min	96	9	8.6 ± 2.9
45°C + 30 min	85	20	19.1 ± 8.7
48°C + 5 min	3	102	97.1 ± 2.9
48°C + 10 min	2	103	98.1 ± 1.7
48°C + 15 min	0	105	100 ± 0
48°C + 20 min	0	105	100 ± 0

[a] Three replicates with five larvae per apple.

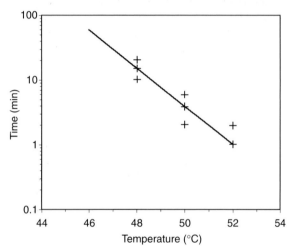

Fig. 12.12. Selected temperature–time combinations used in determining a treatment for controlling Mediterranean fruit fly (+) in oranges based on experimentally determined minimum time and temperature combinations required to achieve complete mortality in a heating block system (from Gazit *et al.*, 2004).

The treatments also included preheating the fruit in 35°C water for 45 min and hydrocooling for 30 min in 3–4°C water. Postharvest quality evaluations after 10 days storage at 4°C revealed changes in weight, firmness, colour, total soluble solids, acidity and volatiles. The results indicated that holding the fruit for 15 min in 48°C water controlled Mediterranean fruit fly with no significant damage to any of the physical quality attributes, although significant changes in volatile flavour profiles were found (Birla *et al.*, 2005).

Pilot-scale RF treatments show that the marketability of fresh fruits may be affected due to the limited range of exposure times and temperatures in obtaining insect control and maintaining fruit quality. The practical implementation of RF treatments becomes difficult for fresh fruits in large-scale RF systems due to the potential for large temperature variations in the treated load. Generally, it is more difficult to develop efficacious heat treatments for fresh fruits than for low-moisture products, such as walnuts.

12.5 Conclusions

Thermal treatments to control quarantine pests are developed most efficiently with a systematic approach using: (i) the thermal death kinetics of the target insect; (ii) knowledge of commodity quality degradation kinetics; and (iii) engineering principles for delivering the thermal energy uniformly in commercial treatments. Commercial applications of heat treatments for several tropical fruits (see Chapter 13, this volume) and research on RF heat treatments for in-shell walnuts demonstrate that these treatments are viable alternatives to fumigation with toxic compounds for some commodities.

However, not all agricultural commodities – especially heat-sensitive fresh produce – can tolerate the thermal conditions required to control insect pests. Systematic studies of commodity and target pest thermal kinetics and pilot-scale testing of candidate treatments for both efficacy and commodity quality are required for each commodity and pest complex. Engineering and economic studies are required to determine the commercial viability of candidate treatments, and confirmatory tests are required for regulatory approval. Ultimately, this overall process of research, development and commercialization will result in the expansion and diversification of agricultural exports and assurance of quarantine security against the spread of unwanted pests through marketing channels.

12.6 References

Aegerter, A.F. and Folwell, R.J. (2001) Selected alternatives to methyl bromide in the postharvest and quarantine treatment of almonds and walnuts: an economic perspective. *Journal of Food Processing and Preservation* 25, 389–410.

Birla, S.L., Wang, S. and Tang, J. (2004) Improving heating uniformity of fresh fruit in radio frequency treatments for pest control. *Postharvest Biology Technology* 33, 205–217.

Birla, S.L., Wang, S., Tang, J., Fellman, J., Mattinson, D. and Lurie, S. (2005) Quality of oranges as affected by potential radio frequency heat treatments against Mediterranean fruit flies. *Postharvest Biology and Technology* 38, 66–79.

Drake, S.R., Hansen, J.D., Elfving, D.C., Tang, J. and Wang, S. (2005) Hot water to control codling moth in sweet cherries: efficacy and quality. *Journal of Food Quality* 28, 361–376.

Feng, X., Hansen, J.D., Biasi, B. and Mitcham, E.J. (2004) Use of hot water treatment to control codling moths in harvested California 'Bing' sweet cherries. *Postharvest Biology and Technology* 31, 41–49.

Gazit, Y., Rossler, Y., Wang, S., Tang, J. and Lurie, S. (2004) Thermal death kinetics of egg and third-instar Mediterranean fruit fly *Ceratitis capitata* (Wiedemann) (Diptera: Tephritidae). *Journal of Economic Entomology* 97, 1540–1546.

Johnson, J.A., Valero, K.A., Wang, S. and Tang, J. (2004) Thermal death kinetics of red flour beetle, *Tribolium castaneum* (Coleoptera: Tenebrionidae). *Journal of Economic Entomology* 97, 1868–1873.

Mitcham, E.J., Veltman, R.H., Feng, X., de Castro, E., Johnson, J.A., Simpson, T.L., Biasi, W.V., Wang, S. and Tang, J. (2004) Application of radio frequency treatments to control insects in in-shell walnuts. *Postharvest Biology and Technology* 33, 93–100.

Mitcham, E.J., Tang, J., Hansen, J.D., Monzon, M.E., Biasi, W.V., Wang, S. and Feng, X. (2005) Radio frequency heating of walnuts and sweet cherries to control insects after harvest. *Acta Horticulture* 682, 2133–2139.

Rao, M.A. and Lund, D.B. (1986) Kinetics of thermal softening of foods – a review. *Journal of Food Processing and Preservation* 10, 311–329.

Tang, J. and Sokhansanj, S. (1993) Effect of drying parameters on the viability of lentil seeds. *Transactions of the ASAE* 36 (3), 855–861.

Tang, J., Ikediala, J.N., Wang, S., Hansen, J.D. and Cavalieri, R.P. (2000) High-temperature, short-time thermal quarantine methods. *Postharvest Biology and Technology* 21, 129–145.

Taoukis, P.S., Labuza, T.P. and Sagus, I.S. (1997) Kinetics of food deterioration and shelf-life prediction. In: Valentas, K.J., Rotstein, E. and Singh, R.P. (eds) *Handbook of Food Engineering Practice.* CRC Press, Boca Raton, Florida, p. 374.

Wang, S., Ikediala, J.N., Tang, J., Hansen, J.D., Mitcham, E., Mao, R. and Swanson, B. (2001a) Radio frequency treatments to control codling moth in in-shell walnuts. *Postharvest Biology Technology* 22, 29–38.

Wang, S., Tang, J. and Cavalieri, R.P. (2001b) Modeling fruit internal heating rates for hot air and hot water treatments. *Post-harvest Biology and Technology* 22, 257–270.

Wang, S., Ikediala, J.N., Tang, J. and Hansen, J.D. (2002a) Thermal death kinetics and heating rate effects for fifth-instar *Cydia pomonella* (L.) (Lepidoptera: Tortricicae). *Journal of Stored Products Research* 38, 441–453.

Wang, S., Tang, J., Johnson, J.A. and Hansen, J.D. (2002b) Thermal death kinetics of fifth-instar *Amyelois transitella* (Walker) (Lepidoptera: Pyralidae) larvae. *Journal of Stored Products Research* 38, 427–440.

Wang, S., Tang, J., Johnson, J.A., Mitcham, E., Hansen, J.D., Cavalieri, R.P., Bower, J. and Biasi, B. (2002c) Process protocols based on radio frequency energy to control field and storage pests in in-shell walnuts. *Postharvest Biology and Technology* 26, 265–273.

Wang, S., Tang, J., Cavalieri, R.P. and Davis, D. (2003a) Differential heating of insects in dried nuts and fruits associated with radio frequency and microwave treatments. *Transactions of the ASAE* 46, 1175–1182.

Wang, S., Tang, J., Johnson, J.A, Mitcham, E., Hansen, J.D., Hallman, G., Drake, S.R. and Wang, Y. (2003b) Dielectric properties of fruits and insect pests as related to radio frequency and microwave treatments. *Biosystems Engineering* 85, 201–212.

Wang, S., Yue, J., Tang, J. and Chen, B. (2005) Mathematical modelling of heating uniformity of in-shell walnuts in radio frequency units with intermittent stirrings. *Postharvest Biology and Technology* 35, 94–104.

Wang, S., Birla, S.L., Tang, J. and Hansen, J.D. (2006a) Postharvest treatment to control codling moth in fresh apples using water-assisted radio frequency heating. *Postharvest Biology and Technology,* 40, 89–96.

Wang, S., Monzon, M., Johnson, J.A., Mitcham, E.J. and Tang, J. (2007) Industrial-scale radio frequency treatments for insect control in walnuts: I. Heating uniformity and energy efficiency. *Postharvest Biology and Technology* {In press}.

Wang, S., Tang, J., Sun, T., Mitcham, E.J., Koral,
    T. and Birla, S.L. (2006b) Considerations in
    design of commercial radio frequency treat-
    ments for postharvest pest control in in-shell
    walnuts. *Journal of Food Engineering* 77,
    304–312.

13 Commercial Quarantine Heat Treatments

J.W. Armstrong[1] and R.L. Mangan[2]

[1]USDA-ARS, US Pacific Basin Agricultural Research Center, Hilo, Hawaii, USA; e-mail: jarmstrong@pbarc.ars.usda.gov; [2]USDA-ARS, Kika de la Garza Subtropical Agricultural Research Center, Weslaco, Texas, USA; e-mail: rmangan@weslaco.ars.usda.gov

13.1 Introduction

World trade in fresh fruits, vegetables and ornamentals is expanding rapidly to meet increasing demands on existing markets and to supply new markets resulting from international trade agreements. For most fresh commodities, postharvest quarantine (or phytosanitary) treatments are required to prevent the spread of exotic pests through marketing channels to areas where they do not normally occur. Without treatments to provide quarantine security, quarantine restrictions limit available markets for fresh commodities. Therefore, effective postharvest quarantine treatments that are not harmful to either the fruit or people coming in contact with or consuming the commodity are essential to the unrestricted trade of fresh fruits, vegetables and ornamentals through domestic and international marketing channels (Paull and Armstrong, 1994).

For many decades, either simple regulatory exclusion or fumigation with a toxic compound (e.g. ethylene dibromide, methyl bromide, hydrogen cyanide) were the primary methods for phytosanitary control of any commodity considered to be a host for one or more quarantine pests. Today, international trade groups and agreements such as the World Trade Organization (WTO), North American Free Trade Agreement and the Association of Southeast Asian Nations Free Trade Agreements have transformed phytosanitary regulations and all but eliminated simple exclusion as a quarantine method. Additionally, fumigation with toxic compounds is no longer the first quarantine treatment option of choice due to human health and environmental concerns. Irradiation and physical treatments using heat or cold are now commonly used instead of methyl bromide.

Most physical treatments in use today have been developed or refined over the past three decades, which at the time of writing, include refrigeration (or cold) treatment, vapour heat and forced hot-air treatment, and hot-water immersion treatment methods. Simply put, cold treatments consist of reducing the temperature of the host commodity below the thermal tolerance limits of the target pest, while heat treatments consist of heating the host commodity beyond the thermal tolerance limits of the target pest. The commercial use of quarantine heat treatments (vapour heat, forced hot air and hot-water immersion) for regulatory purposes has been primarily to disinfest fresh fruits of potential fruit fly (Diptera: Tephritidae) egg and larval infestations before export. Therefore, most of the examples given in this chapter will refer to fruit flies. Recently, however, the US Department of Agriculture (USDA) approved a hot water immersion quarantine treatment method to disinfest a variety of phytophagous insects from tropical ornamental exports from Hawaii (Hara *et al.*, 1996).

13.2 Definitions and Concepts

Before proceeding further, some basic concepts of quarantine specific to the purpose of this chapter need to be defined:

- Artificial infestation: obtaining life stages from target pests, usually under laboratory rearing conditions, and placing them on or inserting them in a commodity.
- Bioassay: using laboratory tests to determine the effects of temperature on the life stages of a target pest (i.e., relative thermotolerance) and select effective treatment parameters for infested commodities.
- Disinfest: to kill the quarantine pest or eliminate it from the host.
- Efficacy data: data gathered by subjecting infested commodities to selected treatment parameters to determine disinfestation success.
- Host: any fresh commodity (fruit, vegetable or ornamental) that, at one or more of its growth stages, can become naturally infested, harbour or otherwise support the development of a quarantine pest and facilitate movement of that pest through marketing channels.
- Load factor: either (i) the number of insects used in each thermal treatment test for identification of relative thermotolerances between target species and life stages; (ii) the total infestation (number of insects) in a commodity used to test the efficacy of a selected candidate treatment; or (iii) the number of a particular commodity treated at one time (e.g. the volume, weight or number of the commodity in the treatment chamber).
- Natural infestation: the direct action of a pest laying or inserting eggs on or in a commodity in the field or laboratory, resulting in the onset of that pest's life cycle.
- Quarantine pest: any organism that is regulated on the basis that entry into a geographical area where it does not occur may result in its establishment and resultant major economic losses.
- Quarantine security: that degree of statistical probability required to ensure that quarantine treatments or systems adequately disinfest host commodities so that, upon transport or shipment of the treated commodities, the targeted pests cannot become established in an area where they do not already exist.

- Quarantine treatment: any individual action or multiple actions that can be used to disinfest a host commodity.
- Thermotolerance: the degree to which a pest species or its life stages are able to survive temperature ranges selected for quarantine heat treatment when compared with other pest species or life stages.

13.3 Quarantine Heat Treatments

Vapour heat, the oldest of the three heat methods, was developed during a Mediterranean fruit fly (*Ceratitis capitata* (Wiedemann)) outbreak in Florida in the mid- to late 1920s (Baker, 1952). Shortly thereafter, the original vapour heat treatment was modified for use against Mexican fruit fly, *Anastrepha ludens* (Loew), as a postharvest quarantine treatment before shipping citrus fruits from the Rio Grande Valley of Texas to US markets elsewhere (Hawkins, 1932).

Vapour heat consists of heating the host fruit by moving hot air saturated with water vapour over the fruit surface. Heat is transferred from the air to the commodity by condensation of the water vapour (heat of condensation) on the relatively cooler surfaces of the fruit being treated. Fruit may be gradually heated over time (approach time) to a target temperature – which may be at the end of the treatment (i.e., all insects have been killed) or held for a specific time (holding time) required to kill all insects.

Gradual heating of fruit is more desirable than rapid heating in order to prevent damage to the fruit (Armstrong, 1994). With the advent of ethylene dibromide and methyl bromide fumigation during the 1940s and 1950s, vapour heat as a quarantine treatment method against fruit flies fell into disuse because the fumigants were more economical and easier to use.

Although American researchers continued to provide new vapour heat treatments for sweet pepper, Chinese peas, cucumber, aubergine, green beans, lima beans, lychee, mango, papaya, tomato and yellow wax beans (Jones, 1940; Balock and Starr, 1946; Balock and Kozuma, 1954; Seo *et al.*, 1974), there were no new commercial applications in the USA until 1984, when ethylene dibromide was discontinued (Ruckleshaus, 1984); fumigation as a quarantine treatment for papaya exported from Hawaii to the US mainland and Japan ended after ethylene dibromide was identified as a carcinogen.

The loss of ethylene dibromide created immediate interest in the use of heat treatments in the USA (Jones, 1940; Seo *et al.*, 1974) and Japan, where commercial vapour heat treatments were developed in the early 1980s with the exporting of sweet peppers, aubergine and other fruit fly hosts from Okinawa, Japan, to markets elsewhere in the country (Sugimoto *et al.*, 1983; Furasawa *et al.*, 1984).

Hot water immersion, the least technologically difficult of the three heat treatment methods to apply commercially, has been used for its effectiveness in killing pathogens for well over 100 years. Jensen (1888) was the first to describe hot-water immersions of seeds to prevent bacterial diseases of plants.

Burditt *et al.* (1963) noted that hot water could be combined with ethylene dibromide and a residual insecticide as a quarantine dip for treatment to kill Japanese beetle (*Popilla japonica*) in nursery plants in 1948. Hot water immersion alone was first used as a quarantine treatment by Armstrong (1982) to kill Mediterranean fruit fly, melon fly (*Bactrocera cucurbitae Coquillett*) and oriental fruit fly (*B. dorsalis Hendel*) eggs and larvae in bananas before export to US mainland markets, followed by quarantine treatments for papayas and mangoes (Armstrong, 1994) and cut tropical flowers (Hara *et al.*, 1996).

Hot water treatment of mangoes is the largest volume heat treatment used for any fresh fruit. Combination hot water treatments to kill fruit flies in mangoes were developed by Seo *et al.* (1972) in Hawaii and by Lin *et al.* (1976) in Taiwan. The first single treatment using hot water for mangoes was carried out by Sharp and Spalding (1984) on Florida-grown mangoes. According to the US Department of Commerce (2003), yearly imports from 2000 to 2003 ranged from 239,000–289,000 t of mangoes. In 2003, the major exporting countries were Mexico (61%), Brazil (13%), Ecuador (9.5%), Peru (7%), Guatemala (3%) and Haiti (2%). A more detailed summary of hot water treatments is presented in Sharp (1994).

Forced hot air, also known as high-temperature forced air, is a modification of the vapour heat treatment developed by Armstrong *et al.* (1989) to kill Mediterranean fruit fly, melon fly and oriental fruit fly eggs and larvae (see Fig. 13.1) in papaya as a replacement for ethylene dibromide fumigation. Forced hot air and vapour heat are essentially the same treatment method, except that fruit surfaces are wet during vapour heat and dry during forced hot air.

Hansen *et al.* (1990) observed that air saturated with water vapour was unnecessary in ensuring quarantine security against fruit fly eggs and larvae, and forced hot-air treatment provided for better fruit quality than either hot water immersion or vapour heat treatment methods (Laidlaw *et al.*, 1996). Forced hot air became the treatment method of choice for the Hawaii papaya

Fig. 13.1. Quarantine pests for which heat treatments are used commercially, from left to right (top row): Mediterranean fruit fly, *Ceratitis capitata* (Wiedemann); melon fly, *Bactrocera cucurbitae* (Coquillett); oriental fruit fly, *B. dorsalis* (Hendel); Mexican fruit fly, *Anastrepha ludens* (Loew) ovipositing in citrus fruit; and Caribbean fruit fly, *A. suspensa* (Loew); (bottom row): fruit fly eggs under surface of tomato skin; fruit fly larva in tomato; white peach scale, *Pseudaulacaspis pentagona* (Targioni-Tozetti); green scale, *Coccus viridis* (Green); and longtail mealy bugs, *Pseudococcus longispinus* (Targioni-Tozetti).

industry for fruit exported to the US mainland; for papaya, aubergine and breadfruit exported from Fiji to New Zealand; and for papaya exported from the Cook Islands, Samoa and Tonga to New Zealand (Waddell *et al.*, 1997; Secretariat for the Pacific Community, 2002).

13.4 Quarantine Treatment Protocols

Quarantine heat treatment protocols (or schedules) are developed by various agricultural research and regulatory agencies, e.g. (i) USDA; (ii) the Animal and Plant Health Inspection Services (APHIS) in cooperation with the USDA Agricultural Research Service (ARS); (iii) the Australia Quarantine Inspection Service in cooperation with the Queensland Department of Primary Industries; (iv) the New Zealand Ministry of Agriculture and Forestry in cooperation with HortResearch New Zealand; and (v) the Japan Ministry of Agriculture, Forestry, and Fisheries.

Research phases consist of identifying the most heat-tolerant life stage(s) of the target pest(s) *in vitro* or *in situ*, testing selected heating protocols (temperatures over time), collecting mortality data in small-scale tests using fruit infested with the most heat-tolerant life stage(s) and, ultimately, demonstrating that the heat treatment provides quarantine security in commercial-scale confirmatory tests. Regulatory phases consist of approving, certifying and regulating quarantine treatments and treatment facilities.

Generally, research organizations provide the efficacy data showing that a quarantine treatment method will provide quarantine security, and regulatory organizations develop the operational standards for implementing and monitoring the quarantine treatment. An approved treatment protocol or schedule consists of all the conditions, parameters and procedures required to ensure the quarantine treatment provides consistent heat control of the target insect under commercial conditions.

Of primary importance for hot water immersion treatment schedules is regulation of water temperature and immersion time, including when to begin timing the treatment after the commodity is immersed and water temperature has reached the required temperature. The weight of individual units being treated may be regulated if unit size is variable for a commodity and size directly affects how quickly heat transfers through the fruit to kill the target pest. Of primary importance in both vapour heat and forced hot-air treatments are: (i) regulation of increasing air temperature over time and internal temperatures; and (ii) final temperatures or holding times required to ensure heat transfer is adequate to satisfy quarantine security conditions.

The practical process of developing heat treatments must also consider changes in product condition during infestation and treatment and guard against revisions in the protocol as the treatment is adapted for commercial use. The large numbers of pests and fruits required to demonstrate effectiveness at a given probit level (see below) usually require tens of thousands of individual pests (Couey and Chew, 1986). The time and costs of laboratory (dose-response) phases and confirmatory phases are reduced when very heavy infestations are made. This is, however, unrealistic in the practical sense because heavily infested fruit would rarely pass through the culling stage.

More importantly, heavy infestation may result in significant changes in heat diffusion and weight loss of the commodity, causing higher mortality than would be experienced under commercial conditions. Additional problems may arise when modifications of the treatment are made, such as cool water dips following heat treatments if they are not used during development of the treatment.

In a review of heat treatment work published in the 1980s and early 1990s, Mangan and Hallman (1998) noted that a number of factors – including deteriorating fruit condition, inaccurate weights and hydrocooling – probably reduced the efficacy of published and approved hot water treatments for mangoes. After live larvae were found in heat-treated mangoes at ports of entry, Shellie and Mangan (2002a, b) demonstrated experimentally that these problems – especially inaccurate estimates of fruit weight – caused under-treatment of commercially treated mangoes. Larger fruit were found to lose significant amounts of weight (c.20%) when heavily infested, and cool water dips shortened total heat dose after removal from the water bath. This combination of inaccurate weights and reduced heat dose allowed larvae to survive. Quarantine security for hot water-dipped mangoes was restored by adding about 12 min treatment for the larger classes (> 500 g) or by eliminating the cool water dip.

13.5 Quarantine Security Statistics

Statistical analyses are used to estimate the probability that a treatment will successfully prevent the target quarantine pest from becoming established in the area into which the commodity is imported. Laboratory bioassays can evaluate the probable success of different treatments. A large-scale confirmatory test follows treatment selection. This sequence of testing resembles the protocol generally used to evaluate the probable success of a treatment to control any pest species on any crop, but is more complex because control must also prevent establishment (Robertson et al., 1994).

Probit-9, which was discussed in Chapter 5, this volume, is reviewed here because the provision of appropriate statistical criteria that will ensure quarantine security is difficult when few statistical standards are available, and even these are subject to debate. For many decades, the USA required that fruit fly disinfestation treatments meet the Probit-9 concept developed by Baker (1939), which assumes insect control at a mortality rate of 99.9968%. This corresponds to no more than three individuals from a treated population of 100,000 surviving a selected quarantine treatment. Using 95% confidence limits, probit-9 could have 29–136 survivors per million, with an average of 32 (Couey and Chew, 1986). Japan often uses a variation of the Probit-9 concept that requires no survivors from a treated population of 30,000 target pests (usually in three treatments of 10,000 target pests per treatment).

Zero tolerance is impossible to demonstrate by treatment efficacy because this predicts that no survivor will ever be found. Additionally, Probit-9 should not be used to predict treatment time due to the high degrees of confidence of high degrees of variation in dose-response required and, in many cases, mortality is curvilinear, not logarithmic. Risks are clearly different for different commodities based on infestation rates. A commodity that averages one pest per unit provides a lesser degree of risk than a commodity that averages 30–40 pests per unit (Paull and Armstrong, 1994).

The Probit-9 concept came under scrutiny for the first time by Landolt et al. (1984), who pointed out that it does not consider several factors that directly impact quarantine situations, such as the actual infestation per volume of treated commodity, the potential for culling infested units of a commodity during processing, the infestation variability between good and poor hosts, or the probability of a potential normal mating pair occurring at the same time at the market destination.

Alternative quarantine statistics that challenge the Probit-9 concept have appeared in the literature (Landolt et al., 1984; Couey and Chew, 1986; Robertson et al., 1994). New Zealand developed a maximum pest limit concept for imported produce (Baker et al., 1990; Harte et al., 1992) that includes a sampling model for the accurate assessment of infestation levels in the host commodity, and a limit on the number of immature pests that may be present in consignments imported during a specified time to a specified location (Baker et al., 1990).

This assumes that the emerging adults will not be able to find a mate because the commercial distribution would scatter the arriving commodity over a large area. The difficulty with this system is the lack of good biological data on infestation indices available, and the limitation of sampling methods for accurate assessment (Paull and Armstrong, 1994). Application of a treatment of known efficiency ensures that the limit is not exceeded if the infestation – determined by the sampling model – is below a predetermined value (Baker et al., 1990).

Mangan et al. (1997) performed an analysis of maximum expected surviving Mexican fruit flies in loads of citrus and mangoes imported into the USA from Mexico. In that analysis, which used actual load numbers of fruit and infestation rates (from field samples), they determined that a maximum pest level of two survivors per load was frequently exceeded even with a Probit-9 level postharvest treatment, and the numbers of fruit currently inspected at port inspection stations (usually < 50 fruit per truckload) is far below that necessary (usually several hundred) to detect infestation rates above this maximum pest limit.

Additional discussions of quarantine security and Probit-9 can be found in Vail et al. (1993) and Follett and McQuate (2001). A biological statistics package for pest risk assessment in commodity quarantine treatments developed by Liquido et al. (1996) is available from USDA-ARS at http://www.pbarr@pbarc.ars.usda.gov.

Although 'quarantine security' was defined for the purpose of this chapter, the term does not have a universally accepted definition. The associated concepts of risk, acceptable risk and affordable risk continue to be debated

(USDA, 1997). Moreover, quarantine security statistics vary by regulatory agency and type of treatment and commodity host status. Therefore, it is imperative to identify the level of quarantine security and associated statistics required by the importing country before the onset of developing any quarantine treatment.

13.6 Developing Quarantine Heat Treatments

Conceptually, heating a commodity to kill a target pest sounds simple, and one could therefore reason that research and development of quarantine heat treatments should be relatively easy and straightforward. Unfortunately, the opposite is true: developing these treatments are usually lengthy processes subject to numerous complexities. Important issues that directly influence the research, development, approval and implementation of all quarantine heat treatments are discussed below.

The Number of Target Pest Species and their Life Stages

The number of such species and their life stages associated with the commodity directly increase the complexity of developing efficacy data. All life stages found in the commodity for each target pest must undergo bioassay for thermotolerance to determine which species and life stage is the most heat-tolerant. However, once the relative thermotolerances are known, efficacy data can be developed using the most thermotolerant species and life stage. For example, four fruit fly species attack fresh fruits and vegetables in Hawaii: Malaysian fruit fly, the Mediterranean fruit fly, melon fly and oriental fruit fly.

The relative thermotolerances determined from bioassays are: Malaysian fruit fly > Mediterranean fruit fly > oriental fruit fly = melon fly, with the late egg stage just prior to hatch = the first-instar larval stage < all other larval stages (Jang et al., 1999). Thereafter, any development of heat treatments against any combination of fruit fly species that included Malaysian fruit fly would need to provide efficacy data against only Malaysian eggs or first-instars; if Malaysian fruit fly was not involved, then only Mediterranean fruit fly eggs or first-instars; if Malaysian and Mediterranean fruit flies were not involved, then either melon fly or oriental fruit fly eggs or first-instars.

The Location of the Target Pest on or in the Commodity

This plays an import role in determining how long heating will be required to sufficiently heat the pest to obtain mortality. Pests found on the commodity surface can be heated directly, while those located inside require the commodity around them to increase in temperature over time until a lethal temperature is reached. It should be noted that the location of the pest in the commodity also can alter which life stage dies first, regardless of relative thermotolerance. If the least tolerant life stage of a pest is found deep inside the commodity, it may be more difficult to kill than more tolerant life stages found on or near the surface. A hypothetical example to illustrate this situation would be the eggs and larvae of a fruit fly that infests watermelon. Although the eggs may be the most thermotolerant stage, the larvae that tunnel deep within the fruit may be more difficult to kill because the surrounding fruit pulp acts as a heat sink and insulates the larvae.

The development of vapour heat and forced hot-air quarantine treatments against fruit fly infestations in mango, *Mangifera indica* L., is an example of a situation where an additional quarantine pest negated the use of a quarantine treatment technology (Armstrong, unpublished data, 1990). Although both of these heat treatments were efficacious in disinfesting fruit flies from mangoes, the treatment was unable to affect adequate mortality against the mango weevil, *Cryptorhynchus mangiferae* (F.), because it was protected by the seed shell. Mango fruits were unable to tolerate the treatment time and temperatures required to penetrate the seed and kill the weevil.

The Host Status of the Commodity in Nature

This usually indicates the relative ease or difficulty of obtaining an infested commodity in the laboratory for efficacy tests, whether by natural or artificial infestation. A very poor host commodity in nature that is also very difficult to infest for use as treatment and control samples directly impedes the development of efficacy data. In this case, artificial infestation methods must be developed to accelerate treatment development.

Armstrong *et al.* (1995b) found that carambola (star fruit), *Averrhoa carambola* L., was a poor host for Mediterranean fruit fly, melon fly and oriental fruit fly. To facilitate adequate infestations of carambola for use in cold-treatment tests, researchers injected eggs into the fruit (the carambola were held until the eggs hatched if first-instars were to be tested). Because the resulting infestations from the injection method lacked uniformity, Armstrong *et al.* (1995b) changed their infestation technique, whereby eggs or first-instars were placed under a flap cut into the surface and lifted to provide space for the eggs or larvae. The flap was then replaced and secured with masking tape. The technique resulted in adequate infestations for developing efficacy data that led to an approved quarantine cold treatment for carambola used by both Hawaii and Taiwan.

Another issue that affects all host commodities, and especially poor hosts, is the survival of the pest during any holding period required before data are gathered. Frequently, techniques that bias towards the survival of the target pest must be developed to ensure adequate numbers result from treated and control commodities held to allow the pest to develop before assessing mortality. Poor or no survival in the test controls yields inadequate data to determine the level of mortality caused by the treatment. For many years, treated and control fruits infested with fruit fly eggs or larvae were held for

1–2 weeks to allow the insect to develop to the pupal stage in order to ascertain the treatment effects on mortality. The treated and control fruits were held on screen platforms in sealed boxes over sand in which the third-instars could pupate after leaving the fruit (see Fig. 13.2).

The pupae were then counted as survivors. Unfortunately, many larvae drowned in the fluids from decomposing fruit that had collected at the bottom of the box, and rapid fungal decay reduced the nutrients available to the larvae for development to the pupal stage. This holding technique resulted in poor pupal recovery from both treatments and controls, thereby prolonging the time required to develop various quarantine treatments. Armstrong *et al.* (1984) developed an improved holding technique for infested commodities by placing the treated and control fruits on dry larval diet on trays in screened cabinets. Fluids from the decaying fruit moistened the diet, from which the larvae could then migrate and complete their development.

Heating Characteristics (Thermal Diffusivity) of the Commodity

These determine the length of time required for heat to move through the commodity until the area around the target pest is heated sufficiently to obtain mortality (heating characteristics are important only for pests found inside the commodity). The subject of thermal diffusivity is covered extensively in Chapter 2, this volume.

Physiological or Morphological Characteristics of the Commodity

Characteristics that protect pests from treatment may occur. Most important among these are physical barriers – such as seed cavities or air pockets – and the seeds themselves, which heat more slowly than the remainder of the

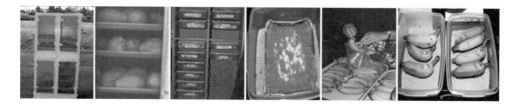

Fig. 13.2. Methods for holding treated and control fruit before observations (left to right): holding cabinet for trays of treated or control fruit on a dry larval diet; papayas on trays of dry larval diet in holding cabinet; plastic containers with aluminum trays containing moist larval diet to rear treated or control eggs or larvae to the pupal stage; fruit fly eggs for pupation on moist larval diet in plastic container with sand at bottom; bananas on a dry larval diet being placed in individual plastic bags; and bananas held on wire mesh platforms over a dry larval diet. These methods provide optimum conditions for rearing treated and control insects by eliminating causes of death (e.g. dehydration, drowning in liquids from decaying fruit, starvation or lack of suitable pupation medium) other than by treatment alone.

commodity and therefore obstruct mortality. For example, research to develop a forced hot-air quarantine treatment against fruit flies attacking atemoya, *Annona squamosa* × *A. cherimola* (hybrid), found that air pockets surrounding the seeds in the fruit flesh insulated larvae from the heat. The treatment times and temperatures were manipulated in an attempt to kill the larvae, but the work had to be abandoned when the larvae were able to survive treatment parameters that the fruit could not tolerate (Armstrong, unpublished data, 1994).

Tolerance of Pests to Treatment

Usually termed thermotolerance, this determines the treatment parameters, viz. the treatment temperature(s) and time required to kill the target pest. If the pest is more thermotolerant than the commodity, heat is not a viable treatment method unless used in combination with other treatments or conditions that reduce the pest's thermotolerance or increase the commodity's thermo-tolerance. One example of reducing target pest thermotolerance is the use of a controlled atmosphere temperature treatment at < 1% oxygen, 15% carbon dioxide and 47°C for 25 min to completely control the most heat-tolerant immature stages of codling moth, *Cydia pomonella* (L.), in sweet cherries, *Prunus avium* (L.) (Neven, 2005). Without the controlled atmosphere, a longer treatment time (45 min) at a lower temperature (45°C) was required to achieve the same level of insect control (Neven, 2005).

Chan and Linse (1989a) developed a preconditioning (preheating a commodity for a period of time to increase its thermotolerance) method to increase thermotolerance in cucumbers for application of quarantine heat treatments. The use of preconditioning is discussed further in Section 13.7, below.

The Economic Feasibility of the Treatment under Commercial Conditions

This is the major factor in overall marketing economics that determines whether a quarantine treatment can be used profitably, and must be considered at the onset of any treatment research and development. Some commodities, such as high-value niche market exotic fruits, have a very high profit margin that will bear the cost of any treatment method necessary. Other commodities have a low profit margin that will bear only the least costly treatment methods, or none at all. Commodity volume must also be factored into the cost of quarantine treatment. Generally, the more volume or throughput, the lower the treatment costs.

Another treatment cost that must be considered is regulatory monitoring. Regulatory agencies may require onsite monitoring and inspections that are charged to the commodity grower or shipper. For several decades, a regulatory inspector of the Japan Ministry of Agriculture, Forestry and Fisheries has been required to live in Hawaii to provide onsite monitoring for all vapour heat treatments of papaya exported to Japan. The *per diem* living expenses for the inspector are paid by the papaya industry. Additionally, exporters pay for any overtime accrued by USDA-APHIS Plant Protection and Quarantine inspectors who monitor vapour heat or forced hot-air treatments of papaya exported to the US mainland.

The Difficulty of Commercial Treatment Applications or Regulatory Monitoring

These issues like economic feasibility, determine whether a treatment will be successful when used under commercial conditions. Quarantine treatments are most successful when they are part of (or easily incorporated into) normal harvest, handling, packing, storing or shipping procedures. Similarly, regulatory monitoring that fits into the normal procedures for a commodity reduces costs and facilitates commodity movement in a timely manner.

An example of a quarantine treatment that was easily adopted for commercial use is the two-stage hot water immersion combined with fruit selection at the mature-green ripeness stage, which was developed for disinfesting papaya of tephritid fruit fly eggs before export from Hawaii (Couey and Hayes, 1986). The hot water immersion phases of the treatment consisted of immersing papaya in 42°C water for 30–40 min, followed immediately by a second immersion in 49°C water for 20 min. The fruit were then hydrocooled to ambient temperatures and dried before sorting and packing.

The treatment was developed around the 49°C, 20-min immersion that was an industry standard for controlling postharvest decay (Couey and Hayes, 1986). The hot water immersion phase of the treatment was easily implemented because the industry was already using both the treatment technology and the 49°C, 20-min treatment; all that was required was the installation of additional immersion tanks to handle the volume of papaya being treated, water circulators and heaters to ensure treatment uniformity and regulatory equipment to monitor the treatments.

The original forced hot-air treatment is an example of a quarantine treatment that was difficult for industry to adopt (Armstrong *et al.*, 1989). The heating profiles were measured precisely during treatment research and reported as 43.0 ± 1.0, 45.0 ± 1.0, 46.5 ± 1.0 and 49.5 ± 0.5°C for the stepped-air temperatures used to heat the fruit; and 41 ± 1.5, 44 ± 1.0, 46.5 ± 0.8 and 47.2 for the fruit centre temperatures at each of the four steps over a 7-h treatment period (Armstrong *et al.*, 1989).

In hindsight, more consideration had been given to the science involved with developing a new treatment technology than to how the papaya industry would implement the treatment, or how the treatment would be monitored by regulatory agencies. The 7-h treatment duration was too long and implementation would completely disrupt harvesting, handling and packing schedules. Moreover, the stepped-air and fruit centre temperatures measured so precisely in the laboratory could not be duplicated under commercial conditions. Hence, the treatment could not meet the regulatory standards that required the same precision that Armstrong *et al.* (1989) reported.

Additional research was able to correct these problems by shortening the treatment duration to 3.5 h, developing a single (rather than stepped) air temperature heating profile, and making the end fruit centre temperature of 47.2°C the primary heating requirement for regulatory monitoring (Armstrong *et al.*, 1995a).

Commodity Tolerance to the Treatment

This is the most important factor in determining whether a treatment method can be used, and must be considered before proceeding too far with any quarantine treatment research and development. It is imperative to develop commodity tolerance data before, or at the same time as, treatment efficacy data are developed. Commodity tolerance must also be considered at both the laboratory and commercial scale (see Fig. 13.3).

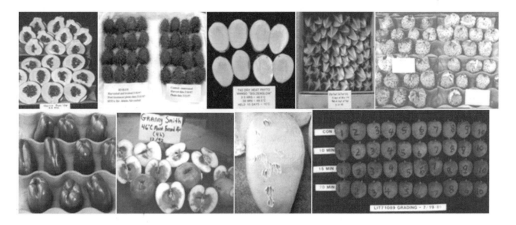

Fig. 13.3. Regardless of the technology used in developing a quarantine treatment, fruit quality tests should coincide with efficacy tests to ensure that the treatment does not adversely affect quality, shelf life or marketability. These photographs show quality observations for a variety of fruits subjected to candidate quarantine treatments. From left to right (top row): papaya subjected to forced hot-air treatment (comparison of control in lower three rows with the treated fruit in the upper three rows indicates a slight heat-induced delay in ripening in the treated fruit); rambutan (a short-duration, forced hot-air treatment caused negligible damage to the skin or spinturns); mango (a two-stage, forced hot-air treatment was used to maintain fruit quality); carambola (or starfruit, which tolerated cold treatment without injury); atemoya (stem splitting, a water content-related issue, occurred in both treatment and control fruit); (bottom row): sweet pepper subjected to a forced hot-air treatment showing no damage; apple showing internal heat-induced damage; mango showing heat-induced fibrous damage that includes both the skin and fruit flesh; and lychee immersed in 49°C water for 0, 10, 15 or 20 min indicate that the number of fruit with heat-induced skin darkening increases with the length of immersion time (top row 0 min, then 5, 10 and 20 min).

The literature is replete with quarantine treatments that have never been used commercially because tolerance studies were not done, or carried out only in the laboratory, and thereafter found to cause unacceptable damage to the commodity in commercial-scale tests. For example, a vapour heat treatment developed against Caribbean fruit fly, *Anastrepha suspensa* (Loew), in Florida grapefruit was never used commercially. Although the efficacy data showed that the treatment would provide quarantine security against Caribbean fruit fly eggs and larvae, later research showed that the grapefruit incurred damage that would make the fruit unacceptable.

Another caveat in developing fruit quality data concurrently with developing efficacy data is to ensure that fruit quality test data are collected under commercial conditions (see Fig. 13.3). Preliminary tests to determine papaya quality in early work by Couey and Hayes (1986) found that fruit treated and held for 2 days at ambient condition showed tolerance to selected heat treatments, whereas fruit treated and held at 10°C for 10 days to simulate storage conditions and shipping time showed the opposite.

The Unexpected

Changes in geographical growing areas, preharvest, harvest or postharvest practices and procedures, development and implementation of new cultivars, differences in annual environmental growing conditions and genetic variability, among many other natural processes, can play havoc with quarantine treatments through either loss of commodity tolerance to the treatment or survival of the target pest.

For example, the two-stage hot water immersion combined with fruit selection at the mature-green ripeness stage was developed to disinfest papaya of tephritid fruit fly eggs before export from Hawaii (Couey and Hayes, 1986). The treatment was used successfully from 1985 to 1991. Fruit selection at the mature-green ripeness stage ensured that, if any infestation was present, it would be only fruit fly eggs and not larvae that would be found in riper fruit. The two-stage hot-water immersion would kill only fruit fly eggs, which would be located within the first 3–5 mm of the fruit surface. Larvae, which tunnelled into the fruit, would not be subjected to enough temperature over time to cause mortality at a level that would ensure quarantine security.

This quarantine treatment worked for over 4 years until two shipments, one in 1989 and another in 1990, were rejected because oriental fruit fly larval infestations were found. The treatment was later rescinded when a genetic aberration called 'blossom end defect', a small opening at the blossom end of the papaya extending to the seed cavity, was discovered (Zee et al., 1989). In rare cases, fruit flies deposited eggs in the blossom end opening, the eggs hatched and the resulting larvae moved from the opening to the seed cavity (see Fig. 13.4). These larvae found deep in the fruit were not killed by the hot water treatment (Liquido, 1990). Hence, the treatment was no longer viable because of an unexpected genetic deformity in the fruit.

Fig. 13.4. Blossom end defect in papaya. Cross-section at left shows the opening from the blossom end to the seed cavity. The larvae from the eggs of the female oriental fruit fly (right) were able to penetrate deeper into the fruit and survive the two-stage, hot water immersion quarantine treatment.

Another unexpected consequence of hot water treatments arose in 1999 when a *Salmonella* outbreak occurred in 13 widely separated states throughout the USA. Using an information network, genetic identification, computer analysis and extensive detective work, a team from the Center for Disease Control and Prevention, state health representatives and scientists were able to trace the infection to mangoes and, using import and marketing records, to a single Brazilian packing house serving one farm (Sivapalsingam *et al.*, 2003). Investigators at the site showed that the problem was caused by contaminated water in the cool water dip tanks, easily remedied by using chlorinated dip water and adequate filtration. Sivapalsingam *et al.* (2003) mentioned that safeguards against these types of contamination should be included as treatments are approved and brought online.

13.7 Commodity Quality

No commodity was ever meant to undergo the rigours of quarantine treatment and, with rare exceptions, most quarantine treatments affect commodities adversely. This is especially the case for heat treatments of fresh horticultural commodities. Horticulturalists and postharvest physiologists emphasize the importance of removing field heat from harvested fruits, vegetables or ornamentals and storing at lower than ambient temperatures in order to reduce postharvest decay and prolong shelf life.

In the case of quarantine heat treatments, however, the commodity is subjected to unusually high temperatures to kill target pests. Heat treatments can cause a variety of physiological damage to commodities, including surface lesions, pitting, scalding, loss of aroma, surface and internal discolorations, wilting or unusual softening, water-soaked flesh, early senescence (loss of Shelf Life), and a predisposition to postharvest decay organisms. The severity of high-temperature stress is primarily determined by the temperature differential and duration of exposure (Paull and McDonald, 1994). Therefore, developing heat treatments for any commodity is an exercise in finding a balance between treatment parameters that will maintain product quality and shelf life and those that will provide quarantine security against the target pest.

Researchers have empirically developed many procedures to reduce the injury caused by heat treatment, most of which seem to rely on the heat shock response (Paull and McDonald, 1994). Bacteria, plants and animals exposed to 34–42°C acquire transient thermotolerance within 30 min, and a unique group of proteins is synthesized (Paull, 1990). For example, the two-stage hot water immersion quarantine treatment for papaya discussed earlier in this chapter caused intermittent fruit quality problems that plagued the Hawaii papaya export industry for the entire time the treatment was used, including fruit that did not ripen or ripened non-uniformly, exhibited surface scalding and unusual fruit softening, as well as postharvest decay issues. Better papaya fruit quality was obtained after replacing hot water immersion with forced hot-air quarantine treatment.

The ideal treatment minimizes damage to fruit and economic costs, while using a maximum temperature the fruit can tolerate for a time period that results in sufficient pest mortality. Determining these parameters requires the experimental finding of the maximum tolerance of various fruit to hot forced air. The experimental machine (Gaffney, 1990) used in these tests requires precise monitoring of fruit, air and dew point temperatures (a function of humidity).

Results depended on precise regulation of forced air temperature: usually the recommended temperature was within 2°C of temperatures that resulted in damage. For example, 'Dancy' tangerine flavour and appearance qualities were indistinguishable from controls when treated for 4 h at 45°C, but significantly damaged at 46°C (Shellie and Mangan, 1995). Valencia oranges in similar tests could be heated for up to 4 h at 46°C, but fruit heated at 47°C were inferior in flavour (Shellie and Mangan, 1994). Navel oranges were also shown to tolerate 46°C hot forced-air treatments, but pretreatments such as hot water shocks at 50°C or treatments in water at 46°C resulted in fruit damage (Shellie and Mangan, 1998). Grapefruit could withstand treatments at 46°C maximum temperatures for 5–7 h but incurred damage at temperatures > 46°C (Shellie and Mangan, 1996).

Although this series of experiments demonstrated that no significant damage to most citrus species or cultivars occurred at certain combinations of temperatures and humidities, the adaptation of laboratory precision to commercial processes has thus far not been successful. The citrus exporters of Nuevo Leon, Mexico, built a unit capable of treating about 10 t of fruit with a computerized system featuring temperature, humidity and dew point sensors through the chamber. Certification testing of this equipment revealed that the unit was unable to maintain uniform temperatures (R.L. Mangan, personal observation). As the load heated, air and fruit surfaces often exceeded 50°C and fruit surfaces were often wet, indicating excessive humidity. Although this unit is not acceptable for later-season fruit, it is adequate for very early-season tangerines and oranges, which are normally excessively damaged by methyl bromide.

Comparison of heating media and heat transmission in various fruit species experiments was carried out to compare fruit heating in water, hot forced air and vapour heat on similar sizes of tropical fruit (Shellie and Mangan, 2000). Temperatures of the media (vapour–air, air and water), fruit surface and interiors, O_2 and CO_2 concentration inside grapefruit were recorded. As expected, thin-skinned fruit (mangoes, papaya) heated more rapidly than did citrus in all media.

Heat transfer to fruit surface and heat stress was most rapid with vapour heat. The fruit surface was actually hotter than the vapour-saturated air due to the release of heat during condensation on the fruit. Oxygen concentration decreased and CO_2 increased to much greater levels in vapour heat and hot water than in hot forced air during the first 2 h of treatment. In another study, Shellie et al. (1997) showed that levels of oxygen < 0.1% in a controlled atmosphere chamber did not cause damage to grapefruit, but higher levels of CO_2 (21%) could induce off-flavours.

Although heat shock response is known to occur in many fresh commodities, researchers have spent several decades investigating the metabolic pathways for inducing thermotolerance. Chan and Linse (1989a) increased thermotolerance in cucumbers, Cucumis sativus L., in order to apply a hot water immersion quarantine treatment against tephritid fruit flies. They found that preconditioning cucumbers at $32.5 \pm 0.5°C$ in air for 24 h increased tolerance to hot water immersions of 30–60 min at 45°C and 30–50 min at 46°C.

Thermodynamic analysis of the heat inactivation of the ethylene-forming enzyme system in conditioned cucumbers showed increased D-values, larger energies of activations and Q10 and larger entropy and enthalpy changes (Chan and Linse, 1989b). The ethylene-forming-enzyme system in conditioned cucumbers was in a more ordered state than in control cucumbers, which was speculated to be due to the formation of hydrophobic bonds during conditioning that had increased the thermostability of the ethylene-forming enzyme system (Chan and Linse, 1989b). Chan (1991) also described the effects of both slow and rapid heating of papaya, showing similar effects in the papaya ethylene-forming enzyme system Chan and Linse (1989b) had previously described for cucumber. Chan (1991) determined that the slower heating of the forced hot-air papaya quarantine treatment produced a preconditioning effect that resulted in little or no damage to the ethylene-forming enzyme system, which provided better fruit quality than the more rapid heating of the two-stage hot water immersion papaya quarantine treatment.

The history of vapour heat traces the general development, complexity and progress in quarantine treatments. As in many pest control programmes, both scientific advances and redirection due to safety and environmental concerns bring about change. Mangan and Hallman (1998) showed that research attention to heat in quarantine treatments greatly expanded following the removal of ethylene dibromide as a fumigant due to health issues and limitations on the use of methyl bromide on account of environmental issues.

According to a review by Baker et al. (1944), Herrera carried out the first studies of tolerance to high temperatures in Mexico from 1900–1901. Crawford expanded this work two decades later, and again in 1927 when he recognized Mexican fruit fly as a threat to Texas citrus production. In response, Darby's research in the USDA Mexico laboratory lead to the development of vapour heat treatments, which were applied during the 1929 outbreak of

Mediterranean fruit fly in Florida. It appears that research attention over 100 years ago was largely elicited by emergency situations, and this pattern continues today.

As previously discussed, vapour heat treatments for citrus listed in the APHIS treatment manual are not currently used. However, following the emergency situations arising from Mediterranean and Mexican fruit fly outbreaks in Florida and Texas during the 1930s, vapour heat or refrigeration treatments were required for Texas citrus shipments originating in infested regions from 1932.

In an unpublished *History of the Mexican Fruit Fly Project 1927–1943*, Berry (1928) listed the volumes of fruit treated in car lots, which are equivalent to 360 boxes, each weighing 80 lbs (total 28,800 lb/13,063 kg). During the period from 1937 (when the infestation was recognized) until 1943 (when Berry stopped keeping data), a total of 6071 carloads (79,305,473 kg) of citrus were shipped after treatment by vapour heat. A cold treatment was also approved for quarantine, but only 78.8 carloads (1,029,364 kg) were treated by this method.

When Berry compared the two treatments, vapour heat required about 14 h and could be carried out in field boxes prior to packing in modified de-greening rooms, while low-temperature treatments required about 20 days after grading and packing in treatment rooms that needed extensive construction. Berry concluded that, for vapour-heated fruit, 'No damage to fruit has been reported'. On the contrary, several packers are convinced that sterilized fruit carries better and arrives at the markets in much better condition than unsterilized fruit.

Shaw *et al.* (1970) described research-leading options to replace vapour heat treatments for citrus. In 1945, Balock and Starr reported that methyl bromide was an effective treatment for larvae and eggs of Mexican fruit fly in citrus, but the treatment was not used at that time because some damage to both fruit colour and flavour occurred. Ethylene dibromide was found to be a better option, but was withdrawn as a treatment due to carcinogenicity (Ruckleshaus, 1984).

However, because methyl bromide fumigation was rapid and economical compared with vapour heat treatment, fumigation with methyl bromide eventually became the preferred treatment method to replace ethylene dibromide (Shaw *et al.*, 1970). Methyl bromide fumigation is, at present, the major treatment of citrus shipped from Texas to the rest of the USA under quarantine conditions, and for most citrus shipped from Mexico to the USA The methyl bromide treatment uses a protocol developed by Williamson *et al.* (1986). During the late 1980s and early 1990s, modifications in treatment practice were made by individual packers to reduce damage by this fumigant, but damage is still present on early-season oranges and tangerines (R.L. Mangan, personal observation).

The hot forced-air treatments discussed above were developed at the request of the Texas and Mexican fruit industries to provide an alternative treatment less damaging to citrus fruits. The Texas fruit industry has not invested in these treatments and continues the use of methyl bromide fumigation as a postharvest treatment. At present, most citrus is shipped from Texas without postharvest treatment under the fly management protocol that involves release of sterilized insects, quarantines and sprays to maintain a pest-free zone (Nilakhe *et al.*, 1991). Under this system, postharvest treatments are seldom required until late in the citrus season, when more mature oranges and grapefruit are less damaged by methyl bromide. Tangerines and early-season oranges continue to be shipped to the USA from Mexico showing varying degrees of methyl bromide burn on the peel, but little internal damage.

13.8 Experimental Heat Treatment Equipment

The equipment used in the laboratory to develop thermotolerance and quarantine heat treatment data for both the target pest and the host commodity – as well as the accompanying efficacy data for candidate quarantine treatments – ranges from simple and inexpensive to complex and costly (see Fig. 13.5).

For example, the original equipment used in the first study of fruit fly thermotolerance consisted of a water bath with a heater and water circulator that maintained the water at constant temperature and provided adequate circulation to prevent thermal stratification (Armstrong, 1982). This system is called a static-temperature hot water bath.

Since that time, both the equipment and methodology used to study fruit fly thermotolerance has become more sophisticated. Transient-temperature hot water baths reduce hot water temperature fluctuations and use computer-driven hot water baths to manipulate temperature profiles (Jang *et al.*, 1999; Thomas and Shellie, 2000). However, several researchers question hot water immersion and its related methodologies (Hallman, 1996, 2000; Waddell *et al.*, 2000; Gazit *et al.*, 2004). Most important among these concerns is the impact of oxygen depletion in heated water on insect mortality, and the number of insects (load factor) tested at one time in hot water immersion devices that could cause increased mortality and result in erroneous data (Mangan and Hallman, 1988; Hansen and Sharp, 1998; Gazit *et al.*, 2004).

To eliminate potential problems caused by using hot water immersion in developing thermotolerance data, a new approach using a heating block system was developed to heat insects directly and uniformly at a given temperature–time combination in the presence of air (Ikediala *et al.*, 2000; Wang *et al.*, 2002a, b; Johnson *et al.*, 2003; Gazit *et al.*, 2004; Fig. 13.5).

At the time of writing, no published comparisons of static-temperature water baths, transient-temperature water baths or heating block systems exist. The best system for determining thermotolerances of target pests will provide consistently accurate data in the shortest period of time with the least number of factors (other than the heating process itself) that influence target pest mortality.

Fig. 13.5. Experimental equipment: (a) hot-water immersion equipment for efficacy and quality tests at USDA-ARS, Weslaco, Texas; (b) computer-controlled water baths for developing fruit fly thermotolerance data at USDA-ARS, Hilo, Hawaii; (c, g) hot-water baths for developing fruit fly thermotolerance data in Vanuatu and Fiji; (d, e, f) heating block systems developed by Washington State University for fruit fly thermotolerance tests; (h, i, j) a simple box and fan system used to develop the first forced hot-air treatments (unit was placed in a heated room and the air drawn up through stacked boxes with open lattice bottoms containing papaya) at USDA-ARS, Hilo; (k) a hydrocooling method to remove heat from fruit after treatment; (l, m) an original forced hot-air treatment unit and later model at USDA-ARS, Hilo; (n) a pilot-scale forced hot-air unit to test horizontal air flow and heating rate parameters developed at the University of Hawaii; and (o, p, q) an experimental prototype forced hot-air and vapor heat treatment facility developed at the University of Hawaii. Treatment parameters were tested on fruit in single layers on trays or bulk bins, and the equipment was also used for thermal mapping of standard papaya bins for bulk treatment of fruit in commercial papaya packing houses before export."

13.9 Heat Treatment Research

Heat treatment research begins with the determination of the most heat-tolerant target pest species and life stage(s) based on relative thermotolerances. Once known, commodities that have been naturally or artificially infested with the most heat-tolerant life stage of the target pest are exposed to candidate heat treatment parameters.

For example, Gazit et al.'s (2004) finding that the Mediterranean fruit fly third-instar was the most heat-tolerant life stage for this target pest of citrus means that treatment will kill all other life stages. Continuing with this example, citrus can be naturally infested with Mediterranean fruit fly third-instars by exposing the fruit to female oviposition, then holding the fruit until larvae hatching from the deposited eggs become third-instars. Infested fruits are then subjected to the candidate heat treatment. Natural infestation is used when the commodity is a good host and the target life stages can be easily reared for use in treatment tests.

Artificial infestation is required when the commodity, target pest life stage or candidate treatment cannot support the use of natural infestation because of commodity physiology, host status or a physical attribute that severely limits the infestation of the commodity and recovery of insects in the controls. Artificial infestation is also used with good hosts to accelerate data acquisition from treatment tests.

For example, Armstrong et al. (1989, 1995a) removed a plug from papaya with a cork borer, inserted fruit fly third-instars, then replaced the plug and held it in place with masking tape. Papayas artificially infested with third-instars in this manner were then treated using forced hot air in efficacy tests. Although this artificial infestation method worked for papaya, it could not be used for citrus because damage to the fruit segments would result in fluid accumulation that eventually could drown the larvae.

Shellie and Mangan (2002a) discussed artificial and cage infestation of fruit for treatment development. In addition to previously mentioned problems with fruit weight loss and breakdown of fruit structure during the larval maturation period, artificial infestation allowed a uniform number of larvae to be inserted and treated. This greatly reduced unknown rates of infestation and skewed distributions of larvae among fruit (Preisler and Robertson, 1992). Mangan et al. (1998) artificially infested citrus for their hot forced-air treatments by carefully inserting a cork borer through the stem end of the peel and manoeuvring the borer though the centre, hollow and core of the fruit to widen this space without damaging sections for insertion of larvae. In that study, mortality in the untreated control fruit averaged 0.92–2.29%, which was considered acceptable.

Fruit size also plays a role in judging the suitability of using artificial infestation. Lychee (Litchi chinensis Sonn.) and longan (Dimocarpus longan Lour.) are too small and fleshy from which to remove plugs but, if the most heat-tolerant life stage were fruit fly eggs or first-instars, artificial infestation by injection (Friend, 1957) would work. Another method for fruit not amenable to either the plug or injection technique is to cut flaps in the surface of the fruit. Armstrong *et al.* (1995b) inserted fruit fly eggs or first-instars into carambola with this process, and secured the flaps with tape prior to cold-treatment tests that resulted in adequate survival in controls, to demonstrate treatment efficacy of their candidate quarantine cold treatment.

The laboratory phase is complete when the candidate treatment has proved efficacious against the most heat-tolerant life stage of the target pest without causing unacceptable damage to the commodity or loss of shelf life. Large-scale confirmatory tests are then required to demonstrate treatment efficacy under commercial conditions, including maintenance of commodity quality. Confirmatory tests are usually monitored by regulatory personnel from the importing country to ensure accurate documentation of equipment, methods, and treatment parameters (see Fig. 13.6).

13.10 Commercial Heat Treatment Equipment and Facilities

Commercial heat treatment equipment in use today is limited to the application of forced hot air and/or vapour heat treatments and hot water immersion treatments (see Fig. 13.6). Hot water immersion can claim the largest amount of treated tonnage based on the importation of mangoes into the US from Latin America. Although hot water immersion was used to export papaya from Hawaii, the quarantine treatment was discontinued in 1991 because of the previously discussed blossom end defect in papaya. Presently, the only other commercial hot water immersion quarantine treatment is used for tropical flowers and foliage exported from Hawaii.

Forced hot air is responsible for the second largest amount of treated tonnage based on the export of papaya, capsicums, aubergine and breadfruit from Fiji and papaya from the Cook Islands to New Zealand. However, the use of forced hot air is rapidly expanding in the Pacific Basin and Pacific Rim, with new treatment facilities now in Tonga, the Republic of Samoa and New Caledonia. At the time of writing, Vanuatu and Vietnam were contemplating the use of forced hot-air treatment technology for exporting papaya and other tropical fruits.

Vapour heat treatment facilities are used by Hawaii for exporting papaya and by Thailand for exporting mango. Whereas hot water immersion quarantine treatment equipment is relatively easy to engineer, forced hot-air and vapour heat treatment equipment requires complicated computer pro-grammes to operate and monitor the treatment parameters and equipment. Companies that produce forced hot-air or vapour heat treatment equipment at the present time include: FoodPro International, San Jose, California, USA; Quarantine Treatments International, Queenstown, New Zealand; Sanshu Sangyo, Kagoshima, Japan; and Takenaka Komuten, Tokyo, Japan. These treatment facilities are relatively expensive compared with forced hot-air treatment equipment (about US$120,000 for a chamber that will treat 8 t of

Fig. 13.6. Commercial heat treatment equipment: (a, b, c) Double hot water immersion treatment for papaya at AMFAC Tropical Fruits, Keaau, Hawaii, USA (c.1986) showing line of immersion tanks, forklift placing bin of papayas in tank and papayas held under heated water by wooden grate; (d, e, f) hot water immersion tanks for treating mangoes for export, Tapachula, Mexico, showing fruit being lowered into tank for treatment, bins of mangoes and empty treatment tank with water circulator openings at end; (g, h, i) a forced hot-air treatment unit at Tropical Hawaiian Products, Keaau, Hawaii (c.1989) showing operators preparing to insert temperature probes while a USDA-APHIS inspector watches, and preparing to close and lock chamber for treatment; (j, k) a double-wide, forced hot air and vapour heat treatment unit at Nature's Way Cooperative, Suva, Fiji, showing loading operations; and (l) loading double-wide, forced hot air and vapour heat treatment chamber at Dole Tropical Fruits, Kahuku, Hawaii.

fruit, such as papaya, at a time). Hot water immersion equipment costs about one-third that of forced hot-air equipment to treat the same quantity of fruit.

Most important in the cost of heat treatments is the amount of commodity treated, or throughput. The more throughput, the lower the treatment costs. In Hawaii, the cost for forced hot-air or vapour heat treatment of papaya ranges from US$0.17 to 0.31 per pound (0.45 kg) of treated fruit. Other cost factors in the use of any heat treatment include facilities, labour, refrigeration, power and shipping.

Regardless of what heat treatment is used, it is a regulatory requirement to protect the treated commodity from reinfestation. This is done most frequently for forced hot-air and vapour heat treatments by engineering treatment equipment to load from the outside of the facility (unprotected), then unload into a protected facility when the treatment is complete. Both ends of the treatment equipment are locked during treatment to prevent tampering. Commodities subjected to hot water immersion can be moved from the treatment equipment to a protected facility in the open because the fruits are too hot for fruit flies to land and reinfest the fruit. However, the time from the end of the treatment to movement into the protected facility is closely regulated.

13.11 Approved Commercial Heat Treatments

Table 13.1 provides a general overview of commercial heat treatments that are presently approved for the export of specific fruits. Some of the treatments are no longer used, but they remain in quarantine treatment manuals as approved treatments simply because they have not yet been removed.

13.12 Summary

Quarantine treatments are needed to ensure that unwanted pests are eliminated from commodities moving through export marketing channels. With the loss of ethylene dibromide and the public demand for less toxic methods, the use of quarantine heat treatments has been revived. Advances in heating- and temperature-monitoring devices, new experimental methods and equipment, a better understanding of thermal dynamics in commodities and thermotolerance among target pests and their life stages, and a concerted research effort beginning in the 1970s has helped to bring about the hot water, vapour heat and forced hot-air quarantine treatments used internationally today.

The forces of economics, new fruits entering markets worldwide and the limits of commodity tolerances to heat will result in future modification of present heat treatment methods and facilitate the advent of new heating technologies, such as radio frequency treatment.

Table 13.1. Commercial heat treatments approved for the export of fruits.

Exporting country	Importing country	Commodity	Treatment and parameters
USA (Hawaii)	USA (mainland)	Sweet pepper, aubergine, pineapple (except smooth cayenne), courgette, squash, tomato	Vapour heat to a pulp temperature of 45°C and a holding period at 45°C for 8.75 h
USA (Hawaii)	USA (mainland)	Papaya, citrus	Forced hot-air or vapour heat to a fruit centre temperature of 47.2°C in > 4 h
USA (Hawaii)	Japan	Papaya	Vapour heat to a fruit centre temperature of 47.2°C; relative humidity must be > 90% during the last 1 h of treatment
USA (Hawaii)	New Zealand	Any fruit	Forced hot air to a fruit centre temperature of 47.2°C in > 4 h
USA (Hawaii)	USA (mainland)	Flower and foliage plants	Hot water immersion
Belize	USA	Papaya	Forced hot air to a fruit centre temperature of 47.2°C followed by a 20-min holding period at 47.2°C
USA (Hawaii)	USA (mainland)	Lychee	Hot water immersion at 49°C for 20 min
Fiji	New Zealand, Australia	Papaya, aubergine, breadfruit, pepper, mango	Forced hot air to a fruit centre temperature of 47.2°C followed by a 20-min holding period at 47.2°C
Republic of Samoa	New Zealand	Breadfruit, papaya	Forced hot air to a fruit centre temperature of 47.2°C followed by a 20-min holding period at 47.2°C
Philippines	Australia, USA, Japan, New Zealand, Korea	Mango	Fruit pulp temperature to 46°C followed by a 10-min holding period at 46°C
Vanuatu	New Zealand	No commodities at this time	Forced hot air to a fruit centre temperature of 47.2°C followed by a 20-min holding period at 47.2°C
Cook Islands	New Zealand	Aubergine, papaya, mango	Forced hot air to a fruit centre or pulp temperature of 47.2°C followed by a 20-min holding period at 47.2°C
Tonga	New Zealand	Breadfruit, peppers, aubergine, mango, papaya, tomato, avocado	Forced hot air to a fruit centre or pulp temperature of 47.2°C followed by a 20-min holding period at 47.2°C
New Caledonia	New Zealand	Peppers, aubergine, mango, lychee	Forced hot air to a fruit centre temperature of 47°C for 20 min (or 43°C for 3.5 h, peppers only)
Mexico	USA	Citrus	Forced hot air to a fruit centre temperature of 44°C in > 90 min followed by a holding period at 44°C for 44 min

Continued

Table 13.1. *Continued.*

Exporting country	Importing country	Commodity	Treatment and parameters
Mexico	USA	Mango	Forced hot air to a pulp temperature of 48°C
Mexico	USA	Clementine	Vapour heat to a fruit centre temperature of 43°C and a holding period at 43°C for 6 h
Taiwan	USA	Mango	Vapour heat to a pulp temperature of 46.5°C with a holding period of 30 min at 46.5°C
Mexico and Central and South America, Puerto Rico, US Virgin Islands, West Indies	USA	Mango	Hot water immersion at 46.1°C for 65–90 min depending on size of fruit based on fruit weight (e.g. 75 min for mangoes that weigh ≤ 500 g)
Chile	USA	Mountain papaya	Forced hot air or vapour heat to a fruit centre temperature of 47.2°C in > 4 h
Thailand	Japan	Mango	Vapour heat to a pulp temperature of 46°C followed by a holding period of 20 min at 46°C
Colombia	USA	Yellow pitaya	Vapour heat at 47°C air temperature to a minimum pulp temperature of 46°C in > 4 h with a holding period of 20 min at 46 °C

13.13 References

Armstrong, J.W. (1982) Development of a hot-water immersion quarantine treatment for Hawaiian-grown 'Brazilian' bananas. *Journal of Economic Entomology* 75 (5), 787–790.

Armstrong, J.W. (1994) Heat and cold disinfestation treatments. In: Paull, R.E. and Armstrong, J.W. (eds) *Insect Pests and Fresh Horticultural Products: Treatments and Responses.* CAB International, Wallingford, UK, pp. 103–119.

Armstrong, J.W., Schneider, E.L., Garcia, D.G., Nakamura, A.N. and Linse, E.S. (1984) An improved holding technique for infested commodities used for Mediterranean fruit fly quarantine treatment research. *Journal of Economic Entomology* 77 (2), 553–555.

Armstrong, J.W., Hansen, J.D., Hu, B.K.S. and

Brown, S.A. (1989) High-temperature forced-air quarantine treatment for papayas infested with tephritid fruit flies (Diptera: Tephritidae). *Journal of Economic Entomology* 82, 1667–1674.

Armstrong, J.W., Hu, B.K.S. and Brown, S.A. (1995a, b) Single-temperature forced hot-air quarantine treatment to control fruit flies (Diptera: Tephritidae) in papaya. *Journal of Economic Entomology* 88 (3), 678–682.

Armstrong, J.W., Shishido, V.M. and Silva, S.T. (1995) Quarantine cold treatment for Hawaiian carambola fruit infested with Mediterranean fruit fly, melon fly, and oriental fruit fly (Diptera: Tephritidae) eggs and larvae. *Journal of Economic Entomology* 88 (3), 683–687.

Baker, A.C. (1939) *The Basis for Treatment of*

Products where Fruitflies are Involved as a Condition for Entry into the USA. USDA Circular 551, Washington, DC.

Baker, A.C. (1952) The vapour heat process. In: *US Department of Agriculture Yearbook.* US Government Printing Office, Washington, DC, pp. 401–404.

Baker, A.C., Stone, W.E., Plummer, C.C. and McPhail, M. (1944) *A Review of Studies on the Mexican Fruitfly and Related Mexican Species.* Miscellaneous Publication 531, US Department of Agriculture, Washington, DC, 155 pp.

Baker, R.T., Cowley, J.W., Harte, D.W. and Frampton, E.R. (1990) Development of a maximum pest limit for fruit flies (Diptera: Tephritidae) in produce imported from New Zealand. *Journal of Economic Entomology* 83, 13–17.

Balock, J.W. and Kozuma, T. (1954) *Sterilization of Papaya by Means of Vaporheat Quick rRn-up.* Special Report No. 7, Fruit Fly Investigations in Hawaii. US Department of Agriculture, Entomology Research Branch, Honolulu, Hawaii.

Balock, J.W. and Starr, D.F. (1946) Mortality of the Mexican fruit fly in mangoes treated by the vapor-heat process. *Journal of Economic Entomology* 38, 646–651.

Burditt, A.K., Balock, J.W., Hinman, F.G. and Seo, S.T. (1963) Ethylene dibromide water dips for destroying fruit fly infestations of quarantine significance in papayas. *Journal of Economic Entomology* 56 (3), 289–292.

Chan, H.T., Jr. (1991) The effects of ripeness and tissue depth on the heat inactivation of the ethylene-forming enzyme system in papayas. *Journal of Food Science* 56, 996–998.

Chan, H.T., Jr. and Linse, E.L. (1989a) Conditioning cucumbers for quarantine heat treatments. *HortScience* 24, 985–989.

Chan, H.T., Jr. and Linse, E.L. (1989b) Conditioning cucumbers to increase heat resistance in the EFE system. *Journal of Food Science* 56, 1375–1376.

Couey, H.M. and Chew, V. (1986) Confidence limits and sample size in quarantine entomology. *Journal of Economic Entomology* 79, 887–890.

Couey, H.M. and Hayes, C.F. (1986) Quarantine procedure for Hawaiian papaya using fruit selection and a two-stage hot-water immersion. *Journal of Economic Entomology* 79 (5), 1307–1314.

Follett, P.A. and McQuate, G.T. (2001) Accelerated development of quarantine treatments for insects on poor hosts. *Journal of Economic Entomology* 94, 1005–1011.

Friend, A.H. (1957) Artificial infestation of oranges with the Queensland fruit fly. *Journal of the Australian Institute of Agricultural Sciences* 23, 77–80.

Furasawa, K., Sugimoto, T. and Gaja, T. (1984) The effectiveness of vapour heat treatment against the melon fly, *Dacus cucurbitae* Coquillett, in eggplant and fruit tolerance to the treatment. *Research Bulletin of the Plant Protection Service Japan* 20, 17–24.

Gaffney, J.J. (1990) *Warm Air/Vapor Heat Research Facility for Heating Fruits for Insect Quarantine Treatments.* Paper No. 906615, American Society of Agricultural Engineering, St Joseph, Michigan.

Gazit, Y., Rossler, Y., Wang, S., Tang, J. and Lurie, S. (2004) Thermal death kinetics of egg and third instar Mediterranean fruit fly (Diptera: Tephritidae). *Journal of Economic Entomology* 97 (5), 1540–1546.

Hallman, G.J. (1996) Mortality of third instar Caribbean fruit fly (Diptera: Tephritidae) reared in diet or grapefruits and immersed in water or grapefruit juice. *Florida Entomologist* 79, 168–172.

Hallman, G.J. (2000) Factors affecting quarantine heat treatment efficacy. *Postharvest Biology and Technology* 21, 95–101.

Hansen, J.D. and Sharp, J.L. (1998) Thermal death studies of third-instar Caribbean fruit fly (Diptera: Tephritidae). *Journal of Economic Entomology* 91 (4), 968–973.

Hansen, J.D., Armstrong, J.W., Hu, B.K.S. and Brown, S.A. (1990) Thermal death of oriental fruit fly (Diptera: Tephritidae) third instars in developing quarantine treatments for papayas. *Journal of Economic Entomology* 83, 160–167.

Hara, H.H., Hata, T.Y., Tenbrink, V.L., Hu, B.K.S. and Kaneko, R.T. (1996) Postharvest heat treatment of red ginger flowers as a possible alternative to chemical insecticidal dip. *Postharvest Biology and Technology* 7, 137–144.

Harte, D.S., Baker, R.T. and Cowley, J.M. (1992) Relationship between pre-entry sample size for quarantine security and variability of estimates of fruit fly (Diptera: Tephritidae) disinfestation treatment efficacy. *Journal of Economic Entomology* 85, 1560–1565.

Hawkins, L.A. (1932) Sterilization of citrus fruit by heat. *Citriculture* 9, 7–8, 21–22.

Ikediala, J.N., Tang, J. and Wig, T. (2000) A heating block system for studying thermal death kinetics of insect pests. *Transactions of the American Society of Agricultural Engineers* 43, 351–358.

Jang, E.B., Nagata, J.T., Chan, H.T., Jr. and Laidlaw, W.G. (1999) Thermal death kinetics in eggs and larvae of *B. latifrons* (Diptera: Tephritidae) and comparative thermotolerance to three other tephritid fruit fly species in Hawaii. *Journal of Economic Entomology* 92 (3), 684–690.

Jensen, J.L. (1888) The propagation of smuts in oats and barley. *Journal of the Royal Agricultural Society of England* 2 (3), 397–415.

Johnson, J.A., Wang, S. and Tang, J. (2003) Thermal death kinetics of fifth instar *Plodia interpunctella* (Lepidoptera: Pyralidae). *Journal of Economic Entomology* 96, 519–524.

Jones, W.W. (1940) *Vapor Heat Treatment for Fruits and Vegetables Grown in Hawaii.* Hawaii Agricultural Experiment Station Circular No. 16, Honolulu, Hawaii.

Laidlaw, W.G., Armstrong, J.W., Chan, H.T., Jr. and Jang, E.B. (1996) The effect of temperature profile in heat-treatment disinfestation on mortality of pests and on fruit quality. *Proceedings of the International Conference on Tropical Fruits: Global Commercialization of Tropical Fruits*, Kuala Lumpur, Malaysia, pp. 343–352.

Landolt, P.J., Chambers, D.L. and Chew, V. (1984) Alternative to the use of probit-9 as a criterion for quarantine treatments of fruit fly (Diptera: Tephritidae)-infested

fruit. *Journal of Economic Entomology* 77, 285–287.

Lin, T.H., Tseng, F.F., Chang, C.R. and Wang, L.Y. (1976) Multiple treatment for disinfesting oriental fruit fly in mangoes. *Plant Protection Bulletin* 18, 231–241.

Liquido, N.J. (1990) Survival of oriental fruit fly and melon fly (Diptera: Tephritidae) eggs oviposited in morphologically defective blossom end of papaya following two-stage hot-water immersion treatment. *Journal of Economic Entomology* 83 (6), 2327–2330.

Liquido, N.J., Barr, P.G. and Chew, V. (1996) *CQT_STATS: Biological Statistics for Pest Risk Assessment in Developing Commodity Quarantine Treatment.* USDA-ARS Publication Series, Washington, DC (available at http://www.pbarr@pbarc.ars.usda.gov).

Mangan, R.L. and Hallman, G.H. (1998) Temperature treatments for quarantine security: new approaches for fresh commodities. In: Hallman, G.H. and Denlinger, D.L. (eds) *Temperature Sensitivity in Insects and Application in Integrated Pest Management.* Westview Press, Boulder, Colorado, pp. 201–234.

Mangan, R.L., Frampton, E.R., Thomas, D.B. and Moreno, D.S. (1997) Application of the maximum pest limit concept to quarantine security standards for the Mexican fruit fly (Diptera: Tephritidae). *Journal of Economic Entomology* 90, 1433–1440.

Mangan, R.L., Shellie, K.C., Ingle, S.J. and Firko, M.J. (1998) High temperature forced-air treatments with fixed time and temperature for 'Dancy' tangerines, 'Valencia' oranges and 'Rio Star' grapefruit. *Journal of Economic Entomology* 91, 933–939.

Neven, L.G. (2005) Combined heat and controlled atmosphere quarantine treatments for control of codling moth in sweet cherries. *Journal of Economic Entomology* 98 (3), 709–715.

Nilakhe, S.S., Worley, J.N., Garcia, R. and Davidson, J.L. (1991) Mexican fruit fly protocol helps export Texas citrus. *Subtropical Plant Science* 44, 49–52.

Paull, R.E. (1990) Postharvest heat treatments and fruit ripening. *Postharvest News and Information* 1 (5), 355–363.

Paull, R.E. and Armstrong, J.W. (1994) Introduction. In: Paull, R.E. and Armstrong, J.W. (eds) *Insect Pests and Fresh Horticultural Products: Treatments and Responses.* CAB International, Wallingford, UK, pp. 1–33.

Paull, R.E. and McDonald, R.E. (1994) Heat and cold treatments. In: Paull, R.E. and Armstrong, J.W. (eds) *Insect Pests and Fresh Horticultural Products: Treatments and Responses.* CAB International, Wallingford, UK, pp. 191–222.

Preisler, H.K. and Robertson, J.L. (1992) Estimation of treatment efficacy when the number of test subjects is unknown. *Journal of Economic Entomology* 85, 1033–1040.

Robertson, J.L., Haiganoush, K.P. and Frampton, E.R. (1994) Statistical concept and minimum threshold concept. In: Paull, R.E. and Armstrong, J.W. (eds) *Insect Pests and Fresh Horticultural Products: Treatments and Responses.* CAB International, Wallingford, UK, pp. 47–65.

Ruckleshaus, W.D. (1984) *Ethylene Dibromide, Amendment of Notice of Intent to Cancel Registration of Pesticide Products Containing Ethylene Dibromide.* Federal Register 49, 14182–14185, US Government Printing Office, Washington, DC.

Secretariat for the Pacific Community (2002) Pacific Fruit Fly Web, http://www.spc.int/Pacifly.

Seo, S.T., Hu, B.K.S., Komura, M., Lee, C.Y.L. and Harris, E.J. (1972) *Dacus dorsalis*: vapour heat treatment in papayas. *Journal of Economic Entomology* 67, 240–242.

Seo, S.T., Hu, B.K.S., Komura, M., Lee, C.Y.L. and Harris, E.J. (1974) *Dacus dorsalis*: vapour heat treatment in papayas. *Journal of Economic Entomology* 65, 1372–1374.

Sharp, J.L. (1994) Hot water immersion. In: Sharp, J. and Hallman, G. (eds) *Quarantine Treatments for Pests and Food Plants.* Westview Studies in Insect Biology, pp. 133–148.

Sharp, J.L. and Spalding, D.H. (1984) Hot water as a quarantine treatment for Florida mangoes infested with Caribbean fruit fly. *Proceedings of the Florida State Horticultural Society* 97, 355–357.

Shaw, J.G., Lopez, D.F. and Chambers, D.L. (1970) A review of research done with the Mexican fruit fly and the citrus blackfly in Mexico by the Entomology Research Division. *Bulletin of the Entomological Society of America* 16, 186–193.

Shellie, K.C. and Mangan, R.L. (1994) Postharvest quality of 'Valencia' orange after exposure to hot, moist, forced air for fruit fly disinfestation. *HortScience* 29 (12), 1524–1527.

Shellie, K.C. and Mangan, R.L. (1995) Heating rate and tolerance of naturally degreened 'Dancy' tangerine to high temperature, forced air for fruit fly disinfestation. *HortTechnology* 5, 40–43.

Shellie, K.C. and Mangan R.L. (1996) Tolerance of red fleshed grapefruit to a constant or stepped temperature forced air quarantine heat treatment. *Postharvest Biology and Technology* 7, 151–159.

Shellie, K.C. and Mangan, R.L. (1998) Navel orange tolerance to heat treatment for disinfesting Mexican fruit fly. *Journal of the American Society of Horticultural Science* 123, 288–293.

Shellie, K.C. and Mangan, R.L. (2000) Postharvest disinfestation heat treatments: Response of fruit and fruit fly larvae to different heating media. *Postharvest Biology and Technology* 21, 51–60.

Shellie, K.C. and Mangan, R.L. (2002a) Hot water immersion as a quarantine treatment for large mangoes: artificial versus cage infestation. *Journal of the American Society of Horticultural Science* 27, 430–434.

Shellie, K.C. and Mangan R.L. (2002b) Cooling method and fruit weight: efficacy of hot water quarantine treatment for control of Mexican fruit fly in mango. *HortScience* 37, 910–913.

Shellie, K.C., Mangan, R.L. and Ingle, S.J. (1997) Tolerance of grapefruit and Mexican fruit fly larvae to heated controlled atmospheres. *Postharvest Biology and Technology* 10, 179–186.

Sivapalasingam, S., Barrett, E., Kimura, A., Van Duyne, S., DeWitt, W., Ying, M., Frisch, A., Phan, Q., Gould, E., Shillam, P., Reddy, V., Cooper, T., Hoekstra, M., Higgins, C., Sanders, J.P., Tauxe, R.V. and Slutsker, L. (2003) A multistate outbreak of *Salmonella enterica* serotype Newport infection linked to mango consumption: impact of water-dip disinfestation technology. *Clinical Infectious Diseases* 37, 1585–1590.

Sugimoto, T., Furasawa, K. and Mizobuchi, M. (1983) The effectiveness of vapour heat treatment against the oriental fruit fly, *Dacus dorsalis* Hendel, in green pepper and fruit tolerance to the treatment. *Research Bulletin of Plant Protection Service Japan* 19, 81–88.

Thomas, D.B. and K.C. Shellie (2000) Heating rate and induced thermotolerance in Mexican fruit fly (Diptera: Tephritidae) larvae, a quarantine pest of citrus and mangoes. *Journal of Economical Entomology* 93 (4), 1373–1379.

USDA (1997) Quarantine security for commodities: current approaches and potential strategies. In: Liquido, N.J., Griffin, R.L. and Vick, K.W. (eds) *Proceedings of Joint Workshops of the Agricultural Research Service and the Animal and Plant Health Inspection Service.* Washington, DC, 56 pp.

US Department of Commerce (2003) FAS online US Trade internet system (http://www.fas.usda.gov/USTrade/).

Vail, P.V., Tebbets, J.S., Mackey, B.E. and Curtis, C.E. (1993) Quarantine treatments: a biological approach to decision making for selected hosts of codling moth. *Journal of Economic Entomology* 86, 70–75.

Waddell, B., Clare, G.K., Petry, R.J.,

Maindonald, J.H., Purea, M., Wigmore, W., Joseph, P., Fullerton, R.A., Batchelor, T.A. and Lay-Yee, M. (1997) Quarantine heat treatments for *Bactrocera melanotus* (Coquillett) and *B. xanthodes* (Broun) (Diptera: Tephritidae) in Waimanalo papaya in the Cook Islands. In: Allwood, A.J. and Drew, R.A.I. (eds) *Management of Fruit Flies in the Pacific (Ecology, Seasonal Abundance).* Australian Center for International Agricultural Research Proceedings No. 76, pp. 251–255.

Waddell, B.C., Jones, V.M., Petry, R.J., Sales, F., Paulaud, D., Maindonald, J.H. and Laidlaw, W.G. (2000) Thermal conditioning in *Bactrocera tryoni* eggs (Diptera: Tephritidae) following hot-water immersion. *Postharvest Biology and Technology* 21, 113–128.

Wang, S., Ikediala, J.N., Tang, J. and Hansen, J.D. (2002a) Thermal death kinetics and heating rate effects for fifth-instar codling moths (*Cydia pomonella* (L.)). *Journal of Stored Products Research* 38, 441–453.

Wang, S., Tang, J., Johnson, J.A. and Hansen, J.D. (2002b) Thermal-death kinetics of fifth-instar *Amyelois transitella* (Walker) (Lepidoptera: Pyralidae). *Journal of Stored Products Research* 38, 427–440.

Williamson, D.L., Summy, K.R., Hart, W.G., Sanchez, R.M., Wolfenbarger, D.A., Bruton, B.D. and Benschoter, C.A. (1986) Efficacy and phytotoxicity of methyl bromide as a fumigant for the Mexican fruit fly (Diptera: Tephritidae) in grapefruit. *Journal of Economic Entomology* 79, 172–175.

Zee, F.T., Nishina, M.S., Chan, H.T. and Nishijima, K.A. (1989) Blossom end defects and fruit fly infestation in papayas following hot water quarantine treatment. *HortScience* 24, 323–325.

Index

ACC oxidase 80–81
ACC synthase 81
Acetaldehyde 255
Acorn 10
Activation energy 154–155
Adenosine triphosphate 252
Aeration 189
　　see also Low temperatures, Physical
　　　　disinfestation
Almond 2, 90, 91
Alternaria 163, 164, 165, 172
Amyelois transitella 133
　　biology 136–138
　　see also Navel orangeworm
Anaesthesia/anaesthetic 253
Anastrepha 141
　　A. ludens (Mexican fruit fly) 141, 142, 256,
　　　　258, 263, 281, (Loew) 314
　　A. oblique 258
　　A. suspensa 281, (Loew) 324
　　see also Caribbean fruit fly, West Indian fruit
　　　　fly
Angoumois grain moth, *Sitotroga cerealella* 9,
　　13
Anoxia 262
Ant 11
Antioxidants 91
Aphid 255
APHIS (Animal Plant Health Inspection
　　Service) 264
Apple 2, 27, 36, 44, 48–49, 80–81, 89, 93,
　　167–169, 172, 263, 277

Gala 263
Granny Smith 263
Apple maggot, *Rhagoletis pomonella* 2–3
Apricot 10–11
Asian longhorned beetle, *Anoplophora
　　glabripennis* 14
Asparagus 82
Aspergillus 164
ATP 252
ATPase 80
Aubergine 6, 10
Avocado 10–11, 80, 84, 86, 95–96, 274–277

β-galactosidase 81
β-mannanase 81
Bactrocera 141
　　B. actrocera tyroni (Queensland fruit fly)
　　　　281
　　B. cucurbitae (melon fly) 141, 142, 143,
　　　　(Coquillett) 314, 318
　　B. dorsalis (Oriental fruit fly) 141, 142, 143,
　　　　(Hendel) 314, 318
　　B. latifrons (Hendel) 318
　　B. oleae (olive fruit fly) 141, 142
Banana 93, 171
Bean 7
Bell pepper 6, 10
Biological 5
Botrytis 163, 164, 165, 166, 168, 170, 171,
　　172
British thermal units (BTU) 225
　　see also Calories, Energy

Broccoli 82
Brownheaded leafroller, *Ctenopseustis obliquana* 253
Browning
  flesh 86
  skin or peel 83–85
Brucid 13
Brussels sprout 82
Bulb 15

Cactus pear 166, 169, 173, 277
Calcium 253
Calories 225
Canola 204
  see also Stored product, quality
Carbon dioxide 80, 87–88, 251, 256, 257, 258
Caribbean fruit fly 240–241, 257
  Anastrepha suspense 3, 14
Cashew 91
Cavitation 261
Cavities 87
Celery 83
Ceratitis 141
  C. capitata (Mediterranean fruit fly) 141, 142, 144, 279–281, (Wiedemann) 314, 318
  distribution map 143
Chalara 164
Chaperonin 262
Cherries 2–4, 12, 27, 49
Cherry fruit fly, *Rhagoletis cingulata* 3
Cherry, sweet 90
Chestnut 169
Chilling injury (low temperature stress) 96, 275, 277
Chinch bug 8
Chlorophyll 83
  breakdown 82
  chlorophyllase activity 82
  peroxidase 82
Chrysmelid beetles 279
Circulating air velocity 39–40
Citrus 6, 9–10, 83, 85–86, 88, 93, 96, 275, 277
Clementine 6
Cnephasia jactatana 260
CO_2 80, 87–88, 251, 256, 257, 258
Codling moth, *Cydia pomonella* 3, 4, 12, 14, 242, 247–248, 252, 253, 256, 257, 260, 261, 263, 264, 321
Cold 5

Coleoptera 183, 185
Coleopteran species 256
Collard 82
Colletotrichum 163, 167, 168, 171, 172
Colour development 81–82, 87, 93
Commercial RF system 301
Commodity 311–313, 315–325, 329–336
  gradual heating 313
  host 311–313, 317–319, 329, 331
    poor host 317, 319
  quality 314, 323–327, 331–332
  status 318–319, 331
  thermotolerance 312–313, 318–319, 321, 326–327, 329–331
  tolerance 16
Complementary log–log transformation 208–209, 211
Computer simulation model 28
Conduction 28–36
  Biot number 36
  Fourier's law 29
  fruit 28–30, 33–34
  plate 29, 32, 36
  sphere 32, 34
  thermal conductivity 29, 34
Conductive heating 216–217, 220
Confused flour beetle, *Tribolium confusum* 9, 14
Controlled atmosphere 251
  arthropod mortality 262
  benefit 254
  effect of temperature 255, 256, 256, 258
  ethylene production 254
  off-flavour 255
  ripening-associated change 254
  tolerance limit 255, 260
Convection 30–33
  convective heat transfer coefficient 30, 32
  empirical approach 30
  forced 30, 32–33
  free 30
  Grashof 31
  laminar 30–33
  natural 30, 32–33
  Nusselt 31–32
  Prandtl 31–32
  sphere 32
  turbulent 30, 32–33
Convective heating 213, 215
  see also Fluid bed, Pneumatic conveyor, Spouted bed

Courgette 6, 10, 83, 275, 277
Cryptorhynchus mangiferae (F.) 319
Ctenopseustis obliquana (brownheaded
 leafroller) 253
Cucumber 275
Cumulative time 127
Cydia pomonella (codling moth) 133, 252,
 253, 256, 257, 260, 261, 263, 264,
 282, 283, 321
 biology 134–136
 distribution map 143

Dacus 141
 D. bivittatus (pumpkin fly) 141
Dacus oleae see olive fruit fly
Data logger 224, 228
Data recorder 75
Date 2
Decay 88, 261
Deploidia 171
Developmental stage 199
Diapausing 260, 261
Dielectric heating 41–52
 Federal Communications Commission
  41–42
 thermal runaway 41
Dielectric loss factor 217–218
Dielectric properties 41–53
 dielectric constant 43
 dielectric loss factor 41, 43, 49
 dipole rotation 43
 ionic conduction 43
 penetration depth 42–44
 permittivity 43
Diet, meridic 243
Disinfestation 183, 189, 190, 193, 217
Dothiorella 163
Dried fruit 7, 8, 11
Drosophila 269, 271, 276, 278, 279, 280, 284
Drosophila melanogaster 240

Economic impact 134, 141
Efficacy of treatment 245–246
Electrolyte leakage 83, 88
Electromagnetic 9, 13–16
Electromagnetic energy 202, 217, 218, 220
 see also Infrared radiation, Microwave,
  Radiowaves
Empirical method 201–202
Energy 190, 217, 225, 226
Energy balance 28, 34

Energy charge 252
Epiphyas postvittana (light brown apple moth)
 253, 256, 258, 261, 262, 263, 264,
 281
Equilibrium relative humidity 90
Equipment 182, 190, 192, 205, 222–225,
 226, 227, 228
 see also Data loggers, Fans, Heaters
Equivalent treatment time 151, 154
Ethanol 255
Ethyl acetate 255
Ethylene 80–81
Experimental method 107–112
 direct immersion 108–109
 glass vial 109
 ideal test condition 107
 infested commodities 108
 metal tube 109
External browning 254, 261, 264
External developer 183, 184, 185

Fans 223–224, 228
Fig 2, 14
Flat grain beetle, *Cryptolestes pusillus* 14
Flavour 89–91
 off-flavour 90–91
Floorboard 12
Flour 9
Flour mills 187, 193, 205, 222
Flower 85
Fluid bed 202, 203, 213–214, 216
Food-processing 220, 221
Food-processing facility 182, 185, 187, 188,
 189, 205, 222, 225
Forced hot air 11
Forced-air heating 86–87
Frankliniella occidentalis (Western flower
 thrips) 253
Fruit 2, 8–11, 13, 15
Fruit fly 6, 10
Fruit ripening 284
Fuller rose beetle, *Asynonychus godmani* 10
Fumigant 5, 7, 187, 193, 229
Fumigation 189, 190, 222, 224, 227
Fundamental kinetic model 144
Fungal infection 88, 94
Fusarium 163, 164, 170, 172

Garlic cloves 83
Gene expression 81–82
Geotrichum 163, 173

Ginger flower 169
Gloeosporium 169, 170
Grain 8, 9, 11–13, 182, 183, 185, 186, 188,
    202, 203, 213, 214, 215, 216, 217,
    218, 229
  quality 216
Grain dryer 216
Grain weevil, *Sitophilus* spp. 9
Granary weevil 9
Grape 2
Grape, table 86
Grapefruit 2, 6, 173, 256, 257, 263
Grapholita molesta (Oriental fruit moth) 261
Grasshopper 8
Greenheaded leafroller, *Planotortix octo* 253
Growth regulator 93–94
Growth in vegetable 82

Harvest season 93–94
Hazelnut 2, 91
Heat 5–16
Heat disinfestation 190–193, 213, 214, 217,
    220, 229
  of stored product 190–191, 212, 220
  of structure 191–193, 202, 220, 221, 222
Heat shock 240, 262
Heat shock protein (HSP) 83, 94, 194,
    269–284
  HSP40 271
  HSP60 271, 272
  HSP70 269, 270, 271, 272, 278, 279, 282,
    284
  HSP90 271, 272
  HSP100 271, 272
  small HSP 271, 272
  subcellular localization 271
Heat stress 227
Heat tolerance
  in equipment 277
  in insect 194, 198, 199
Heat transfer 27–53, 212–213, 226
  theory 27–28
Heat treatment 190, 191, 193, 199, 205, 209,
    214, 220–222, 223, 224, 225, 226,
    227–228, 311–316, 318–319,
    321–322, 324–325, 327–328,
    331–332, 334–335
  approved 315, 334
  bioassay 312, 316, 318
  calculator 225–227
  checklist 227–228

commercial 311, 322, 332–335
confirmatory tests 315, 332
dose–response 315–316
efficacy data 312, 315, 318–319, 323–324,
    329
heating profiles 322
load factor 312, 329
problem 227, 228
thermal diffusivity 320
thermotolerance 312–313, 318–319, 321,
    326–327, 329–331, 334
transient-temperature 329
Heaters 222, 224, 227, 228
  electrical heater 191, 202, 222
  gas heater 191, 202, 222, 224, 225
  steam heater 191, 202, 222, 223
Heating block system 110–111, 127
Heating curve 47–52
  apple 48
  cherry 50
  fruit size 49
  orange 51
  walnut 47
Heating rate 157–158, 247–248, 261, 262,
    263
Heating system 202
Heating uniformity 297–301, 304–305
High temperature 193–194
  effects on insect 193
  damage 227
High-temperature controlled atmosphere 12
Hot air 27, 30, 33, 37–40, 47–52
Hot water 10, 27, 37–39, 48–51, 263
Hot water heating 85–88, 95–96
Hot water treatment 244
Huhu, *Prionoplus reticularis* 256
Hydrocooling 246
Hypercarbia 253
Hypoxia 253, 261

Impaired ripening 254, 261, 264
Indianmeal moth 13
Induction temperature 269, 270, 280
Infestation 241–245, 312, 315–317, 319, 324,
    328, 331–332, 334
  artificial 312, 319, 330–332
  natural 312, 319, 330–331
Infrared 13
Infrared radiation 202, 218–220
Infrared thermometer 64–67
  electromagnetic spectrum 64, 66

emissivities 67
thermal imaging 66–68, 75
Insect biology 134–145
codling moth 134–136
Indianmeal moth 138–139
navel orangeworm 136–138
red flour beetle 139–141
tephritid fruit flies 141–145
Insect development 199
Insect distribution 134, 143, 145
in fruit 263
Insect–free product 144
Insect mortality model 112–124
activation energy 114, 116–117
Arrhenius plot 115
CLL 118, 122, 125
degree-minute model 123–124
kinetic model 112, 114, 116, 118, 126–127
logarithmic model 115
logit 118, 120, 122, 125
modified first-order model 119, 122, 125
order of reaction 113, 115–116
Insect pest 182, 183–187, 217, 228
see also Stored product insects
Insecticide (pesticide) 188, 189, 229
residue 188, 229
Internal browning 254, 261
Internal developer 183, 184
Irradiation 5

Kale 82
Kamani nut 4
Khapra beetle, Trogoderma granarium 9, 11
Kiwifruit 81, 88, 261

Lactate 255
Larger grain borer 13
Lemon 2, 167, 173
Lepidoptera 183, 185
Lepidopteran 257
Leptinotarsa decemlineata (Colorado potato
beetle) 281
Lesser grain weevil 13
Lethal temperature 92
Lethal time 118–120, 125, 127
Lettuce 83, 283
Life stage 253, 258
Light brown apple moth, Epiphyas postvittana
238, 253, 256, 258, 261, 262, 263,
264
Lime 2, 16

Lipoxygenase 91
Little brown apple moth, Epiphyas postvittana
12
Locust 8
Locusta migratoria (African migratory locust)
281
Logit transformation 208–209, 211
Longan, Dimocarpus longan Lour. 6
Low temperature 189
see also Aeration, Physical disinfestation
Lychee, Litchi chinensis Sonn. 6, 86, 93

Macadamia 91
Maize 13
Malaysian fruit fly 142, 318
Malting barley 204
see also Stored products, quality
Mango 2, 6, 14, 80–81, 83–87, 89–90, 93,
95–96, 171, 172, 174, 246, 263,
274, 275
Mango weevil, Cryptorhynchus mangiferae
14, 319
Mathematical model 298–300
Maturity 95
Mealiness 90
Mealybug 6
Mechanical method 189
see also Physical disinfestation
Mediterranean fruit fly, Ceratitis capitata 4, 6,
9, 238–239
Meditterranean flour moth, Anagast kuehniella
9
Melon 86, 93, 170, 172
Melon fly, Bactrocera cucurbitae 239, 314,
318–319
Membrane
chloroplast omega-3 fatty acid desaturase
89
integrity 89
permeability 252
phospholipid saturation 89, 91
thermostability 88
Merchant grain beetle 13
Metabolic arrest 252
Metabolic heat rate 252
Metabolism 252, 255, 256
Methyl bromide 187, 188, 190, 193, 224, 227
Mexican fruit fly, Anastrepha ludens (Loew) 6,
9, 239, 242, 245–247, 256, 263,
313–314, 327–328
Microorganism 90

Microwave 13–14, 27, 41, 44, 202, 207, 217, 220
Millet 13
Mite 253
Mixing number 298–299
Model application 126
Model comparison 124
Modified atmosphere 188
Moist forced air 85
Moisture 196–197, 200
   moisture content 90–91, 203, 208, 211
   see also Relative humidity (RH)
Monilinia 163, 169, 170, 172
Montreal Protocol 7
Mortality 190, 193, 195, 198–200, 208, 209, 210, 216, 220
   due to moisture 200
   due to rate of heating 199
   in developmental stage 199
   in species 198–199
Mortality data 195–197
   Psocids (Psocoptera) 195, 196, 209
   R. dominica 195, 196
   S. oryzae 195, 197
   T. castaneum 195, 197
   T. confusum 195, 197
   T. variabile 195, 197
Mortality model 209–212
Mung bean 275
MW 41–44, 46–47

Navel orangeworm 240, 247
Nectarine 2, 10, 12, 86, 90, 172
Nectria 170
New Zealand flower thrips, *Thrips obscuratus* 10
Nitidulid 13
Nut 2, 7–8, 11, 15
Nysius huttoni (wheatbug) 264

O_2 80, 87–88, 90, 256, 257, 258, 259
Oatmeal 13
Obscure mealybug, *Pseudococcus affinis* 257, 258, 259
Olive 2
Olive fruit fly 240
Omnivorous leafroller, *Platynota stultana* 252, 255
Onions, green 82
Orange 2, 6, 167, 168, 169, 170
Oriental fruit fly, *Bactrocera dorsalis* 6, 10,

239, 244, 314, 318–319, 324–325
Oriental fruit moth, *Grapholita molesta* 261, 264
Ornamental 15
Oxygen 80, 87–88, 90, 256, 257, 258, 259

Pacific spider mite, *Tetranychus pacificus* 253
Papaya 2, 6, 10, 80, 85, 87, 93, 96, 168, 244–245, 274
Peach 2, 10, 12, 85, 90, 93, 169, 172
Peanut 91
Pear 93
   Bartlett 260, 264
   Packham's Triumph 261, 264
Pecan 14, 90, 91
Penicillium 163, 167, 168, 170, 171, 172, 173, 174
Pepper 172, 275, 277
Pepper, bell 93
Persimmon 275, 277
Pesticide *see* Insecticides
Phaeomoniella 163
Phosphine 187, 189, 190, 192
Physical disinfestation 188
Phytophthora 168
Phytosanitation 3
Pilot-scale 293–296, 303, 307
Pineapple 2, 6, 10, 169
Pistachio 2, 91
Pitting 83–84, 254, 260
Planotortix octo (greenheaded leafroller) 253
Platynota stultana (omnivorous leafroller) 252, 256, 258
Plodia interpunctella (Indianmeal moth) 133
   biology 138–139
Plum 2, 169
Pneumatic conveyor 202, 206, 215–216
Polygalacturonase 81
Polyscytalum 163
Pomelo 2
Population 186, 193
Potato 83
Powder post beetle 11
Preconditioning 96
Preconditioning effect 155
Pre-harvest factor 93–95
Pre-treatment 96
Prionoplus reticularis (Huhu) 256
Probit analysis 118, 120–121, 208–209, 211, 216
Probit-9 4

Processed food 7
Processing facilities 7, 9, 11
Protein
  denaturation 83
  stabilization 272
  synthesis 83
Protocol development 291–300
  quality curve 291–292, 294
  regulatory agencies 293–294
  strategies 291
  systematic development 293, 295
Prune 2, 90
Pseudococcus affinis (obscure mealybug) 257,
    258, 259
Psocids (Psocoptera) 185, 198, 199
  mortality data 195, 196
Pulse 182, 184, 185
  pest of 184, 185

Q_{10} 80, 91
Quarantine regulation 2–4
Quarantine treatment 274, 311–334
  approach time 313
  cold 276
  economic feasibility 321–322
  ethylene dibromide 311, 313–314,
      327–328, 334
  forced hot-air 311, 314–315, 319,
      321–323, 326–327, 331–334
  high-temperature forced air 314
  holding time 313, 315
  hot air 277
  hot water 274, 277
  hot-water immersion 311–313, 324
  load factor 312, 329
  methyl bromide 311, 313, 326–329
  phytosanitary 311
  protocol 315–316, 328–329
  schedules 315, 322
  vapour heat 312–315, 319, 322, 324,
      326–328, 331–334

Radio frequency 15–16, 27, 41–42, 44, 47,
    49–50
Radio frequency heating 91, 242, 247
Radiowave 202, 217, 218
Raisin 2, 90
Rambutan 6
Rancidity 90–91
  hydrolytic rancidity 92
  peroxide value 91–92

Rangeland 8
Raspberry 93
Rate of heating 200–201, 202, 220
Recovery 157–158
Red flour beetle, *Tribolium castaneum* 9, 13
Relative humidity (RH) 195, 196–197, 200,
    208
Resistance
  to contact insecticide (pesticide) 188, 229
  to fumigant 187
Respiration 79–80, 83
  alternative pathway 80
  climacteric 80, 82
  cyanide-insensitive pathway 80
  preclimacteric 80
Respiratory control ratio 80
RF 41–51
  apple 305–307
  cherry 305
  fringing 46
  generator 45–46
  orange 306–307
  treatment 295–304
  walnut 295–304
Rhagoletis 141
  R. indifferens (western cherry fruit fly) 141
  R. pomonella (apple maggot) 141
Rhizopus 164, 165, 166
Rhyzopertha dominica (lesser grain borer)
      183, 184, 198, 199, 200, 201, 209,
      213, 219, 220
  mortality data 195, 196
  see also Internal developer
Rice weevil 13
Ripening 80–82, 93
Roach 11

Saccharomyces cerevisiae 272
Safety 190, 227, 228
Sapote mammey 86
Sarcophagi crassipalpis (flesh fly) 281, 282
Sawtoothed grain weevil 13
Scalding 83–84, 86, 93, 96
Scale-up 293
Security 311–312, 314–318, 324–325
  maximum pest limit 317
  probability 312, 316–317
  probit 315
  probit-9 316–317
  regulatory monitoring 321–323
  statistic 316–318

Seed 13, 15
Senescence 82
Sensor application 68–76
    calibration 69–70, 75
    measurement error 74
    precision 68–70, 75
    response time 68–70
Sensory quality 89–90
Simulation model 208–212
Sitophilus oryzae (rice weevil) 183, 184, 198,
        209, 217, 218, 219
    mortality data 195, 197
    see also Internal developer
Softening, abnormal 87
Soil 13
Solar 12
Spices 12
Spouted bed 202, 204, 205, 214–215, 216
Squash 6
Statistical transformation 208–209
Stone fruit 264
Stored product 182, 183, 184, 189, 190, 212
    quality 203–204
        canola 204
        malting barley 204
        wheat 203
Stored-product insect 183, 186, 187, 188,
        189, 190, 193, 194, 219, 228
    see also Insect pest
Strawberry 2, 82, 86, 166, 168, 169
Stress, dual 263
Structural treatment 225
    energy requirement 225
    see also Heat disinfestation of structure
Structure 11–12, 15, 182, 183, 184, 189, 192,
        193, 202, 205
Sweet cherry 171, 260
Synergistic effect 263
System approach 8

Table grape 166, 170, 171
Tangerine 2
Tarsonemid mite 9
TDT curve 115–117, 296, 299, 302–303
Temperature coefficient (Q_{10}) 80
Temperature control 76
Temperature distribution 34, 42, 48, 50, 52
Temperature measurement 56–77
    fibre-optic sensor 62, 73
    magnetic resonance imaging 67–68
    principle 56–68

resistance temperature device (RTD) 57, 70
resistance thermometer 57
sonic thermometer 63, 70
temperature scale 57
thermistor 59, 71
thermocouple 60, 61, 70–71
type T thermocouple 60–61, 75
Termite 11, 14
    Anastrepha oblique 6
    Anastrepha serpentine 6
    Anastrepha spp. 6
    Bactrocera cucurbitae 6
Tetranychus pacificus (Pacific spider mite) 253
Tetranychus urticae (two-spotted spider mite)
        253, 256, 260, 261
Tephritid fruit flies (Tephritidae) 133
    biology 141–144
Thermal conductivity 35–37, 47
Thermal death data 144–158
    LT 147–149, 156
    most heat resistant life stage 146, 150, 152
    TDT curve 146, 150, 153
    thermal death constant 146
Thermal death kinetic 105–115
Thermal diffusivity 32, 34–40
Thermal radiation 28–33
    heating rate 28, 33
    surface emissivity 33
    thermal energy 34
Thermal tolerance 105–106, 108–109
Thermotolerance 94, 96, 241, 243, 247, 262,
        270, 272, 274, 275, 277, 281, 282,
        284
Thrips 255
Titratable acidity 80
Tocopherol, alpha-tocopherol 91
Tomato 6, 10, 80–81, 88, 90, 93, 170, 172,
        274–277
Transcription 269, 273, 277, 282
Translation 269, 273, 277, 278,
Trehalose 283
Tribolium castaneum (red flour beetle) 133,
        184, 185, 194, 199, 200, 209, 211
    biology 139–141
    heat shock protein 194
    mortality data 195, 197
    see also External developer
Tribolium confusum (confused flour beetle)
        194, 195, 197, 209, 211, 212
    mortality data 195, 197
    see also External developer

Trogoderma variabile (warehouse beetle) 184, 185, 195, 197
  see also External developer
Two-spotted spider mite, *Tetranychus urticae* 253, 256, 260, 261

Uniformity index 299, 301–302, 304

Vapour heat 9, 10
Vapour pressure 86
Vapour pressure-deficit heating 85
Vapour-forced air heating 85–87, 95–96
Vegetable 8, 9, 15
Volatile 81, 89
  acetaldehyde 90
  ethanol 90

Walnut 2, 14, 90–91, 256
Walnut quality 296–304
  FA 296–297, 302–304

kernel colour 302, 304
  PV 296–297, 302–304
Water content 90
Water loss 86
Water vapour-saturated air 85
West Indian fruit fly 242, 245
Western cherry fruit fly, *Rhagoletis indifferens* 141
Western flower thrips, *Frankliniella occidentalis* 253
Wheat 9, 13, 182, 203, 204, 217, 219, 220
  see also Stored products, quality
Wheatbug, *Nysius huttoni* 264
Wood product 11, 14

Yellow pitaya 6
Yellowing 82–83

Z value 115–117, 126